Applying Mathematics

Applying Mathematics

Immersion, Inference, Interpretation

Otávio Bueno and Steven French

OXFORD
UNIVERSITY PRESS

Great Clarendon Street, Oxford, OX2 6DP,
United Kingdom

Oxford University Press is a department of the University of Oxford.
It furthers the University's objective of excellence in research, scholarship,
and education by publishing worldwide. Oxford is a registered trade mark of
Oxford University Press in the UK and in certain other countries

First Edition published in 2018

Impression: 1

Published in the United States of America by Oxford University Press
198 Madison Avenue, New York, NY 10016, United States of America

British Library Cataloguing in Publication Data
Data available

Library of Congress Control Number: 2017959950

ISBN 978-0-19-881504-4

Printed and bound by
CPI Group (UK) Ltd, Croydon, CR0 4YY

For Patrícia, Julia, and Olivia,
minhas meninas
For Dena and Morgan

Preface

This book had a somewhat more prolonged gestation than most. Some of the work on which it is based was undertaken in the mid-to-late 1990s and was shaped through the discussions of a small reading group in the philosophy of mathematics at the University of Leeds. And many of those discussions centred around the reaction to Mark Steiner's now very well-known book, *The Applicability of Mathematics as a Philosophical Problem* (Steiner 1998). It seemed to us then that his analysis rendered the applicability of mathematics mysterious, or at least, more mysterious than it really is, and that close attention to the historical details would help to dispel that air of mystery. It is to this claim, namely that we should look more carefully at how mathematics has actually been applied in physics, that we returned in early 2010 in Miami when the bones of the book were laid down. In the meantime, we had also further developed the partial structures variant of the so-called Semantic Approach as a philosophical device that could be deployed in the analysis of a number of issues and views, from inter-theory and theory–data relations in general to forms of structural empiricism and realism more specifically. Coupling that historical sensitivity with the flexibility of this framework seemed to us to be an obvious way to proceed in examining how mathematics is actually applied to science—specifically quantum physics—in practice, rather than in the minds of philosophers.

Still, despite the clear road ahead of us, it took a further six years and an intense few days in Leeds before we had something that we felt could be submitted to the withering gaze of our colleagues! The book as a whole can be seen as a further extended paean to the myriad virtues of the partial structures approach but even if you, the reader, are sceptical about these, we hope you will find something interesting in our historical and philosophical discussions.

It is set out as follows: Chapter 1 tackles Steiner's claims and suggests that the mystery that he sees in the applicability of mathematics can be dispelled by adopting a kind of optimistic attitude with regard to the variety of mathematical structures that are typically made available in any given context. This might suggest that applying mathematics is simply a matter of finding a mathematical structure to fit the phenomena in question. Nevertheless, a cautionary note can still be sounded, represented by Mark Wilson's concern that this attitude does not in fact capture the manner in which mathematics is applied, not least because mathematics is more 'rigid' than this form of optimism assumes. What has to happen for mathematics to be brought into contact with physics is for certain 'special circumstances' to be in place. We suggest a related approach that builds on the central role of idealizations in enabling mathematics to be applied—it is via such idealizations that these circumstances are effectively constructed.

Much of the rest of the book consists of a series of case studies illustrating that approach and the manner of the construction of these special circumstances. We think that it is important to present the details of such case studies, rather than toy examples, in order to illustrate how mathematics was applied in the context of actual scientific research rather than in its regimentation through textbooks. The case studies also exemplify three different roles associated with the application of mathematics: a representational role, a unificatory role, and an (alleged) explanatory role.

Before we present these case studies, however, we need to set out the framework in which these various moves and roles will be examined. Thus, in Chapter 2, we consider, at one end of the spectrum, the structuralist account associated with Joseph Sneed, Wolfgang Stegmüller, and others and, at the other, the models-focused account of Ron Giere. The former, we claim, is heavy on formalism at the expense of a consideration of practice, whereas the latter draws on a number of interesting case studies but omits the crucial formal framework. We suggest that both these extremes should be rejected in favour of an approach that is appropriately formal, while retaining the ability to represent science as it is actually practised.

And such an approach of course is represented by the partial structures account that we also present in Chapter 2. Before we do so, however, we highlight and represent the heuristic role of surplus mathematical structure, as emphasized by Michael Redhead. The exploitation of such structure is a crucial feature of scientific practice that we shall return to again and again throughout the book and we maintain that it is also nicely accommodated within the partial structures framework, extended to include a variety of partial morphisms (such as partial isomorphism, partial homomorphism, etc.).

In Chapter 3, we then show how this framework may be used to give an account of how scientific models *represent* systems and in a way that pays due regard to the relationship between the relevant mathematics and such models. In particular, the open-ended nature of these developments can be accommodated within such a framework and the use of partial homomorphisms holding between structures allows us to capture the way in which relations may be transferred from one such structure to another. Finally, the expansion of the framework to include partial homomorphisms also allows us to accommodate the fact that typically not all of the mathematics is used in a particular application, leaving surplus structure that can be subsequently exploited.

The pay-off begins in Chapter 4, where we present the introduction of group theory into quantum mechanics in the later 1920s and early 1930s. Here one can make a somewhat arbitrary distinction between the 'Weyl' and 'Wigner' programmes, where the former is concerned with using group theory to set the emerging quantum physics onto secure foundations and the latter has as its focus a more practical use as a means of solving complex dynamical problems. Here we emphasize that the application of the mathematics to the physics depended on certain structural 'bridges' within the mathematics itself and also that both this mathematics and the physics were in a state of flux at the time. Given that and the fact that what we had was a partial importation of group theory into physics, we argue that the partial structures approach offers a suitable framework for representing these developments. More importantly, perhaps, we show how one can resist Steiner's claim that it is the mathematics that is doing all the work in these cases, creating a mystery as to how it can thus capture the physical—rather, it is only because of prior idealizing moves on the physics side that the mathematics can be brought into play to begin with. This will be a recurrent theme throughout our work: granted the significance of the mathematics, the crucial work is undertaken on the empirical side of things!

This analysis is then extended in Chapter 5 to include not only the 'top-down' application of group theory but also the 'bottom-up' construction of models of the phenomena, with London's explanation of the superfluid behaviour of liquid helium in terms of Bose–Einstein statistics as our case study. Thus, here we have not only the introduction of 'high-level' mathematics in the form of group theory but also a degree of modelling at the phenomenological level. We claim that in moving from top to bottom, from the mathematics to what is observed in the laboratory, the models involved and the relationships between them can be accommodated by the partial structures approach, coupled with an appreciation of the heuristic moves involved in scientific work. Furthermore, as in the previous examples, this case fits with the 'immersion, inference, and interpretation' account of the application of mathematics (the inferential conception), whereby immersion of the phenomena into the relevant mathematics allows for the drawing-down of structure and the derivation of certain results that can then be interpreted at the phenomenological level.

In Chapter 6, we turn to a different form of application where the aim was to unify apparently unrelated domains, such as quantum states, probability assignments, and logical inference. Here our case study is John von Neumann's development of an alternative to the Hilbert space formalism that he pioneered—one that is articulated in terms of his theory of operators and what we now call von Neumann algebras. This allowed him to accommodate probabilities in the context of systems with infinite degrees of freedom and here too, as well as 'top-down' moves from the mathematics to the physics, we also find 'bottom-up' developments from empirical features, to a particular logic and thence to mathematical structures. Through a combination of such moves, crucially involving exploration of the structural relations that hold between the mathematical and the physical domains, von Neumann was able to articulate the kind of unification across such domains that represents a further important aspect of the application of mathematics.

In Chapter 7, we move on to consider whether, in bringing the mathematics into application in this manner, there are grounds for taking the relevant mathematical entities to be *indispensable*. Our case study here is Dirac's use of the delta function in quantum mechanics and we point out, first, that Dirac himself was very clear about the function's dispensability and, second, even if certain mathematical theories were indispensable, this wouldn't justify the commitment to the existence of mathematical entities. To illustrate this second point we use a further example, that of Dirac's prediction of the existence of antimatter via the exploitation of surplus mathematical structure. We maintain here that the commitment to the existence of certain objects requires the satisfaction of certain criteria of existence and it is unclear whether mathematical entities meet these criteria. Let us be explicit: we agree that mathematics is indispensable to science but this does not imply that we should be committed to the existence of mathematical entities.

The putative explanatory role of mathematics is further pursued in Chapter 8, again in the context of the so-called indispensability argument and the claim that certain scientific features have a hybrid mathematico-physical nature. Our conclusion here is that, with regard to the former, the possibility of mathematical entities acquiring some explanatory role is not well motivated, even within the framework of an account of explanation that might be sympathetic to such a role. Second, the example of such a hybrid property that has been given is that of spin, but we argue that the assertion of hybridity also lacks strong motivation and comes with associated metaphysical costs.

In Chapter 9, we pursue these themes further by examining both the role of idealization with specific regard to explanation, and the broad criteria of acceptability that, we argue, any such explanatory account should meet. Our case study here is that of the phenomenon of universality, and the role of the renormalization group, which has been held up as another example of mathematics playing a significant explanatory role. Again, we press our argument that once we have a clear framework for understanding representation, and an equally clear understanding of what is required of any explanation, such claims simply do not hold up. Here the immersion, inference, and interpretation framework comes to the fore once again and contrary to Bob Batterman's proposal we maintain that, on the one hand, this framework can accommodate the relevant cases he presents but, on the other, the relevant mathematics does not itself play an explanatory role. Nevertheless, we agree with Batterman that his examples shed new light on the practice of science. By articulating the above framework within the partial structures approach, we can account for the nature and significance of the phenomena involved, and thereby offer an understanding of them within a unitary account of scientific practice.

Our final chapter (Chapter 10) then considers two sets of criticisms of our overall approach. The first is that there may be cases where the special circumstances constructed via idealizations simply do not arise and thus our account cannot be applied to them. We examine one such possible case, the discovery of the Ω^- particle, and conclude that it presents no particular obstacle to our approach. The second concerns the role of the partial structures framework in capturing scientific practice, and here we emphasize that we see it as a meta-level representational device that, we argue, suits the purposes of philosophers of science when considering this practice. In particular, when it comes to the various different aspects of the applicability of mathematics, it does the job!

Contents

Acknowledgements

We owe a huge debt to many people, across many years and different countries. We won't be able to name all of them, but we would like in particular to highlight: Jody Azzouni, Manuel Barrantes, Mark Colyvan, Newton da Costa, Sarah Kattau, Décio Krause, James Ladyman, Peter Lewis, Kerry McKenzie, Joseph Melia, Chris Pincock, Juha Saatsi, Scott Shalkowski, Harvey Siegel, Amie Thomasson, Bas van Fraassen, and Pete Vickers, among so many friends and colleagues with whom we discussed and corresponded about the issues examined in this work. To all of them, including those not individually named above, our sincere thanks.

For extensive comments on the manuscript, we are grateful to Manuel Barrantes and two anonymous reviewers for Oxford University Press (OUP). Their feedback led to substantial improvements in the final version of the work. We'd also like to thank Peter Momtchiloff at OUP for all his tireless support and patience throughout the long gestation of this work. It has been a pleasure to work with him and his colleagues at OUP.

Steven would also like to thank, as always, Dena, Morgan, and The Ruffian.

Otávio would like to thank Patrícia, Julia, and Olivia for showing him the world, and for making life such a great adventure.

On a Sunday afternoon in June 2016, as we worked together on this project in Leeds, Olivia (Otávio's then 9-year-old daughter) called to ask what we were up to. We told her that we were working on the book, and she promptly asked whether it would be the biggest book in the world. We explained to her that it wouldn't. Unfazed, she then proclaimed: "Okay, then make it the best one!" Clearly, it won't be easy to live up to her expectations. But at least we did the best we could.

A few months later, as we were putting the final touches in the work, Julia (Otávio's then 11-year-old daughter) browsed through the manuscript and complained that the titles weren't colourful enough. If she had her way, this book would be called *The Awesomeness of Partial Structures*!

Finally, we would like to thank Cambridge University Press, Chicago University Press, Elsevier, Oxford University Press, and Springer, for permission to reproduce material from the following papers: Bueno, O. (1999): "Empiricism, Conservativeness and Quasi-Truth", *Philosophy of Science 66*: S474–S485. Bueno, O. (2000): "Empiricism, Mathematical Change and Scientific Change", *Studies in History and Philosophy of Science 31*: 269–96. Bueno, O. (2005): "Dirac and the Dispensability of Mathematics", *Studies in History and Philosophy of Modern Physics 36*: 465–90. Bueno, O. (2016): "Belief Systems and Partial Spaces", *Foundations of Science 21*: 225–36. Bueno, O. (2017): "Von Neumann, Empiricism, and the Foundations of Quantum Mechanics", in Aerts, D., de Ronde, C., Freytes, H., and Giuntini, R. (eds.), *Probing the Meaning and Structure of Quantum Mechanics: Superpositions, Semantics,*

Dynamics and Identity (Singapore: World Scientific), pp. 192–230. Bueno, O., and French, S. (1999): "Infestation or Pest Control: The Introduction of Group Theory into Quantum Mechanics", *Manuscrito 22*: 37–86. Bueno, O., and French, S. (2011): "How Theories Represent", *British Journal for the Philosophy of Science 62*: 857–94. Bueno, O., and French, S. (2012): "Can Mathematics Explain Physical Phenomena?", *The British Journal for the Philosophy of Science 63*: 85–113. Bueno, O., French, S., and Ladyman, J. (2002): "On Representing the Relationship between the Mathematical and the Empirical", *Philosophy of Science 69*: 452–73. Bueno, O., French, S., and Ladyman, J. (2012): "Empirical Factors and Structural Transference: Returning to the London Account", *Studies in History and Philosophy of Modern Physics 43*: 95–103. French, S. (1999): "Models and Mathematics in Physics: The Role of Group Theory", in Butterfield, J., and Pagonis, C. (eds.), *From Physics to Philosophy* (Cambridge: Cambridge University Press), pp. 187–207. French, S. (2000): "The Reasonable Effectiveness of Mathematics: Partial Structures and the Application of Group Theory to Physics", *Synthese 125*: 103–20. French, S. (2015): "Between Weasels and Hybrids: What Does the Applicability of Mathematics Tell Us about Ontology?", in Beziau, J.-Y., Krause, D., and Arenhart, J. (eds.), *Conceptual Clarifications: Tributes to Patrick Suppes* (London: College Publications), pp. 63–86.

List of Illustrations

1

Just How Unreasonable is the Effectiveness of Mathematics?

1.1 Introduction

Eugene Wigner's famous description of the 'unreasonable effectiveness' of mathematics has, explicitly or implicitly, structured much of the discussion of applicability in recent years (Wigner 1960). Those two simple words of his phrase have laid down the parameters of the debate: on the one hand, mathematics is clearly effective as, at the very least, the language of science, but on the other, given the apparently stark differences between mathematics and science,[1] this effectiveness appears unreasonable. Put bluntly, the question is: how can something apparently so different from science be so useful in scientific practice? This question invites comparison of the mathematical and scientific domains, and in order to effect such a comparison one must choose a suitable mode of representation. It further invites us to pay close attention to the elements of practice, in order to discern the intricate workings of the mathematical in the domain of the physical.

Of course, certain notable philosophies of mathematics do present highly articulated representational frameworks. However, they fail to appropriately consider the practice itself. Looking in the other direction, as it were, one can identify several attempts to redress the balance, to present the nitty-gritty of both mathematical and scientific practice. Nevertheless, admirable as these attempts might be, they are deficient when it comes to the other side of the coin: they fail to present these details within an appropriate mode of representation. Even worse, in one notable case, the details are presented in a reconstruction that is so artificial as to give the most Whiggish of historians pause. In this opening chapter, we shall argue that this detaching of practice from the relevant history actually perpetuates the mystery of applicability.[2] In subsequent chapters we shall attempt to dispel the air of mystery by offering what we consider to be a closer reading of the historical circumstances.

---

[1] These differences are both epistemic, in terms of the manner in which the respective truths in mathematics and science are established, and ontological, in terms of the apparent causal efficacy of the relevant objects.

[2] This sense of mystery was perpetuated by Wigner himself, who wrote, '...the enormous usefulness of mathematics in the natural sciences is something bordering on the mysterious and...there is no rational explanation for it' (1960, p. 2).

1.2 Mystery Mongering

In a programme of work leading up to his renowned analysis, *The Applicability of Mathematics as a Philosophical Problem*, Mark Steiner has articulated his own understanding of Wigner's phrase in terms of the unreasonable effectiveness of mathematics *in scientific discovery* (Steiner 1998).[3] The view he sets out is both radical and regressive: it is that the very concept of mathematics is *anthropocentric*, just as the physics of the Middle Ages was.[4] What he means by this is that the practice of mathematics can be understood in terms of the twin criteria of beauty and convenience, both of which are species-specific (1998, p. 7). Relying on mathematics in scientific discovery therefore amounts to relying on human standards of what counts as beautiful and convenient and hence is an anthropocentric policy (ibid.). His claim, then, is that such a policy 'was a necessary factor (*not* the only factor) in discovering today's fundamental physics' (ibid., p. 8; italics in the original).[5]

The argument in support of this claim runs as follows (ibid., pp. 3–5): quantum physicists have been extraordinarily successful at 'guessing' the laws of the atomic and sub-atomic world(s). They have arrived at these laws by using mathematical analogies in addition to experimental work. In many cases, these analogies were 'Pythagorean' in the sense that *at the time they were established*, they could only be expressed in the language of pure mathematics (ibid., pp. 3, 54).[6] In certain 'remarkable instances', the analogy was based on the *form*, rather than the meaning, of the relevant equations (ibid., p. 54). Steiner gives an example of what he calls a 'formal argument' from analogy, in an earlier paper:

Suppose we have effected a successful classification of a family of 'objects' on the basis of a mathematical structure S. Then we project that this structure, or some related mathematical structure T, should be useful in classifying other families of objects, even if (a) structure S is not

---

[3] As a reader of an earlier draft of our book noted, Wigner was primarily concerned with the role of mathematics in describing or *representing* nature, whereas Steiner is concerned with its *heuristic* role in scientific discoveries. Although these concerns are distinct, Steiner takes Wigner's as motivating his, casting it as '...an *epistemological* question about the relation between Mind and the Cosmos' (1998, p. 45; italics in the original). In this chapter we shall focus primarily on the latter role although subsequently we shall touch on the former as well.

[4] Umberto Eco (1998, pp. 64–5) once wrote that we still live in the Middle Ages, in the sense that, for many of us and in particular those who have not come much into contact with modern physics, the furniture of the world is the same as it was then. On Steiner's view, even though some of us have refurnished our conceptual houses, we cannot escape anthropocentrism in both the styling and description of this new furniture.

[5] Azzouni offers a response in terms of the empirical projectibility of mathematical predicates (Azzouni 2000, p. 226, note 20). However, this is obviously a rather general response, with the details left to be filled in for specific cases. Here we will be concerned both with those details in the context of specific historical episodes and with framing the relevant embeddings within a particular formal framework, namely the partial structures variant of the Semantic Approach to theories.

[6] This is not to say that the applicability of mathematics involves its expressive power only. It is also, and fundamentally, applied with respect to the relevant *derivations*. However, it is mathematics' expressive power that Steiner is focusing on here.

known to be equivalent to any physical property, and (b) the relationship between structures S and T is not known to be reducible to a physical relation.   (1989, p. 460; see his 1998, p. 84)

But, 'such formal analogies appear to be irrelevant analogies and irrelevant analogies should not work at all' (1989, p. 454). So, the success of such analogies is puzzling and 'unreasonable'. It is so, however, only from a 'naturalist', anti-anthropocentric perspective that regards the universe as indifferent to human values as expressed in the above criteria.

As a result, in Steiner's view:

The strategy physicists pursued, then, to guess at the laws of nature, was a Pythagorean strategy: they used the relations between the structures and even the notations of mathematics to frame analogies and guess according to those analogies. The strategy succeeded.   (1998, pp. 4–5)

And a Pythagorean strategy cannot avoid being anthropocentric, because (again) 'the concept of mathematics itself is species-specific' (ibid., p. 6).[7] In particular, the criteria in terms of which mathematicians regard a structure as *mathematical* are internal, two such criteria being beauty and convenience, as already noted (ibid., pp. 6–7).[8]

Before we go on to consider the examples Steiner draws upon to flesh out this picture, let us pause to consider the view of scientific discovery that it is predicated upon. The central motif is that of physicists 'guessing' the laws of nature.[9] Thus, he writes:

by the end of the nineteenth century, physicists began to suspect that the alien laws of the atom could not be deduced from those governing macroscopic bodies. Nor, of course, could they be determined by direct observation. Atomic physics seemed reduced to blind guessing, with an uncertain future.   (1998, p. 48)

We are given a dichotomy between deduction and induction, and with the failure of both, blind guessing seems the only other option. However, there is a considerable amount of work that has been done in the philosophy of science on the topic of heuristics, or theory construction, which provides an alternative and very rich description of what is involved in scientific discovery (see, in particular, Post 1971; Nickles 1980; Nersessian 1992). Post, for example, has emphasized the importance of what he calls the 'General Correspondence Principle', according to which a new theory should recover significant empirical and theoretical elements of its predecessors. In other words, we never lose the best of what we have, where empirical success ultimately determines what is retained. Furthermore, the Correspondence Principle

---

[7] For a response that tackles Steiner's claim of anthropocentrism head-on, see Bangu (2006). He argues that mathematicians' rejection of the maxim that all mathematical concepts should be explicitly definable in a uniform way, which he takes to be an anthropocentric strategy, indicates their reluctance to embrace anthropocentrism. Our criticism of Steiner proceeds along different lines.

[8] Thus, with regard to note 5 above, Steiner notes that 'in the Middle Ages, physics *was*, arguably, anthropocentric—at least covertly—because it classified events into heavenly and terrestrial, a distinction reflecting our own parochial point of view' (1998, p. 6; italics in the original).

[9] The word is used repeatedly throughout the book.

also plays a heuristic role in delimiting the field of possible successors to a theory and thus acts as a kind of 'projectibility principle' (Rueger 1990), which constrains theory construction by 'projecting' into the new domain those aspects of the old theory which are judged to be relatively independent from theory change. Thus—just to foreshadow our subsequent discussion—in developing an early form of 'atomic physics', Bohr deployed his version of the Correspondence Principle in order to derive reliable empirical results that were independent of the details of certain physical processes. More significantly, perhaps, he developed the notion of a 'stationary state' in such a way that it was immune to changes in the theory of radiation (Rueger 1990, p. 210), and it was this notion that formed part of the core of the 'new' quantum mechanics of Heisenberg, Schrödinger, and Dirac.[10]

In addition, Post, and others, also highlighted the heuristic role played in modern physics by symmetry principles, for example, something we shall be returning to in subsequent chapters. In certain significant cases such principles, together with various mathematical techniques, have been imported from one domain, where they are empirically supported (albeit indirectly perhaps), into another, relevantly similar to the first. In the case of what Steiner calls the 'alien laws of the atom',[11] Kuhn, and before him Klein, have mapped out in detail the moves made by Planck, Einstein, and Ehrenfest in constructing the basis of the 'old' quantum theory. In particular, the whole edifice is founded on Planck's effective importation into the quantum domain of statistical techniques developed by Boltzmann as part—the crucial, central part—of classical statistical mechanics. The significant difference, of course, lay in the different *counting* which Planck used, and Einstein and, in particular, Ehrenfest, devoted considerable effort to understanding just what Planck had done.[12]

Leaving aside the details for now, our point is just to point out that there were numerous connections between classical and quantum physics which do not fit into either the deductive or inductive moulds, yet can be, and have been, accommodated within recent discussions of theory construction.[13] Now, to be fair, Steiner does cover Peirce's thoughts on scientific discovery, but the latter's notion of abduction is, again, taken to be equivalent to guessing (1998, p. 49). There has also been considerable discussion of abduction (a useful overview can be found in Douven 2011), but it is worth noting here Rescher's point that the notion can be detached from Peirce's own idiosyncratic underpinnings of instinct (Steiner 1998, pp. 49–50) and understood in terms of heuristic 'principles of method', where the focus is on the exportation into

---

[10] Similarly, and more recently, Fraser has argued that renormalization theory should be understood as isolating features of current forms of quantum field theory which are independent of future theory change (Fraser 2016).

[11] And the rhetoric here is significant, with the word 'alien' used to dramatic effect.

[12] In large part this understanding was only completed with the developments we shall be considering in subsequent chapters.

[13] See, for example, Saunders on the 'heuristic plasticity' of mathematical techniques and devices, such as Fourier analysis and the Poisson bracket as they are exported from the classical to quantum domains (Saunders 1993).

new domains of appropriately supported methodologies, as represented by symmetry principles, for example (Rescher 1978, pp. 61–3; Curd 1980).

To be even fairer, Steiner's own presentation of the role of analogies can be seen as an attempt to fill in some of the heuristic details. However, the view of this role as 'formal' or 'Pythagorean', in the above senses, is based on a somewhat skewed account of the development of the physics in question. Jumping forward to the later development of quantum theory, let us consider another example, namely Schrödinger's discovery of wave mechanics (Steiner 1998, pp. 79–82). Steiner states that Schrödinger assumed that a particle of constant energy 'corresponds' to a wave of appropriate frequency (given by Planck's famous equation $E = h\nu$ with Planck's constant h set to 1) and satisfies a wave equation that is '*formally* identical to the equation for a *monochromatic* light wave in a *nonhomogeneous* medium' (1998, p. 79; italics in the original). In order to deal with problems such as the interaction between atoms and light, Steiner writes that Schrödinger had to eliminate the energy term, rewriting it in terms of the classical separation between kinetic and potential energy. By appropriate differentiation, Schrödinger then obtained his famous wave equation.

The crucial point for Steiner is that Schrödinger then 'forgot' his crucial assumption above and allowed solutions where the energy was not constant. But this leads to bizarre consequences such as the superposition of waves of different frequency/energy and Steiner asks: 'how can *one* particle be assigned to two different energies?' (ibid., p. 80; italics in the original). Furthermore, the wave function in Schrödinger's equation is 'irredeemably complex; it has no physical interpretation' (ibid., p. 81). As is well known, only the square of the modulus has 'physical meaning' (via Born's famous probabilistic interpretation[14]). Despite these profound interpretational problems, Steiner continues, 'Schrödinger conjectured the equation with all its solutions' (ibid., p. 81). These problems illuminate the dramatic divergence from classical physics and underpin the claim that the reasoning involved in Schrödinger's 'conjecture' was not merely Pythagorean but 'formalist', in the sense of 'allowing the notation to lead us by the nose' (ibid., p. 80, note 11).

Now this is a curious but extremely telling vignette. We shall not dwell on the full details of its historical deficiencies here, referring the reader instead to the much more comprehensive account given in Moore's biography, for example (Moore 1989).[15] Instead, we simply wish to emphasize that Steiner reaches his conclusion by conjoining *heuristic* considerations with *interpretational* ones *that emerged later*. Schrödinger himself perceived the non-classical import of superposition in his famous response to the EPR argument (see Moore 1989, pp. 306–13) and, similarly, had recognized the

---

[14] Although this was not Born's own original view; see Wessels (1980).

[15] It is not clear from which of Schrödinger's 1926 papers Steiner is getting this distorted picture. The citation given is to 'Quantisation as a Problem of Proper Values', but both of Schrödinger's first two papers of 1926 were given this title. Furthermore, the summary Steiner gives is that of the 'derivation', if it can be so dignified, of the time-dependent wave equation which Schrödinger presented in his *fourth* paper in the series.

complex aspects of the wave function in his fourth paper on wave mechanics published in 1926. At the beginning of the series, however, his metaphysical worldview was explicitly that of a classical continuum with particles as merely the epiphenomena of an underlying wave-like reality[16]—a view that emerged in his 1925 paper on quantum statistics. The connection with wave mechanics lies in Einstein's earlier introduction of de Broglie's form of wave–particle duality in his account of what came to be known as Bose–Einstein statistics.[17] Having published on and lectured about quantum statistics, Schrödinger was certainly aware of Einstein's account (Wessels 1979) and subsequently claimed that 'wave mechanics was born in statistics' (Moore 1989, p. 188).

The classical worldview also formed the background to the development of the wave equation itself. As is well known, Schrödinger first obtained the *relativistic* form of the equation (which subsequently came to be known as the Klein–Gordon equation), following de Broglie's own relativistic treatment (Moore 1989, pp. 194–200). His notes reveal that he simply substituted de Broglie's relationship between the momentum p and wavelength λ ($p = h/\lambda$) into the steady state equation of classical physics.[18] Unfortunately, this relativistic form of the equation appeared to be ruled out by experiment, as it did not fit the data regarding the fine structure of the hydrogen spectrum. It turned out that this was because the equation did not include a component representing the spin, which was, of course, only introduced later.

Nevertheless, from the relativistic equation one could straightforwardly obtain the non-relativistic form and this was the subject of Schrödinger's first paper setting out wave mechanics. Again, the framework is supplied by the above metaphysical worldview and the 'derivation'[19] is explicitly based on the Hamilton–Jacobi equation from classical mechanics (Moore 1989, pp. 200–5). Using standard variational techniques,

---

[16] There are also Schrödinger's broader philosophical views to take into account. As is well known, he was greatly influenced by both Schopenhauer and Indian philosophy (which he was introduced to through reading Schopenhauer; see Moore 1989, pp. 111–14). In the autumn of 1925, just prior to the development of wave mechanics, he wrote a large part of what was subsequently published as *Meine Weltansicht* (ibid., pp. 168–77). This covered a broad range of related themes, from the decline of traditional metaphysics to the significance of the philosophy of Vedanta and thoughts on evolutionary biology and heredity. As Moore notes, 'it would be simplistic to suggest that there is a direct causal link between [Schrödinger's] religious beliefs and his discoveries in theoretical physics' (ibid., p. 173), but nevertheless, he suggests, the unity and continuity inherent in Vedantic philosophy can be seen as reflected in the unity and continuity of wave mechanics (ibid.).

[17] The central idea was to apply the statistics of photons to indistinguishable gas molecules, using de Broglie's claim that molecules also had wave-like characteristics as the bridge.

[18] As Moore reminds us:

Of course, one cannot actually *derive* the Schrödinger equation from classical physics. The so-called "derivations" are therefore never logically rigorous—they are justifications rather than derivations. It would be reasonable to simply write down the final equation as a *postulate* of wave mechanics, and then to show that it gives correct results when applied to various calculations. But before you can write it down, you have to have it, and it did not spring fully fledged from Erwin's mind like Venus from the scallop shell. The paths leading to the discovery are thus of interest, since they show the process by which the discovery was made. (Moore 1989, p. 197; italics in the original)

[19] We recall Moore's point above.

the non-relativistic wave equation more or less 'drops out', and Schrödinger was able to show that the relevant eigenvalues gave the standard integral values of the principal quantum number, thus capturing the quantization of the energy levels of the hydrogen atom.[20] Interestingly, Schrödinger doesn't mention the role of de Broglie's work until almost the end of the paper.[21] Again, he adverts to the importance for his discovery of his earlier work in quantum statistics. The paper concludes with a representation of the emission and absorption frequencies of Bohr's theory in terms of the phenomenon of beats.

In the second paper, submitted four weeks after the first, Schrödinger presented a new 'derivation' of the wave equation, based on the well-known analogy between the Hamiltonian formulation of mechanics and *geometrical* optics (Moore 1989, pp. 207–9).[22] What is particularly interesting is that he then draws a further similarity between the failure of classical mechanics at the atomic level and the failure of geometrical optics in order to motivate an extension of the above analogy into wave mechanics:

we know today in fact that our classical mechanics fails for very small dimensions of the path and for very great curvatures. This failure is so similar to that of geometrical optics that it becomes a question of searching for an undulatory mechanics by further working-out of the Hamiltonian analogy.   (translated in Moore 1989, p. 207)

What is also noteworthy is Schrödinger's explicit insistence in this work, deriving from his philosophical worldview, that the structure of the atom *had* to be understood within the framework of space and time, 'for we cannot avoid our thinking in terms of space and time, and what we cannot comprehend within it, we cannot comprehend at all' (ibid., p. 208).[23] It is in this paper that he realizes that the wave function is a function in many-dimensional configuration space, rather than 'real', three-dimensional space. (He will eventually note, in a later paper, the difficulties that this brings for his wave picture; see below.) And, finally, he applies the wave equation to the one-dimensional harmonic oscillator, the rigid rotor, and the spectra of diatomic molecules.

However, it is in the third paper in the series that Schrödinger presented what he considered to be the first major application of his theory (Moore 1989, pp. 213–14). This was to the Stark effect for hydrogen, in which the lines of the spectrum are split by an applied electric field. Here he applied standard perturbation techniques, again

---

[20] Schrödinger insists that this 'mysterious whole number requirement' has its basis in the requirement that the wave function be finite and single-valued (Moore 1989, p. 205).

[21] It is relevant that de Broglie's theory involved travelling waves, whereas Schrödinger's featured *standing* waves.

[22] See Steiner 1998, p. 79, note 9.

[23] Moore writes: 'It is evident in this paper that Schrödinger believed that a classical wave picture based upon continuous matter waves would provide the most satisfactory foundation for atomic physics' (ibid., p. 208).

taken over from classical mechanics, to give good agreement with experiment. Finally, in his fourth paper, submitted only six months after the first, Schrödinger shifted his attention from stationary systems and eventually obtained the time-dependent wave equation that is the central focus of Steiner's considerations. It is here, in this paper, that Schrödinger begins to perceive the difficulties for his wave picture that arise from the complex nature of the wave function and, again, its dependence upon configuration space (ibid., pp. 217–20).[24]

Thus, it simply was not the case that Schrödinger 'conjectured' his equation in the face of these problems with superpositions and complex numbers, as Steiner suggests. What was crucial for him was (a) his underlying classical worldview that underpinned, in particular, the Hamiltonian analogy between mechanics and optics, which Steiner admits was not Pythagorean for Schrödinger, and (b) the agreement with experimental results, particularly in the last two works. It is these factors which 'led him by the nose', not the notation. It is only with Whiggish hindsight that Schrödinger's strategy appears Pythagorean, as we are invited to ask ourselves how he could possibly have obtained the results he did, given that these problems showed that the wave function had no physical meaning.[25] However, at the time that Schrödinger obtained his equations, it was *not* the case that the wave function had no physical meaning for him. On the contrary, it had a very definite meaning in terms of a classical continuum theory. It was only as the theory was further explored that the interpretational problems became apparent (we shall return to these problems later).

Thus, casting the moves that Schrödinger made as a 'Pythagorean strategy' seems wholly inappropriate. This confusing misrepresentation of the historical situation in light of subsequent interpretational difficulties more generally features in Steiner's account—which is central to his whole project—of what he calls 'perhaps the most blatant use of formalist reasoning in physics'. This was the 'successful attempt by physicists to "guess" the laws of quantum systems using a strategy known as "quantization"' (1998, p. 136). Explicitly sacrificing historical accuracy for clarity, Steiner presents the details of the quantization strategy in mathematical terms that would have been, as he admits, unfamiliar to Schrödinger. It is no surprise then that the conclusion he arrives at is that 'the discoveries made this way relied on symbolic manipulations that border on the magical' (ibid.).

---

[24] Einstein is attributed with pointing this out to Schrödinger. And, of course, the disparities, both metaphysical and epistemic, between this configuration space and 'ordinary' space continue to be emphasized today, in discussions of the interpretation known as *wave function realism*, for example (see Ney and Albert 2013).

[25] Of course, it may be claimed that what Steiner is presenting is a 'rational reconstruction' of the historical situation. While we are unwilling to become too embroiled in a discussion over the usefulness or otherwise of such reconstructions and a Whig attitude to history in general, we think it is worth noting that some reconstructions are more rational than others. If too much relevant detail is simply omitted, the reconstruction may come to seem very irrational indeed!

He begins with the wave equation again, presented in the context of Hilbert space[26] and in terms of the Hamiltonian. Classically the latter involves position and momentum but, notes Steiner, 'in quantum mechanics, of course, the very idea that the electron has both a momentum and a position is meaningless' (ibid., p. 140). Hence the classical Hamiltonian, although mathematically well defined, is physically meaningless, and the application of Schrödinger's equation, although *formally* justified, has no *physical* basis.[27] This is Steiner's overall point, as he considers the extension of the theory from the hydrogen atom to the helium atom and on to the uranium atom. However, the reasoning that is involved here is not that which was used in the relevant *discoveries*. The non-commutativity of position and momentum operators is a result that comes from what was, in 1926–7, the alternative tradition, namely the matrix mechanics of Heisenberg, Born, and Jordan. That it undermines the classical basis of Schrödinger's theory is well known and was a crucial feature that von Neumann, for example, had to confront in demonstrating that wave and matrix mechanics were two mathematical representations that were underpinned by the framework of the Hilbert space formalism.[28] Von Neumann's approach also faces problems of interpretation, of course, particularly with regard to the celebrated 'projection postulate' which formally represents the so-called 'collapse' of the wave function. But when one moves away from the latter formalism in, for example, the standard exercises in undergraduate pedagogy,[29] the problem of obtaining a coherent interpretation of quantum mechanics becomes even more acute. However, this is not a problem to do with *discovery*, and the 'magical' reasoning, which has been artificially reconstructed to appear 'formalist', does not feature in the historical developments themselves, contrary to Steiner's claims.[30]

Steiner goes on to support his case with the example of the application of group theory to quantum physics. As we shall consider this example in some detail in Chapter 4, we shall postpone further consideration until then. For the moment, we simply insist that it is only by paying due attention to the historical situation that one can appreciate the heuristic moves that were made.[31] Steiner, by contrast, presents a

---

[26] In an appendix (1998, pp. 177–96), Steiner shows that the Hilbert space formalism is applicable to quantum mechanics (inspired by a paper in the *American Journal of Physics* for undergraduate physics students) and points out that no explanation for this is currently known (see also p. 117). It is unclear what sort of explanation he might be looking for here. A historically based explanation will be given in Chapter 6.

[27] Since Schrödinger 'eliminated' energy in his derivation by rewriting it in these terms, Steiner takes this to be a further reason why this derivation should be regarded as based on a Pythagorean strategy.

[28] Muller, for example, has argued that to demonstrate this equivalence, von Neumann effectively had to chop off Schrödinger's ontology (Muller 1997). We will return to this point in Chapter 6.

[29] Steiner repeatedly cites Landau and Lifshitz's classic textbook (Landau and Lifshitz 1958), used by generations of undergraduate physics students the world over, including one of the authors.

[30] And if one were to insist (again) that what Steiner is giving is a rational reconstruction, then at best the claim as to the Pythagorean nature of the discovery makes sense only in the context of the Copenhagen interpretation.

[31] Simons suggests a similar point in his review of Steiner's book: '...while professing to give us the history of discoveries by physicists Steiner gives us a history largely sanitized of the physical motivations

series of highly contrived case studies that support his anthropocentric account only by virtue of obscuring these moves and confusing historical developments with subsequent interpretational difficulties.

If Steiner might be accused of 'mystery mongering', there are others who can be situated at the other extreme of the spectrum of attitudes towards this issue of the applicability of mathematics. Such folk see no mystery to applicability whatsoever, but are guilty of a different misdemeanour, namely *mathematical optimism*.

1.3 Mathematical Optimism

This is the phrase coined by Mark Wilson to describe the attitude that 'for every physical occurrence there is a mathematical process that copies its structure isomorphically' (Wilson 2000, p. 297). He is careful to distinguish 'lazy' or 'trivial' optimism, from a more respectable form. Lazy optimism is the view that for any given physical structure[32] a corresponding mathematical surrogate can always be constructed in the following way: (1) take all the possible properties of the world and apply times and positions to them in all possible combinations; (2) arbitrarily designate which of these sentences are true and which false. This gives a range of mathematical structures, one of which perfectly captures the physical world (ibid., pp. 301–2).[33]

However, this assumes that, on the physical side, all the points of space-time can be *simultaneously* assigned unique coordinates and this can only be done for certain curved surfaces (ibid., p. 303). What actually happens is that, when describing the Earth, say, the mathematician provides a collection of two-dimensional charts, together with an 'accompanying filigree' (ibid.) which effectively tells us how to

and correctives that drove the physicists themselves' (Simons 2001, p. 183). As he acknowledges, Simons received comments on his review from members of a Leeds reading group that included one of the authors.

[32] Throughout this book we shall understand 'physical structure' in a very general sense that is, hopefully, compatible with both realist and empiricist sensibilities: we mean by 'structure' those components and features of a system, including, significantly, the relations between the components, that can be represented set-theoretically (via the framework set out in Chapters 2 and 3). What makes it physical, as opposed to mathematical, is of course a profound issue, having to do with the distinction between the concrete and the abstract, among other aspects. We shall not go into this any further here, but shall take the view that the distinction can be cashed out empirically, in the sense that physical structures, but not mathematical ones, yield empirical or (broadly) observable results. (This can also be represented via empirical sub-structures; if one prefers one can further explicate this 'yielding' of empirical results in causal terms.) At a certain level of abstraction, then two or more physical systems may be represented as 'having' the same structure, but at the concrete level they cannot. We are grateful to Manuel Barrantes for raising this point.

[33] Whether we can make sense of physical structure without using (some) mathematics is a further issue that we shall not pursue here. Husserl famously complained that mathematics comes between us and the conceptual roots of science in the 'lifeworld' (Husserl 1954), a complaint that has been echoed more recently by Maudlin (2014, p. 9). Here we shall set such concerns to one side and take as the basis of our approach the historical fact that mathematics has been so used and the issue, of course, is how it has come to be used so effectively.

piece together the various bits of information contained in those charts. According to Wilson, there is an important and general lesson to be drawn here: not just any old collection is worthy of the name 'structure'. What is required in addition is some explanation or interpretive story (ibid., p. 304).

This lazy optimism should be contrasted with a philosophically and historically more respectable kind which draws on developments in mathematics itself. There are two aspects of such developments that are relevant here. The first has to do with theory change and progress in mathematics itself. Phillip Davis, for example, has noted that 'in mathematics there is a long and vitally important record of impossibilities being broken by the introduction of structural changes' (Davis 1987, pp. 176–7). Thus, Euler extended the notion of function from a dependence on explicit mathematical formulae to any mapping between ordinate and abscissa (Wilson 2000, p. 301). Moving forward to the nineteenth century, Wilson draws on Davis's point to explore the impact on geometry of a change in mathematical setting:

It was discovered that a wide variety of traditional mathematical endeavors could be greatly illuminated if their basic questions were reset within richer domains of objects than had previously seemed appropriate.   (Wilson 1992, p. 150)

In other words, mathematics itself is not static, of course, but undergoes structural changes (this is something we shall also emphasize in the discussion of our case studies below), and thus the optimistic hope is that, even in cases where the mathematics suitable for capturing a physical process is not currently available, such changes will generate the appropriate mathematical structures. An example of this might be found in the well-known case of Dirac's introduction of the delta function, which we shall consider in Chapter 7. In this case, interestingly enough, questions arise as to whether the function is suitable mathematically (since it is inconsistent). Von Neumann noted this but acknowledged that if an alternative were not to be found, the relevant physics would have to be changed to accommodate this new piece of mathematics. We shall return to this example in subsequent chapters, but it does raise the further question whether the (respectable) mathematical optimist would be happy with such a drastic change (namely to an inconsistent piece of mathematics).

The second aspect that supports an optimistic attitude concerns the role of 'surplus structure' at the mathematical level. This is the phrase coined by Michael Redhead in his consideration of the relationship between mathematics and physics as part of an analysis of the role of symmetry in (scientific) theory change (Redhead 1975). We shall consider it in more detail in Chapter 2, but the core idea is as follows: when a relationship is established between a scientific theory T and a mathematical structure M' only part of M' (strictly, a sub-structure M) is applied to T. The 'surplus structure' of M' may then itself be open to physical interpretation, yielding certain further theoretical developments. The heuristic advantages of re-expressing a theory within such a mathematical structure can be seen in the example that we shall explore in much greater detail in Chapter 7: namely, Dirac's prediction of the positron. Here the 'surplus structure' is exemplified by the negative energy solutions of his relativistic wave equation and giving this a physical interpretation via a new form of matter proved to be a major theoretical advance.

A similar idea is expressed by Scheibe in his discussion of the 'mathematical overdetermination of physics' (Scheibe 1992). He considers the example of a gas law, where the values of the physical quantities involved are described by real numbers. This gives a 'wealth of 3-termed relations between numbers' (ibid., p. 143) in terms of which the law can be formulated. It also gives further operations and relationships on the mathematical side that have no physical counterpart. Scheibe is right, of course. Nevertheless, some of this richer surplus structure can subsequently be found to have such counterparts. We will clarify this notion in Chapter 2 by showing how it can be represented formally, and examples of the role of such mathematical surplus structure will be given as part of our case studies. (Of course, the status of the counterparts will ultimately depend on whether they are interpreted realistically or not, which is an additional layer of complexity in these issues.)

This gives us two reasons to be optimistic then: (1) developments and progress within mathematics itself; and (2) the existence of surplus structure on the mathematical side.

A good representative of the optimistic attitude is Ivor Grattan-Guinness, who has identified three broad cases covered by the applicability of mathematics, in that a mathematical theory can be applied (a) to another piece of mathematics; (b) to a scientific theory, and (c) 'as part of a mathematico-scientific theory to reality' (Grattan-Guinness 1992, p. 92).[34] His analysis is articulated in terms of the similarity and non-similarity of structures, where these may be mathematical, scientific, or the 'empirical interpretations', in 'reality', of the latter, corresponding to (a), (b), and (c), respectively.[35] An unproblematic example of 'intra-mathematical similarity' is the interpretation of $a + b$ as number and lengths (ibid., p. 94), where the non-similarity appears with regard to negative numbers. A more interesting example concerns Descartes' claim that 'all curved lines which can be described by any regular motion, should be received in Geometry' (ibid.) which Grattan-Guinness describes as 'a typical example of an isomorphism between one branch of mathematics and another' (ibid., p. 94). Here too, of course, non-similarity is present between types of algebraic expressions and fundamental types of geometrical curves, leading, Grattan-Guinness claims, to difficulties regarding classifications in algebraic geometry. Recalling Steiner's claims above, it is interesting to note Grattan-Guinness's suggestion that in some

---

[34] With regard to (b) and (c), Grattan-Guinness suggests that instead of talking of 'applied' mathematics, one might use the older term 'mixed mathematics'; for further on this distinction, see Kattau (2001).

[35] With regard to these 'empirical interpretations', Grattan-Guinness refers here to the 'ontological similarities' between mathematical structures and these interpretations.

cases of intra-mathematical structural similarity a crucial role is played by the mathematical notation. He gives the example of Schröder who carried over to algebraic logic the practice in projective geometry of presenting theorems in pairs, related by specific structure-preserving rules (ibid., p. 95). In such cases, the focus is on the formalism itself, and not its referents (or, at least, not in the first place).

An example of structure similarity holding between mathematics and science is that of the use of the differential and integral calculi to model continua (ibid., pp. 96–7). Laplace, for instance, represented cumulative intermolecular forces, on the physical side, by integrals, on the mathematical. Poisson subsequently shifted to a representation in terms of sums, as did Cauchy who argued that integrals could not capture the sensitivity of the effect of such forces on a molecule to the location of other molecules. Here the construction of an appropriate model on the physical side is crucially important for establishing the relationship with the mathematics. Moving in the opposite direction, non-similarity may appear again when physical phenomena are represented by mathematical structures not all of whose terms have a physical interpretation. Here Grattan-Guinness gives the example of the development of the Fourier series to represent, linearly, what are known to be non-linear phenomena (ibid., pp. 97–100). In effecting such a representation, idealizations have to be made on the physical side in order to cast the phenomena into linear form (we shall return to this shortly). Furthermore, such series consist of time terms, multiplied by exponential decay terms and trigonometric terms, involving sine, cosine, etc. The interpretation of the time terms is straightforward but what of the trigonometric terms? If heat diffusion is modelled by such a series, for example, must heat be understood in terms of periodic waves (ibid., pp. 97–8)? Fourier himself refused to make that ontological move, but in the field of acoustics, Bernoulli proposed a 'structure-similarity' between the periodicities and pitch level, an idea that was later developed mathematically by Ohm. Another example is Kelvin's application of the series to the tides, in which the tidal level was calculated as the sum of various components, but 'the trigonometric terms as such were not interpreted as a planetary effect' (ibid., p. 99). These terms can be viewed as aspects of 'surplus structure', in the above sense, where the question as to their physical interpretation is decided on the basis of the relevant context.

In classical physics, this decision is comparatively easy to make, since, it can be argued, we have a good grip on the state of nature we are trying to model to begin with (or, at least, a better grip than we do on the corresponding notion in quantum physics). In the example above, we have a notion of 'the tides' to which the Fourier series can be applied. This notion may not be 'intuitive', of course, having been obtained by means of already mathematized, 'phenomenological' models, perhaps. But underlying it is a comparatively clear idea of the 'state' to which the mathematics is being applied. This idea is then refined 'using a mathematical structure whose very development is motivated by the desire to have a mathematical formalism adapted to expressing the refined intuitive concepts of state' (Sklar 1998, p. 245). It is only by having such a clear idea that we can even talk of the relevant context in terms of which one can decide which terms are surplus and which not.[36]

Thus, consider the example of atoms. In the second half of the nineteenth century, the relevant 'state of nature' was represented by thermodynamical phenomena. The mathematics deployed by Maxwell–Boltzmann statistical mechanics (in particular, the statistical techniques) can be seen as the 'mathematical formalism adopted to express the refined intuitive concepts of state' represented in Clausius's work, for example, and the expressions for entropy in the latter in particular. As far as the 'energeticists' were concerned, the terms referring to atoms within this formalism were representative of just so much surplus structure, devoid of physical reference. It wasn't until the extension of this formalism by Einstein and Smoluchowski to cover Brownian motion that physicists in general came to acknowledge these terms as referring and therefore as 'surplus' no longer.

The problem, however, is that in quantum mechanics we lose that grip on the notion of state (Sklar 1998, pp. 246–9). Recalling our discussion of Schrödinger's wave mechanics above, as is well known, Schrödinger effectively carried over a concept of state derived from classical wave theories into the quantum realm. However, this concept was then undermined by the application of the mathematics itself as it came to be realized that the spatial dispersion of the wave packets could not be reconciled with the spatial localizability of the particles they were describing. This led to a radical reconceptualization of the nature of the physical states that the mathematics was supposed to represent (ibid., p. 246). The example of matrix mechanics fits nicely with Grattan-Guinness's vignettes above, since here the Fourier series was deployed again, in a generalized form (Sklar 1998, p. 247; see also Saunders 1993, again).[37] Driven by the empirical fact that the lines of atomic spectra needed two numbers to characterize them, rather than just one as in the classical case, Heisenberg constructed—via analogy—a two-dimensional form of Fourier analysis and, famously, eschewed consideration of the underlying 'state' of the atom altogether. With the embedding of both representations in a more abstract formalism—to be discussed in Chapter 6—the issue becomes even more acute: what 'states of nature' are the mathematical state vectors of Hilbert space supposed to represent? As Sklar notes, 'this is the notorious interpretation problem of quantum mechanics' (1998, p. 247). With this lack of referential clarity on the physical side, it may become equally unclear what structure is actually surplus and what is not. We may put this point another way: with a firm (or, at least, firmer) grip on the physical

---

[36] This broadly relates to a concern that has been raised: there may be many such structures that would count as 'surplus', so the question arises as to why one would be chosen over the others and subsequently interpreted. In general, this would be decided on a contextual basis. We shall touch on this again in Chapter 2 (and we are grateful to Manuel Barrantes for bringing the concern to our attention).

[37] The Fourier series is the principal example given by Saunders to represent his notion of heuristic plasticity (1993). It is particularly interesting from our perspective because it is used to illustrate a fundamental commonality between classical and quantum mechanics.

states in classical mechanics, it is comparatively straightforward to delineate surplus from non-surplus structure, as in the case of the tides and the trigonometric components of the Fourier series. In the case of quantum mechanics, however, the grip is much looser and elements that, classically, would surely be regarded as surplus, come to be regarded as referring to physical states. A nice example would be the interpretation of the negative energy states in Dirac's relativistic equation as, first, 'holes' in space-time and subsequently, positrons. (We will discuss this case in detail in Chapter 7.)

Returning to Grattan-Guinness's analysis, we can see how the structure-similarity/ non-similarity framework allows for a more gradated understanding of ontological commitment or, as Grattan-Guinness himself puts it, more generally, of a given theory exhibiting different 'levels of content'. It is by paying attention to historical episodes such as the above that:

one can understand how the 'unreasonable effectiveness of mathematics in the natural sciences' occurs: there is no need to share the perplexity of [Wigner] on this point if one looks carefully to see what is happening. [...] The *genuine* source of perplexity that the mathematician-philosopher should consider is the *variety* of structures and of levels of content that can obtain within one mathematico-scientific context.

(1992, p. 105; italics in the original)

The mystery that has so exercised Steiner, and others, following Wigner, is dispelled by a wave of optimism fuelled by the variety of mathematical structures available.

Grattan-Guinness's similarity/non-similarity dichotomy offers the beginnings of an appropriately fine-grained framework for getting to grips with the applicability issue, and we wholeheartedly agree with him that the air of mystery that surrounds this issue dissipates in light of detailed historical considerations. However, this focus on practice has left the more formal treatment behind. In Chapter 2, we shall suggest a way of combining the two which not only formally represents Grattan-Guinness's notions of similarity and non-similarity but also captures the openness of developments on both the physical and, in particular, the mathematical sides as well as Redhead's notion of surplus structure. Before we get there, however, we need to move away from this optimistic stance and acknowledge a cautionary note sounded by Wilson.

1.4 Mathematical Opportunism

The suggestion that applicability can be dealt with simply in terms of relating structures crucially depends on the assumptions that for *any* physical structure there is a corresponding mathematical structure and therefore the relating of these structures will be straightforward. According to Wilson, the optimistic attitude that 'for every physical occurrence there is a mathematical process that copies its structure isomorphically' (2000, p. 297) does not, in fact, adequately capture the manner in which mathematics is actually brought to bear upon the physical sciences.

The obstacles to adopting such an attitude are illustrated via examples of physical propagation processes such as the spreading of spilt milk across a surface (Wilson 2002), or the transmission of waves through an upright post when a weight is placed on top of it. The 'deep' question that has to be faced is whether mathematics can capture the nature of such processes by devices natural to mathematics itself (ibid.).[38] Thus, considering first the example of spilt milk, insofar as the boundary of the spill is expected to stay continuous, it can be modelled by an analytic function which, Poincaré insisted, was all the mathematician needed to model physical phenomena (analytic functions are expandable in power series; ibid., p. 149).[39] However, Hadamard pointed out that analytic functions are:

overly articulated upon a global scale and [...] seem unsuited as models for the more freely distributed spreads of quantities we encounter in the physical world (thus the behavior of an analytic function everywhere can be reconstructed from what any small piece of it is doing, but—presuming that the fluid does not spread signals infinitely fast—the boundary of spilt milk will not manifest such a hidden correlation).   (ibid., p. 150)

Hence, the idea is that an analytic function is globally 'articulated' in a way that the physical process being represented is not—we cannot reconstruct the behaviour of the fluid boundary everywhere on the basis of what a small segment of it is doing. The mathematics, Wilson suggests, displays a kind of 'rigidity', in the sense that it is forced to 'exploit features contained within its descriptions that cannot be plausibly regarded as invariably present in the physical world' (2000, p. 298).

In other words, the claim is that the *mathematics itself* is inflexible and not so easily adaptable to give a representation of the physical situation as people might think. This rigidity might be seen as the source of Grattan-Guinness's 'structural non-similarity', if the phenomenon in question can be unproblematically modelled by a suitable mathematical structure at a low, or 'phenomenological', level to begin with (we shall return to this later). Higher-level structures may then display the kind of rigidity Wilson refers to, leading to the exploitation of features at the higher level that are not invariably present in the phenomenological structure. Of course, one might question whether the phenomena can be unproblematically modelled by

---

[38] One might ask the (meta-)question of just how deep this question is! Have there *ever* been grounds for the optimistic attitude expressed here? How did the initial plausibility of the claim that mathematics can capture the nature of a physical process by devices *natural to mathematics* ever arise in the first place? If one subscribes to the view that much of mathematics is deeply *unnatural*, then it is surprising that such devices could ever work. In other words, why should we be surprised that devices not natural to mathematics have to be brought into play in these sorts of situations?

[39] This example is used to press the deficiencies of lazy optimism: one can cobble together a set-theoretical 'structure' which models the milk boundary, but 'the mere fact that such a set falls within the ontology of set theory does not entail that such a set becomes amenable to *mathematical study* in any interesting sense of the term, any more than the fact that all of the *Uncle Scrooge* comics can be assigned a colossal Gödel number renders Carl Barks scholarship a subdiscipline of recursion theory' (2002, p. 147). However, this may be thought unfair: the point is not whether or not a given set-theoretical structure is amenable to mathematical study, but rather, for the optimist, the issue is whether there is such a structure to be used in applied mathematics.

phenomenological structures to begin with; that is, these lower-level structures may also be 'rigid'. In response, one might insist that there has to be *some* unproblematic and non-rigid modelling at *some* level, albeit very low perhaps, for us to even begin to bring a phenomenon within the purview of science. What needs to be acknowledged here is the multi-levelled kind of modelling that goes on in bringing mathematics to bear on a phenomenon (and again we shall spell this out in Chapters 4–7).

Consider now the example of placing a weight on top of an upright post. This will generate a series of waves whose evolution can be described by a set of 'core' differential equations, together with appropriate boundary conditions (Wilson 2000, pp. 306–7). As it is generally too complicated to include the specific dissipative factors that lead to the waves eventually dying away, scientists typically bundle all these factors under the umbrella of 'frictional effects' and move to the equilibrium situation in which the kinetic energy terms can be dropped from the reformulated core equations. In other words:

No longer do we attempt to predict the final state of the post by mathematically tracking its ongoing temporal evolution, watching its kinetic energy gradually ebb away. Instead we demurely turn our gaze away from the post for a decent interval and calculate what its shape might possibly be *after* the messy process of energy loss has been completed.

(ibid., p. 307; italics in the original)

Now, Wilson continues, this giving up on the attempt to mathematically track the physical process may not put a damper on mathematical optimism so long as the reformulated equations can be regarded as logically subservient to the original core. Unfortunately, however, as he notes, situations exist for which this relationship of subservience is reversed. Suppose the weight is dropped on the post with such force that the material actually fractures (ibid., p. 312). In such a case, we have a clear disparity between the discontinuity of the topology of the post and the smooth evolution mapped out by the core equations.[40] The standard methodology in such cases is to continually monitor the situation, mathematically speaking, to see if unstable cracks are about to form, and when they are, overrule the core equations and restart the process of tracking the behaviour of the post after the onset of fracture. In other words, we must '*stop watching* the bar mathematically' and 'patiently loiter until nature supplies us with a *new set* of boundary opportunities relative to which we can track the smooth evolution of the interior' (ibid., p. 312; italics in the original). According to Wilson, the optimists' hope that all the physical changes in the post can be tracked mathematically has had to be abandoned. Instead, the application of the mathematics must be 'opportunistic':

Applied mathematics has retreated from providing a direct description of nature and replaced it by a methodology subtly tagged to the opportunistic appearance of structures that can be temporarily exploited as boundaries.   (ibid., pp. 312–13)

---

[40] This is another example of 'rigidity' in the mathematics.

Mathematical opportunism is then understood in the following terms: 'it is the job of the applied mathematician to look out for the special circumstances that allow mathematics to say something useful about physical behavior' (ibid., p. 297). This attitude, in contrast with the optimism delineated above, acknowledges that mathematics itself may be too rigid to 'snugly fit' the physical world, and that the mismatch between the two is typically overcome by introducing an 'alien element' on the mathematical side that is not present in the physical world itself (ibid., p. 299). The moral that, according to Wilson, should be drawn is that physical phenomena are modelled by bringing a variety of pieces together, where these various pieces of mathematics may not harmonize sufficiently to serve as a direct representation of any physical structure. Thus he offers a form of 'honest' optimism, warranted by the successful history and practice of applied mathematics itself. From this optimistic perspective, mathematics appears to be 'unreasonably uncooperative' and has to be wrestled into application.[41]

Wilson makes a very useful point here. However, there are two criticisms that can be levelled at the above claims. First of all, with regard to the apparent abandonment of the optimists' fond hopes, the question arises as to why the optimist couldn't just return to the situation of the fractured bar and describe it with the new complex mathematical structure (involving the new set of boundary conditions). From the fact that current practice does not appear to do this it does not follow that it cannot be done this way; that is, by introducing *new* mathematical structures. This is effectively what von Neumann did with his theory of operators and von Neumann algebras, as we shall see: if you need more mathematics, you just make it!

Second, it should be noted that in many cases, and particularly those of importance in theoretical physics, the mathematics is not applied directly to the phenomena, but to *representations* of these phenomena; indeed, as we have already noted, in some cases, such as those we shall be considering, there is a whole nexus of structures lying between the phenomenon itself and the very top level of mathematics that is applied, ranging from models of the data, through models of phenomena to empirical substructures and up to theoretical superstructures.[42] As in the case of the Fourier series

---

[41] For the reasons outlined here, Wilson (2002) finds unattractive the argument from the indispensability of certain mathematical entities to their reality, since it assumes a nice, smooth relationship between science and mathematics. 'On the contrary', he writes, 'it seems to me more likely that mathematics is characterized by its own internal methods of investigation, which can be adapted *only with difficulty* to physical circumstances' (ibid., p. 144; italics in the original). We shall consider this 'Indispensability Argument' in Chapter 8.

[42] Of course, the question then arises: what is the representational relationship between the (mathematized) data models and the (non-mathematized, or 'purely physical') phenomena? Strictly speaking, of course, there can be no relationship of isomorphism (partial or otherwise) between a mathematical structure and something non-mathematical, such as a physical system. Van Fraassen, for example, appeals to the essential use of indexicals within our linguistic practices in order to establish the relevant relationship with the world (2008, p. 3; interestingly he cites Weyl's famous claim that such indexicals are the 'ineliminable residue of the annihilation of the ego' (ibid., p. 89)). This is another profound issue that we cannot respond to in detail here; suffice it to say that it is one that has to be answered by any scheme that

considered above, what is modelled by the analytic functions is not the milk itself but a representation of it. Hence the mismatch between, at the bottom, the physical phenomena and, at the top, the mathematics is overcome by a nexus of intervening structures, beginning with the model of the data itself. It is in the context of such a nexus that the wrestling together of the mathematics and physics takes place.

In response, it might be argued, as we noted above, that even the very lowest-level representation of the phenomena will typically be mathematical in nature, and thus the problem can arise even here, with no opportunity for intervening 'matching'. However, even at this very lowest, 'phenomenological', level fundamental idealizations are introduced, which effectively 'smooth' the matching process. Consider again the example of the spilt milk: to treat it as a fluid with, crucially for Wilson's argument, a continuous boundary, its molecular constitution has to be ignored, *at least in part*. The nature of the idealization introduced even at this stage is nicely illustrated in a standard text on fluid dynamics (Tritton 1977, pp. 42–4). In order to apply the mathematics of continuously varying quantities, the notion of a 'fluid particle' has to be introduced, and the fluid regarded as a continuous aggregate of such particles, each having a certain velocity, temperature, etc. The fluid particle must be large enough to contain many actual molecules yet still be effectively a point with respect to the fluid flow as a whole (ibid., p. 43). Furthermore, although molecules may enter and leave such a 'particle', those within it must be regarded as strongly interacting with each other, else there would be little point in taking them collectively to form such a particle in the first place (ibid., p. 44). Therefore, the mean free path of the molecules must be significantly smaller than the size of the fluid particle, and it is in these terms that the idealization of regarding the fluid in terms of a continuum can be introduced.[43] Only then can the equations of motion of the fluid be formulated on a continuum basis. It is through such idealizing moves that the mathematics and physics are brought together—albeit opportunistically—as we shall again emphasize in later chapters.[44]

What this simple example supports is a view of applicability that offers an alternative account to both 'lazy' optimism as well as opportunism, and which, although in the spirit of Wilson's 'honest' account, also differs significantly from it.[45] It suggests

involves some symbolic system acting in a representational capacity (see French and Ladyman 1999). Again we should thank Manuel Barrantes for raising this question.

[43] Tritton actually refers to this as the 'continuum hypothesis'. We have avoided this phrase for obvious reasons.

[44] Pursuing his concern about what it is that the mathematics is supposed to represent, Sklar notes that in the case of statistical mechanics the fundamental role played by certain idealizations again brings to the fore the issue of what exactly is the 'state of nature' that is being represented (1998, pp. 254–9).

[45] See Stöltzner (2004), who suggests that 'mathematical optimism and mathematical opportunism appear as two modes of a single strategy whose relative weight is determined by the present status of the field to be investigated' (ibid., p. 123). Thus, if a given scientific theory is already 'mature', conceptually or otherwise, then one might well adopt an optimistic stance, availing oneself of the range of mathematical structures on offer to further develop the theory; on the other hand, if the status of the theory is still

that mathematics is not applied in a lazily optimistic manner, in the sense of simply finding a mathematical structure to fit the phenomena in question, since these phenomena themselves first have to be represented by means of some structure and this will typically involve significant idealizations.[46] But neither is it applied quite in the manner that Wilson suggests: the 'unreasonable uncooperativeness' of mathematics is effectively managed via the physical representations of the phenomena. Thus our approach is not dissonant with Wilson's but where he sees uncooperativeness from the standpoint of the history and practice of mathematics, we see effectiveness achieved by means of appropriate physical representations. In a sense then, we are approaching the same issue from different perspectives[47] but the point is: with the introduction of these idealizations, the mathematics doesn't have to 'look out for the special circumstances that allow mathematics to say something useful about physical behaviour'. The special circumstances are effectively *constructed* by means of these idealizations.

Our own more detailed case studies can be seen as further exemplifying this view as 'special circumstances' are effectively created via the introduction of idealizations to allow the mathematics of group theory and Hilbert spaces to be applied to physical systems, for example.[48] Of course, as Wilson himself would be the first to admit, this does not rule out the kind of development that gives succour to a kind of optimism either. Just as developments in classical physics encouraged Euler to generalize the notion of function, so the requirement of constructing an adequate mathematical framework for quantum mechanics led Dirac to come up with his delta function and also von Neumann to create his operator theory. Nevertheless, we agree that the wrestling together of the mathematics and physics into some sort of contact does not result in something that can be regarded as a 'direct portrait' of physical systems and the *philosophical* optimists' reliance on isomorphism to represent the relationship is

provisional, if, as Stöltzner puts it, 'experience with models is fragmentary' (ibid.), then one may find it being axiomatized in an opportunistic manner. Stöltzner calls this approach, following von Neumann, 'opportunistic axiomatics'. We shall return to von Neumann's overall approach in Chapter 6.

[46]   For a response to Steiner that also emphasizes the importance of idealizations in allowing descriptions of empirical phenomena to be embedded in mathematical models, see Azzouni (2000).

[47]   And here again we are grateful to a reader of an early draft for pressing us to be clear on this.

[48]   According to Azzouni (2000, pp. 210–14), Wigner's apparent surprise over the applicability of mathematics evaporates once one appreciates what Azzouni calls the 'inferential opacity' associated with many mathematical systems; that is, we are typically unable to comprehend, in one fell swoop as it were, how a conclusion follows from the premises, even when we have all the steps of the proof. Thus, we are surprised when highly abstract mathematical structures come to be applied, even though they are related to much less abstract systems. A similar point about surprise in this context was made by Wittgenstein (Simons unpublished). However, we would suggest that one aspect of the 'inferential opacity' is effectively dissolved by the kinds of moves we are describing here—it is only when one ignores such moves, or simply takes the starting point of the process (namely the complex mathematical structure) and the end (the mathematically informed theory) without paying attention to the intervening moves on both sides, that one may be surprised as to how the two can be connected. Whether Wigner really was surprised in this sense is another matter.

entirely naïve.[49] What are required, instead, are more sophisticated representational devices, and it is to these that we shall turn in Chapter 2.

Let us sum up what we consider to be the most important aspects of discussions of applicability:

(1) There is, on the one hand, a considerable body of work offering formal approaches to the relationship between mathematics and science, but which fails to adequately consider the relevant practice.

(2) On the other hand, there are also several studies which focus on the practice, but which lack a formal framework.

(3) Within such studies an appropriate historical sensitivity is required if an aura of mystery surrounding applicability is not to be created.

(4) Nevertheless, even the historically sensitive may fall prey to an unwarranted optimism, and the moves that are undertaken in practice *on both the mathematical and scientific sides* need to be taken into account.

In what follows we shall offer a formal framework that is capable of accommodating:

(i) the openness to further developments of both mathematical and scientific theories;

(ii) the nature and role of surplus structure on both the mathematical and physics sides; and

(iii) the kinds of idealizing moves that are made and intervening representations that are introduced in order to wrestle the mathematics and science together.

Furthermore, in terms of this framework, we shall examine, in what follows, three roles associated with the application of mathematics: a representational role, a unificatory role, and an explanatory role. Let us briefly consider each in turn.

(a) *Representational role*: this involves, in general, the use of mathematics for the representation of physical phenomena. We will focus in particular on four categories, as follows:

(a.1) *Applying genuinely new mathematics*: The application of genuinely new mathematics ('new' in the sense of being new to that particular context of application), and the kinds of heuristic moves, idealizations, and analogies that are required to bring the relevant mathematical and physical structures together, as in the case of the introduction of group theory into quantum mechanics, will be examined in Chapter 4.

---

[49] Of course, these representational devices are intended to accommodate the *practice* of the applied mathematician and not to solve the problems of applied mathematics themselves. We will return to the issue of the status of philosophical representational frameworks vis-à-vis the understanding of scientific practice in Chapter 10.

(a.2) *Appropriately representing physical phenomena*: Applying new mathematics from the 'top down', as it were, may not be sufficient, as the relevant physical phenomena need to be represented in a certain 'bottom-up' way in order for the mathematics to be appropriately applied, as in the case of the use of Bose–Einstein statistics to account for the behaviour of liquid helium (Chapter 5).

(a.3) *Applying problematic mathematics*: Sometimes, the application of new mathematics is suspicious, in the sense that it involves elements whose consistency is not transparent, as in Dirac's introduction of the delta function (an earlier example can be found in the use of infinitesimals in initial formulations of the calculus). In such cases, certain dispensability strategies will be deployed (as we shall investigate in Chapter 7).

(a.4) *Interpreting mathematics*: The introduction of some mathematics is typically not enough to account for physical phenomena, and the mathematics needs to be properly interpreted in order for the phenomena to be accommodated (as we shall examine also in Chapter 7 in the context of Dirac's equation).

(b) *Unifying role of mathematics*: the unifying use of mathematics, whereby it is used to bring different domains together, as in the case of John von Neumann's search for a unified mathematical formulation of quantum mechanics, in which the representation of probability, quantum states, and their dynamics together with an appropriate underlying logic could all hang together (Chapter 6).

(c) *Explanatory role of mathematics*: the alleged explanatory role of mathematics according to which certain parts of mathematics are taken to be explanatory of physical phenomena will be examined, with examples ranging from the life-cycles of cicadas, to the bridges of Königsberg, to certain universality phenomena in statistical physics. We shall argue that no such explanatory role can be accorded to the mathematics itself (Chapters 8 and 9).

It is in terms of this framework that we shall consider our examples of mathematical applications in what we hope is a historically sensitive manner. Before we do so, however, we need to introduce our formal approach in the context of various alternatives that might also be considered. We shall begin by sketching the origins of the so-called 'semantic' or model-theoretic approach. This provides the general framework within which our own approach can be situated. We would like to clearly delineate our version of the semantic approach from what we see as two extremes. At one, we have the work of the structuralists who, we maintain, have emphasized highly elaborate formal considerations at the expense of an adequate account of the scientific practice that is supposedly being represented. At the other extreme, we have Ron Giere's approach, which has offered a detailed and interesting account of certain practices, but remains informal when it comes to their representation.

In between these extremes, one can situate Bas van Fraassen's work, which formally represents those aspects of practice it is interested in using the so-called 'state space' approach. The states of systems are represented in a (model-theoretic) state space and laws of succession and coexistence for the system can then be conveniently represented. We see our approach as also situated in the middle ground, but we have different axes to grind: rather than laws of certain theories, we are interested in relationships between theories and, of course, between those of mathematics and science. To that end, we prefer Newton da Costa's partial structures approach, which, we maintain, offers an appropriate formal framework for representing the details of practice we shall be concerned with. Finally, we shall take the above points (i)–(iii) as representing fundamental desiderata that any such formal framework must satisfy to be adequate. In turn, (a)–(c), we take it, are key desiderata for an account of the application of mathematics.

2

Approaching Models
Formal and Informal

2.1 Introduction

As we indicated in Chapter 1, in order to analyse the relationship between science and mathematics, we need an appropriate framework in terms of which this relationship—and the moves made on both sides—can be represented. The so-called 'Received View' offers one such framework, according to which theories are viewed as axiomatic calculi, the language of which contains a sharp distinction within its vocabulary between observational and theoretical terms. Whereas the former are interpreted directly, the latter receive a partial observational interpretation by means of correspondence rules. These function as a kind of dictionary in relating terms of the theoretical vocabulary with terms of the observational vocabulary. A theory is then represented as the conjunction of theoretical postulates within the calculus and the correspondence rules.

According to the usual story that is told within the history of the philosophy of science, it was principally this sharp distinction in the vocabulary that led to the demise of this account, as critics pointed to problems in maintaining the distinction in the context of scientific practice (see e.g. Putnam 1962). However, Frederick Suppe (1989), for instance, has questioned such stories, noting that the Received View does in fact have the resources to overcome these objections. The cause of the decline must therefore be sought elsewhere, and Suppe himself has pointed the finger at the correspondence rules—among other things—which, by virtue of (a) being regarded as a constituent of the theory itself and (b) encoding experimental techniques and practices, have the consequence that a change in such techniques and practices, leading to the addition to or modification of the correspondence rules, implies a change in the theory itself, effectively generating a new theory. What is required, according to Suppe, is an alternative approach to the relationship between theory and—broadly speaking—'observation' and, he claims, this can be found in Patrick Suppes' model-theoretic analysis.

Before we discuss Suppes' work, however, it is worth noting the broader context in which Suppe's point can be situated: the 1960s saw an increased emphasis on aspects of scientific practice which apparently could not be accommodated within the Received View. This can be seen, most obviously perhaps, in Thomas Kuhn's work which not only focused on the overall context in which science is practised, as represented by that sense of the word 'paradigm' that he subsequently labelled 'the disciplinary matrix', but also brought to the forefront the role of the paradigm in its other guise of 'exemplar' or model. It is through the exemplar-as-model that scientists are inducted into their chosen field, and it is through the exemplar-as-model that science 'progresses' in the absence of clear rules. However, Kuhn's work has perhaps overshadowed other critiques of the Received View that also emphasized the heterogeneous nature of the various elements of scientific practice. Peter Achinstein, for example, insisted that the Received View was incapable of accommodating the nature and role of *models* in science (Achinstein 1968).

From the perspective of this framework, a model for some theory is obtained by substituting (partially) interpreted predicates for certain predicate symbols, interpreted individual constants for certain individual constants, etc., in the underlying formal calculus. In this way, the model supplies the 'flesh' for the formal skeletal structure (Nagel 1961, p. 90). Since a *theory* on the Received View is just such a partially interpreted, formal calculus, a model of a given theory T is effectively just another theory that shares the same deductive structure, but differs in interpretation (Braithwaite 1962, p. 225). Typically, this new interpretation supplied by the model will be expressed in terms that are, in some sense, more familiar than those used by the original interpretation of the theory itself, and in this manner the model fulfils what was taken by some to be one of its fundamental roles, namely, that of providing understanding (see e.g. Hesse 1963). Such a claim was vigorously resisted by certain proponents of the Received View, who insisted that models have, at best, a didactic or heuristic function (Carnap 1939, p. 68), and that any understanding they conveyed was, at best, parasitic on the original interpretation of the theory and, at worst, illusory and misleading (Braithwaite 1962). However, this attitude towards models couldn't stand up to the sustained assault directed against it by Achinstein, who pointed out: (a) that not all models are linguistic—some are physical objects; (b) that in some cases the relevant denotation of terms in a given theory and model is not different; (c) that some models are not developed as alternative interpretations of any theory; and (d) that, in some cases, there is no identity of formal structure between theory and model—the relationship typically involves a 'partial' aspect.

With the decline of the Received View, a kind of formal space was opened up for the model-theoretic or 'Semantic' representation of theories, originally developed in the 1940s and 1950s. An outline of the relevant history can again be found in Suppe (1989, pp. 5–15), but, putting things crudely and without going into a lot of detail, it began with E.W. Beth who was concerned to provide 'a logical analysis—in the broadest sense of this phrase—of the theories which form the actual content of the various sciences' (Beth 1949, p. 180). As far as Beth was concerned, the appropriate formalization could be obtained from Alfred Tarski's semantic method, and in a sense this constitutes the heart of the approach:

> The Semantic Conception gets its name from the fact that it construes theories as what their formulations refer to when the formulations are given a (formal) *semantic* interpretation. Thus 'semantic' is used here in the sense of formal semantics or model theory in mathematical logic.
>
> On the Semantic Conception, the heart of a theory is an extralinguistic *theory structure*. Theory structures variously are characterized as set-theoretical predicates (Suppes and Sneed), state spaces (Beth and van Fraassen), and relational systems (Suppe). Regardless which sort of mathematical entity the theory structures are identified with, they do pretty much the same thing—they specify the admissible behaviors of state transition systems.
>
> (Suppe 1989, p. 4; italics in the original)

By the mid-to-late 1970s, Suppe was able to claim that: 'The semantic conception of theories [...] is the only serious contender to emerge as a replacement for the Received View analysis of theories' (1977, p. 709). In the late 1980s, he insisted that 'The Semantic Conception of Theories today probably is the philosophical analysis of the nature of theories most widely held among philosophers of science' (Suppe 1989, p. 3), an opinion that is also shared by a recent critic of the conception (Halvorson 2012, 2013; for a response, see van Fraassen 2014). As a reaction to this hegemonic development, a series of criticisms have been levelled against it, akin to those that Achinstein levelled at the Received View. Thus, for example, it was argued that neither the relation between models and the objects they model, nor certain kinds of scientific models themselves can be captured set-theoretically (Downes 1992); that actual models in science are not related to theories in the manner apparently presupposed by the semantic approach; and, relatedly, that scientific models are in some sense 'autonomous' from theory in a way that cannot be captured by the semantic approach. Each of these objections, together with Achinstein's original concerns, have been responded to using a form of this approach which represents models set-theoretically in terms of 'partial structures' (see e.g. da Costa and French 2003). We shall elaborate this characterization shortly, as we believe it provides a significant framework within which we can represent the relationship between mathematics and the sciences.

Before we do, let us return our attention to the passage from Beth, given above. There are two aspects of this quote that we would like to emphasize here: first, the phrase 'logical analysis' expresses the desire for an appropriate formalization of scientific theories, and, second, the emphasis on the 'actual content' of science reflects a concern with how they are regarded in actual practice. These two phrases effectively encapsulate the parameters of our discussion to follow, in the sense that the variants of the model-theoretic approach we shall consider can be located within a conceptual space delineated by these two aspects. As we shall see, certain forms of this approach present highly formalized accounts of what a theory is, but at the expense of appropriately capturing the 'actual content' of science; whereas others focus on the content, but fail to provide an adequate formal characterization. In our view, both these extremes should be rejected in favour of approaches that are appropriately formal, while retaining the ability to represent science as it is actually practised.

2.2 One Extreme: The Structuralists

The structuralist programme originated with the work of Wolfgang Stegmüller and Joseph Sneed in the 1970s. Curiously, its initial concern lay with the formalization of Kuhn's theory of science.[1] However, over the past several decades, it has effectively extended its remit to encompass not only the formalization of particular branches of science but also many of the classical problems in general philosophy of science, such as: the notion of scientific explanation, the hypothetic-deductive method, the notion of approximation to the truth, and the confirmation and testing of scientific theories, among others.

The heart of this programme is constituted by the claim that the notion of 'theory' can be structurally analysed into distinct concepts, each lying at a different structural level. At the lowest level are 'theory-elements', at a medium level there are 'theory-nets', and at the highest level we have 'theory-holons'. These concepts can be defined in model-theoretic terms as follows. We begin with the class of potential models of the theory, Mp, which satisfy so-called 'frame conditions'. These are general conditions that effectively define the relevant scientific concepts that feature in the theory. Those models that in addition satisfy the laws of the theory are called simply the 'actual models', M. In both cases, the relevant axioms can be summarized in terms of a Suppesian set-theoretical predicate. Furthermore, some of the concepts expressed by a theory are internal to that theory, whereas others are determined from 'outside', as it were. The models satisfying the axioms for these 'outside' concepts are called 'partial potential models', Mpp. In addition, there exist (second-order) relationships between both models of the same theory and models of different theories; the former are called 'constraints', C, and the latter 'links', L. These various components constitute the 'core' of the theory, K:

$$\langle M_p, M, M_{pp}, C, L \rangle$$

However, this is not sufficient to identify a theory. We also need to introduce its 'domain of intended applications', I. Since this will be delimited by 'outside' concepts, the intended applications are conceived of as a subclass of the partial potential models of the theory. If $C_n(K)$ represents the theoretical content of the core of the theory, then the relationship between the theory and its domain of intended

---

[1] After being presented with this formalization at the International Congress of Logic, Methodology and Philosophy of Science in 1975, Kuhn himself declared that it precisely captured his intentions with regard to the notions of paradigm, incommensurability, etc., and in his farewell address as President of the Philosophy of Science Association, he pointed to this relationship with the structuralists as evidence that, like the incoming President Bas van Fraassen, he had long had an interest in, and approved of, the model-theoretic approach in general.

applications can be expressed via the claim that the latter can be subsumed under the theoretical content, or:

$$I \in C_n(K).$$

This claim is, of course, truth-apt, and will typically be false, although one may be able to pick out a subclass of I for which it is true.

Returning to our tri-level notions above, a theory element can now be identified with the pair $\langle K, I \rangle$. However, on the structuralist view, a scientific theory is usually more complex than this, and has further internal structure. It may be more convenient to represent a theory as an aggregate of theory elements, called a theory net. What holds the aggregate together, as it were, is the existence of a common M_p and M_{pp}, together with an ordering relation s between theory elements, which expresses the internal structure of the theory. Typically, according to the structuralists, this structure is strongly hierarchical, with a single fundamental law at the top, as it were, and various 'specializations' of that law underneath, each determining a new theory-element. Hence, a theory-net is a finite set of theory elements ordered by s. This represents the structure of scientific theories *synchronically* (see Moulines 1996, p. 11). Adopting a diachronic perspective, the evolution of theory-nets is represented by the concept of—no surprise here—a 'theory-evolution', which is a sequence of theory-nets, such that (a) the theory-elements of each theory-net in the sequence are specializations of some theory-elements of the previous net, and (b) the domains of application of each net overlap at least partially with the domains of the predecessor net (ibid., pp. 11–12). In addition, theory-nets which are different, in the sense of differing in their classes of potential models M_p, may also be interrelated in such a way that they effectively 'work' together and form what is called a 'theory-holon', representing the most complex unit in the structuralist programme.

We have only outlined the principal formal elements of the programme here. But it should already be apparent that these elements, taken by themselves, do not seem to be particularly 'intuitive', and the connection to science and to scientific practice is far from being obvious. This is not a particularly original remark, of course. As long ago as 1978, Mario Bunge in his review of Stegmüller's book *The Structure and Dynamics of Theories* (Stegmüller 1976) wrote that:

the new philosophy of science advanced in this work is at least as remote from living science as any of the rival views criticized in it.   (Bunge 1978, p. 333)

Likewise, Suppe has remarked that the structuralists display 'a cavalier disrespect for actual scientific practice' (1989, p. 20). And we shall refrain from diving into the vitriol discharged by Truesdell (1984) in his scathing dismissal of the whole programme.[2]

---

[2] Truesdell levels a number of serious, not to say outrageous, accusations against the structuralists in the course of his essay (they are summarized on p. 568), but his two principal complaints concern their

Of course, the structuralists will insist that the 'intuitive' features of scientific practice will be captured through the application of this framework. However, closer analysis reveals that even then the formal presentation of certain notions does not seem to grasp the intuitive content that one usually assigns to them. Let us consider, for instance, the concept of theory links, as presented above. As formally presented, a link is simply an ordered pair of two different potential models. As we indicated, this is taken to capture the 'intuitive' notion of conditions connecting models of different theories. However, one might wonder as to how the association, in an ordered pair, of two different models supplies any 'conditions' connecting them. Surely this is not enough to formally represent a theory link. Similar questions arise in connection to the other concepts presented. The structuralists' analysis of theoretical reduction, for example, has been criticized by Raimo Tuomela for proposing a relationship that is merely formal:

I would not be surprised if some arbitrary and entirely disconnected theories could be proven to stand in this reduction relationship to each other. (Tuomela 1978, p. 220)[3]

There are also concerns as to the extent to which the structuralist approach can be deemed a 'non-statement' view. The 'empirical claim' of a theory, according to structuralism, is expressed by a sentence, the Ramsey-Sneed sentence, which is an existential statement to the effect that certain theoretical and non-theoretical properties of the entities within the theory's domain are related in terms of the prescribed structure (Diederich 1996a, p. 16). Theoretical terms appear only as variables in such a sentence and, it is claimed, 'theoreticity' is thus relativized to the relevant theory itself (p. 16). The Ramsey-Sneed sentence, which may obviously be quite long and complicated, expresses the theory's claim regarding its various applications. Thus, theories are initially thought of in terms of a class of structures, but when they are applied—and it is in this context that one expects such structures to be especially important—the emphasis apparently shifts to the Ramsey-Sneed sentence. In application, at least, it seems as if we cannot get away from linguistic formulations.

The extent to which formalization is taken too far is revealed by the structuralists' attempt to accommodate 'social' or 'pragmatic' considerations by literally introducing the whole scientific community into a theory element! Thus, 'a pragmatic theory-element is a pair $\langle K, SC \rangle$, where K is a theory-element core and SC a scientific

---

misunderstanding and misuse of the history of science and their disregard of formal axiomatizations already presented in the literature of mathematical physics. With regard to the last point, at least, one might attempt a defence of the structuralists on the grounds that their axiomatizations are designed to serve a different—and more explicitly philosophical—purpose than those of mathematical physicists.

[3] Interestingly, Tuomela (1973, chapter 3) also argued that in practice the 'domain of intended applications' was highly idealized and, crucially, 'open' in a manner that required something other than ordinary extensional set theory to capture it.

community' (Diederich 1996b, p. 79). This attempt to represent the community set-theoretically is clearly problematic and surprising. In particular, it is far from clear in what respect taking a scientific community as a set solves any philosophical, methodological, or historiographical problems. On the contrary, there seems to be a category mistake being made here, in which pragmatics is confused with syntax (or, at least, pragmatics is reduced to purely extensional considerations). A similar move was made earlier in Balzer, Moulines, and Sneed (1987, p. 216), where the notion of a diachronic (idealized) theory element was introduced. This is a set-theoretic structure $\langle K, I, G \rangle$, where K is a core, I is the domain of intended applications, and G is a scientific generation; that is, the set of those members of a scientific community who are active within a certain historical period (Diederich 1996b, p. 79). Again, the same point holds: What gains are there to be made in taking the set of a scientific community? What kind of understanding do we acquire with such a manoeuvre?

These moves not only reveal the extent to which the structuralist programme can be judged extreme, but also illuminate an important issue to do with the purpose of formalization in philosophy of science. In an important paper, Suppes (1968) sets out the various reasons why formalization in general, and set-theoretic formalization in particular, might be desirable: the clarification of conceptual problems and the exposure of foundational assumptions; the explicit revelation of the meaning of concepts; the standardization of terminology and methods of conceptual analysis; the elimination of 'provincial' and non-essential features of the representation of theories; the provision of a degree of 'objectivity'; the formulation of a theory in terms of self-contained assumptions; and the establishment of the minimal set of such assumptions. Now, all these reasons can be thought of as having to do with foundational analysis broadly conceived. The general idea is to use set-theoretical formalization to expose the various elements constituting the structure of a theory, and we agree that in this respect it can be extremely effective. However, our discussion here suggests that it might be less effective when it comes to accommodating social or pragmatic factors, at least when it is used in the manner indicated above. Fixated on formalization, the structuralists appear to have lost the sense of when such formalization is appropriate and when it is more illuminating to relax somewhat and shift attention to scientific practice. Of course, such a shift can also go too far, as we shall suggest in our next section.

2.3 The Other Extreme: Giere's Account of Models

Recalling Beth's emphasis on the 'actual content' of science, Ron Giere's intention to capture this aspect is clear:

When viewing the content of a science, we find the models occupying center stage. The particular linguistic resources used to characterise those models are of at most secondary interest. There is no need rationally to reconstruct scientific theories in some preferred language for the purposes of metascientific analysis.   (Giere 1988, p. 79)[4]

This sounds very much like the Semantic Approach and, indeed, Giere himself remarks that his view is very similar to van Fraassen's (ibid.).[5] However, as we shall indicate, it features two elements that create unnecessary problems: his reluctance to characterize the models themselves and the introduction of a linguistic element via his 'theoretical hypothesis'. Before we engage in any criticism, however, we should note the positive aspect of Giere's account.

This consists in not only his attention to scientific practice, through his quite extensive case studies of nuclear physics and continental drift, for example, but more particularly, in his attempt to effectively 'loosen' the relationship between the models and the systems modelled. Typically (see e.g. van Fraassen 1980), this relationship is characterized within the semantic approach in terms of an isomorphism holding between the model and system. However, as Giere notes, in actual practice not only are scientific models not isomorphic to the relevant systems, but the ways in which they are not are explicitly highlighted. Thus an 'introductory' model of the simple pendulum would typically not account for friction or air resistance and even more sophisticated models would remain idealized with regard to the relevant functions representing these factors. Hence, 'if we are to do justice to the scientists' own presentations of theory, we must find a weaker interpretation of the relationship between model and real system' (Giere 1988, p. 81).[6]

Giere's own suggestion for this weaker interpretation is to turn to the notion of 'similarity' that, as he notes, can be regarded as holding in certain relevant respects and degrees. The 'respects' pick out the relevant parameters, such as the length of the string and period of oscillation in the case of the simple pendulum, and the 'degrees' encapsulate the extent of the idealizations involved (and here he appears to favour a purely qualitative expression of these degrees; ibid., p. 81). Acknowledging the standard criticism of similarity as too vague, he then points out that work in cognitive science and neuroscience appears to demonstrate that human cognition operates according to some sort of 'similarity metric' that meshes nicely with his naturalistic philosophy in general (ibid.). Furthermore, it is not difficult to see that this vagueness can be turned to Giere's advantage in allowing him to capture various aspects of scientific practice. Putting this point more specifically, the respects and degrees in which the similarity holds between model and system will be determined by the relevant context, thus allowing other factors—social as well as epistemic—to come

---

[4] Giere also refers to the 'methodological tactic of staying as close to scientific usage as possible' (ibid., p. 84).

[5] Earlier (ibid., p. 62) he cites—favourably—not only van Fraassen but also Suppes and Suppe.

[6] The significance of idealizations is sometimes presented as a recent realization but, of course, it has been appreciated for many years.

into play. As we shall see in Section 2.5, this 'contextual' aspect of similarity can in fact be captured in more formal and less vague terms via partial structures.[7]

The extent to which a model is similar to the relevant system is expressed by means of a 'theoretical hypothesis', which is, of course, a linguistic entity, unlike the models or systems themselves (ibid., p. 80). And unlike the models or systems the hypotheses can be true or false—depending on whether the asserted relationship holds or not— but to focus on truth would be to miss the point:

> To claim a hypothesis is true is to claim no more or less than that an indicated type and degree of similarity exists between a model and a real system. We can therefore forget about truth and focus on the details of the similarity.   (ibid., p. 81)[8]

Thus, a theory on this account, as Giere admits, is a rather heterogeneous entity as it comprises both a family of models, defined by the relevant laws of the theory, and a set of theoretical hypotheses claiming that various of these models are similar to certain real systems. As he says, this implies that a theory is not a well-defined entity in the sense that we cannot give necessary or sufficient conditions for what is included and what is not. At best all we can say is that for a given model to be part of the theory of classical mechanics, say, it must bear a family resemblance to some family of models already in the theory (ibid., p. 86). And it is the collective judgement of scientists that determines whether or not that resemblance is sufficient.

Now, as we have indicated, although this approach has a certain attractive quality, it incorporates two rather 'awkward' features. The first concerns the issue of how models are to be understood; and the second has to do with the introduction of theoretical hypotheses into this framework. Let us take each of these issues in turn.

For Giere, a model is an 'abstract entity', possessing the properties assigned to it in the relevant scientific text. He goes on to claim that this understanding meshes with scientists' use of the term and, furthermore, 'overlaps' with that of the logician (ibid., p. 79). This raises an ontological problem: If a model is an *abstract* entity, how can it be an object in either the scientists' or logicians' sense? Consider the Crick and Watson model of DNA, for example—an example explicitly used by Giere in his introductory text (1979). From Giere's perspective there is 'the' model of DNA, understood as an abstract entity, and then there is the 'actual' physical model, as displayed in that famous photograph with Crick and Watson standing proudly beside it. Presumably the relationship between the two will then be one of instantiation. From the point of view of the semantic approach, however, the term 'model' is defined set-theoretically in the well-known way, and this definition is then taken to

---

[7] Giere also indicates, without going into much detail, that the relations between the models themselves can also be represented in terms of 'similarity' (ibid., p. 86). For a more formal account of this relation, see Bueno (1997), French and Ladyman (1999), and da Costa and French (2003).

[8] In particular, at most all we need is the redundancy theory of truth (ibid., p. 83).

represent the kinds of models we find in science, physical and otherwise (see da Costa and French 2003).

The ambiguity in Giere's understanding of 'model' is revealed by a kind of ambiguity in the literature as to where he should be situated with regard to the semantic approach. He himself has argued that the analysis of modelling given by cognitive science—which he draws upon (Giere 1994)[9]—gives further aid and comfort to the semantic approach and van Fraassen, for example, clearly regards him as a fellow traveller, yet other commentators, such as Herfel, for example, set him apart (Herfel 1995). Herfel's criticisms, although confused in themselves, are illustrative of the difficulties generated by the vagueness over what a model is within the above scheme. Thus, Herfel suggests—bizarrely—that if 'real', physical objects are 'made of' matter, then models, as abstract entities, must be made of 'abstract stuff' (ibid., p. 71).[10] But then, he notes, Giere acknowledges both that 'textbook models', with all their idealizations, do not actually exist and that they are socially constructed. Missing the point with regard to this last phrase, Herfel declares that material objects, such as the Sydney Harbour Bridge, are also 'socially constructed' and hence this offers no clear distinction between models as abstract entities and material objects. By this point in the discussion the reader might well feel a rising sense of frustration, manifested in the thought that if only Giere had been explicit in defining 'model' set-theoretically this perverse miscomprehension might have been avoided!

Having said that, Herfel's principal claim has to do with the explanatory power-lessness of models-as-abstract-entities. Indeed, he claims that Giere's analysis is simply empirically false, since when we look beyond the kinds of textbook presentations Giere himself draws on,[11] we can easily find examples of 'concrete'—that is non-abstract—models in science, such as Lorenz's model of the strange attractor in chaos theory (ibid., pp. 72–4). Of course, as Herfel notes, this model is also mathematical, so 'concrete' is not to be taken as synonymous with 'material'[12]; rather the abstract–concrete distinction is to be regarded as analogous to that between analytically obtained results and numerically obtained results in mathematics (ibid., p. 77). And, of course, the Lorenz example is well chosen, since the relevant (non-linear) equations must be solved iteratively using computer simulations (ibid., p. 79). This latter aspect represents a form of 'numerical experimentation' that, Herfel claims, renders this aspect of scientific practice more amenable to the sort of analysis that Hacking for example gives of experiment, than the one Giere gives of modelling.[13]

---

[9] Autobiographical note: it was in the context of a discussion of this work with students on the HPS MA programme at Leeds, that James Ladyman and Steven French began to think about introducing partial isomorphisms as a means of capturing the relationships between models.

[10] Thus, he repeatedly refers to Giere's account as 'Platonistic' (ibid., pp. 71, 77).

[11] Here Herfel seems to have overlooked Giere's own painstakingly constructed case studies.

[12] Although he later refers to 'concrete material (but nevertheless mathematical) models' (ibid., p. 83).

[13] This provides a kind of 'existential' aspect to the notion of 'concrete' model which Herfel does not make explicit. Thus, he records Hacking's 'entity' realism, which holds that manipulating an entity, in order to create new phenomena, gives one grounds for believing that the entity exists (ibid., p. 75).

Now it is not immediately clear to us that the notion of 'abstract' model *should* be identified with that of 'analytic' solutions to the relevant equations, nor is it clear that the above account is incapable of accommodating the kinds of models that one finds in chaos theory, for example. Giere might be inclined to respond, *à la* Suppes, that what Herfel is pointing to is simply one of the many *roles* of models in scientific practice, rather than a different ontological status. Of course, Herfel would insist that abstract entities cannot be manipulated in the way that these concrete models can but Giere would then argue that what is manipulated is a concrete *instantiation* of the relevant abstract model on a computer (and Herfel does acknowledge the role of instantiation in Giere's account, when it comes to concrete, *material* models (ibid., p. 70)).[14] As far as we are concerned, however, this whole issue of the supposed abstract nature of models is a red herring (but one we have had fun fishing for!); if Giere is embraced back within the semantic approach, then his sense of model can be understood as set-theoretic and representative of various kinds of models in scientific practice—material, mathematical, whatever.

There is now the second issue to be dealt with, namely the introduction of the 'theoretical hypothesis' that asserts a relationship of similarity between the system of interest and one of the models specified in the theoretical definition. It is interesting to note that van Fraassen has also adopted this schema in his works since *The Scientific Image* (see van Fraassen 1989, pp. 6–7, and 1991), although, not being a realist, he is careful to avoid Giere's commitment to similarity per se, merely asserting that there is some kind of relationship between model and system. However, there is a minor problem to be faced by this schema when we focus on highly theoretical models, for example: this is that typically there exists a wide variety and range of other models which intervene or are inserted between the phenomena and the most theoretical structures of a theory.[15] Hence a relationship of straightforward similarity is too simplistic and cannot capture the complexity of these relationships. The solution is pretty obvious, of course: we merely rewrite the theoretical hypothesis as the claim that a part of the world as described in terms of one of these

Presumably, if we understand the kind of manipulation involved in dealing with Lorenz's model as leading to the creation of new phenomena, then Herfel would concede that this gives us good grounds for believing that the model actually exists, as a concrete non-abstract entity. In the text, however, he refers to the ineliminable use of digital computers (ibid., pp. 75, 80), which he also calls 'concrete', leaving the reader to wonder if it is the model which exists, from this Hackingite perspective, or the actual computer in which/on which/where the modelling takes place. We might speculate that the kinds of criticisms levelled at Hacking's claims with regard to the 'manipulation' of entities such as electrons—to the effect that there is no such direct manipulation but rather a reliance on the causal properties of the entities as represented in low-level causal theories, and hence we do not have theory-independent grounds for believing they exist—could be carried over to Herfel's analysis!

[14] See note 13. Herfel's reasoning here can be paraphrased as follows: abstract entities cannot be manipulated; the Lorenz model can, hence it is not abstract. Giere would presumably agree with the premise but deny that what is being manipulated is the Lorenz model itself, but rather a particular instantiation of it.

[15] Morrison (1999) refers to these models as 'mediating' between high-level theory and the phenomena.

'intermediate' models of the phenomena can be regarded as a system of the kind specified in the theoretical definition (Hughes 2010). Nevertheless, there appear to be two further, and more acute, problems that this schema must face.

The first is that this schema in particular, and the semantic approach in general, cannot capture the practice of model building in science since, it is claimed, much of this proceeds in a way which is relevantly unrelated to theory. This latter claim has developed from the assertion that there is a 'phenomenological' level at which this model building takes place, which is independent from theory in method and aims (Cartwight, Shomar, and Suárez 1995), to the more general insistence that certain models are 'functionally autonomous' from theory (Morrison 1999).[16] The assumption behind these claims is that the semantic approach is tied to a view that regards all models as proceeding from, or being constructed from, theory (see Cartwright et al. 1995, p. 139). However, this assumption is false. The semantic view offers a set-theoretic way of representing theories in terms of (formal) models and, further, can similarly represent scientific models in this manner (Suppes 1961; da Costa and French 2003). Likewise, from the perspective of Giere's schema, we shouldn't be misled by the name '*theoretical* hypothesis', since we could easily extend this notion to cover the relationship between a system and some low-level, 'phenomenological' model. How scientific models are constructed in practice is another issue entirely and in order to handle it appropriately one would have to pay due consideration to the relevant heuristic moves involved. Indeed, Cartwright et al.'s own example—the famous London and London model of superconductivity—has been the subject of a detailed case study which has concluded that its construction was *not* independent of theory in methods or aims (French and Ladyman 1997).[17] Similarly, Morrison's examples—of Prandtl's model of viscous flow and the shell model of the nucleus—fail to be autonomous in any sense which presents a problem for the semantic approach (da Costa and French 2000; we shall return to this issue of the supposed 'autonomy' of models in Chapter 4).

The second problem is potentially more serious in that it suggests that this idea of a 'theoretical hypothesis' introduces a linguistic element into the semantic approach and leaves it open to the same sorts of criticisms that were levelled against the Received View's correspondence rules. The point is clearly made in the following passage:

If theories are identified with, or even through, families of models, then the non-linguistic approach is committed to some distinction between theories themselves, and the claims that they can be used to make when applied to real-worldly systems. Classical mechanics, according to Giere's elegant analysis (1988, Chapter 3) consists of hierarchically arranged clusters of models, picked out and ordered by Newton's laws of motion plus the various force-functions. On Giere's view (also endorsed by van Fraassen 1989, p. 222), the relationship between a law-statement and a model is a definitional one: a model is an abstract entity that satisfies the

---

[16] Subsequently, both Cartwright and Suárez expressed a shift in their allegiance to Morrison's account.
[17] The debate has continued with a response by Suárez and Cartwright (2008) and a counter-response by Bueno, French, and Ladyman (2012a) and (2012b).

definition. But neither the definition itself nor the resulting model tells us which physical systems, if any, the model represents, or how. Construed as Giere suggests, classical mechanics makes *no* claims about physical systems. It only identifies a cluster of models: abstract mathematical entities that may or may not have physical counterparts. Giere reinstates the empirical content by means of what he calls "theoretical hypotheses," linguistic items express-ing representational relationships, in specified respects and to specified degrees, between abstract structures and given classes of real systems.... Although Giere insists that we have models "occupying center stage" (1988, 79), in the analysis of theories considered in isolation from their applications, he has to admit that a linguistic element is indispensable if models are to do any representational work.   (Hendry and Psillos 2007, 139–40; italics in the original)

Of course, theoretical hypotheses are not like correspondence rules (ibid.) insofar as they do not incorporate the various experimental procedures involved in relating the theory to the data and hence are not open to the charge that if any of these procedures are changed or new ones added, then we have new correspondence rules and hence a new theory. Nevertheless, Hendry and Psillos do seem to have a point: the theoretical hypotheses are linguistic and they are a constituent of the theory on Giere's view. Giere himself needs to have such a linguistic element in his overall philosophy of science, since he wants to talk about changes in scientists' beliefs in terms of a non-Bayesian 'satisficing' account of belief change. From the perspective of representing the relationships between theories/models or between the latter and the phenomena, however, the theoretical hypothesis is redundant since these relationships can be represented set-theoretically, as we shall shortly see.

Giere's account, then, represents the further extreme of our spectrum, as incorp-orating some nicely detailed case studies (which we haven't the space to go into here) but forgoing the relevant set-theoretic formalism and, as a consequence, getting into trouble. What we shall now do is move to a middle way, that is appropriately formal while also capable of perspicuously accommodating certain episodes from scientific practice. However, before we get there, we need to consider a notion, already mentioned, that will prove crucial to our account, namely that of 'surplus structure'; that is, additional structures that may not initially seem to play any role in repre-senting physical systems, but which subsequently turn out to be heuristically fruitful.

2.4 Surplus Structure

Surplus structure is a significant source of potential developments in scientific practice, and, as we have already had opportunity to note, this notion plays a major role in our analysis. It was introduced by Michael Redhead (1975) via his alternative to the syntactic approach, as an account of theories that represented them in terms of function spaces.[18] A further core element of our approach has to do with

---

[18] This can be understood as a further variant of the semantic approach that bears interesting comparison to van Fraassen's state-space approach. However, to go into details here would take us too far afield.

the openness and partial nature of our theories, and, as we shall see, this, together with Redhead's notion, can be appropriately captured via the partial structures variant of the semantic approach. The importance of the notion of surplus structure is to highlight the significance of further mathematical structure and how, properly interpreted, it can be effectively mapped into the physical world. Indeed, this partial transportation of structure is one of the key features of the application of mathematics to physics and central to this process is the choice of which are the relevant parts to be mapped and how they should be properly interpreted so that the mapping is adequate. As we shall also see, it requires careful attention to the details and appropriate interpretive work to figure that out since sometimes the same structure can be compatible with entirely different physical set-ups.

As the title suggests, Redhead's (1975) important paper, 'Symmetry in Intertheory Relations', is concerned with the relations between the symmetries embodied in various theories in physics. In a certain respect, it represents a direct continuation of the work of Post (1971), who, as we have already noted, attempted to resist the anti-cumulativist philosophies of Kuhn and Feyerabend by emphasizing the structural continuities in theory change.[19] Certain of these continuities were expressed through the role of fundamental symmetry principles—such as Lorentz invariance—which were not only passed on, as it were, from a given theory to its successor, but also transferred between domains. In both these respects symmetry principles are seen as playing an important heuristic role and Redhead's concern is to capture this heuristic role as well as the way in which the symmetries of two theories may be interrelated (Redhead 1975, p. 77).

In meeting this concern Redhead distinguishes physical symmetries from mathematical ones; the former are regarded as characteristic of a theory independently of any particular mathematical formulation, whereas the latter depend on the particular formulation (ibid., p. 85). And in order to explain what is meant by a mathematical symmetry, he needs to address the issue of how mathematics is related to theoretical physics. In order to do *that*, he needs to outline 'a scheme for relating a mathematical structure to a physical theory' (ibid., p. 86). In elaborating this scheme Redhead begins with what he subsequently acknowledges is Tarski's representation of a scientific theory as a set of sentences closed under logical consequence. Hence, he begins his discussion of 'The Relation of Mathematics to Physics' (the title of section 4 of his paper) with the claim that 'we may think of a theory as consisting of axioms and deductive chains flowing from these axioms to produce theorems and finally empirical generalizations to be confronted with experiment as expressed in the form of singular observation statements' (ibid., pp. 86–7). However, Redhead appears to acknowledge the Suppesian concern with expressing scientific theories via a particular formalized language and takes it to be an 'empirical-historical fact' that theories in

---

[19] Feyerabend described it as a 'brilliant essay' and a 'partial antidote' to his own view (Feyerabend 1975, p. 61 note 17).

physics can be represented as mathematical structures.[20] He then immediately considers the possibility of *embedding* a theory T in a mathematical structure M', in the usual set-theoretic sense of there existing an isomorphism between T and a substructure M of M'. M' is then taken to be a non-simple conservative extension of M.

There is an immediate question regarding the nature of T. To be embedded in M' it must already be 'mathematized' in some form or other. Thus, the issue here is not so much Wigner's original understanding of the inexplicable utility of mathematics in science, in the descriptive sense of its being the indispensable language in which theories are expressed, but rather, in line with Steiner's interpretation of Wigner's idea, the way in which new theoretical structure can be generated via this embedding of a theory, *which is already mathematized*, into a mathematical structure. As we shall see, this framework lends itself particularly well to the consideration of the role within physics of symmetries, as represented by group theory.

An uninterpreted calculus C can be introduced of which T and M can be taken as isomorphic models, and likewise a calculus C' for M' can be introduced which drives the introduction of a new theory T' which in turn is partially interpreted via the structure T. This is an immensely fruitful way of considering the heuristic advantages of re-expressing a theory within a new mathematical structure (as we briefly indicated in Chapter 1 with the example of the negative energy solutions of Dirac's equation). In particular, it is here that the notion of surplus structure is introduced for the first time. This is presented as the relative complement of T in T', suitably axiomatized, although it might be more appropriately identified as the extension of the (sub-)structure of the mathematical representation of T' which is isomorphic to T. This meshes nicely with the central idea of T being *embedded* into the 'larger' structure T'.[21] However, the mathematical surplus that, it is claimed, drives these heuristic developments cannot be captured by the formal notion of a structure extension since this simply involves the addition of new elements to the relevant domain with the concomitant new relations. What is required is 'new' structure which is genuinely surplus and which supplies new mathematical resources. We shall return to this point below.

At this point the realist-empiricist debate intrudes: for the positivist, the surplus structure will be the mathematical representation of theoretical terms and relations and for the constructive empiricist it will be the theoretical models lying beyond the empirical substructures, where this theoretical superstructure incorporates physical as well as mathematical elements. For the realist, aspects of this structure may cease to be regarded as surplus as the relevant terms come to be accorded ontological

---

[20] There is a slight ambiguity here as shortly afterwards Redhead refers to the empirical-historical fact that 'physical theories are always *related* to mathematical structures' in the way he has subsequently spelt out (1975, p. 89; italics added).

[21] The framework adopted by Redhead is explicitly a structuralist one, articulated from the perspective of his 'function space' approach, within which laws are taken to establish correlations between events (and here Redhead states that he is following Wigner; see, in particular, Wigner 1964).

reference. Thus, to use again the example of atoms from Chapter 1 (see also Redhead 1975, p. 88), terms referring to atoms come out of the cold and into the 'essential' structure following Smoluchowski's and Einstein's theoretical analyses of Brownian motion—at least as far as the realist is concerned, for the anti-realist may still regard them as 'ontologically surplus', albeit pragmatically useful devices. This is one way of capturing Fine's later idea of a differentiated ontological attitude towards theoretical entities: electrons are definitely in, quarks came over in the early 1970s, whereas magnetic monopoles are still surplus. Presumably it may also go the other way, in the sense that a term may subsequently come to be seen as having been part of the surplus structure all along and hence unworthy of ontological commitment. This may be one way of capturing the idea that certain theoretical terms may come to be regarded as 'idle', in the sense that they are regarded as not doing any useful work when it comes to the derivation of empirical consequences and hence they are seen as ontologically surplus (see e.g. Psillos 1999, chapter 5).

In other cases, as Redhead notes, there may be no possibility of ontological reference, yet the surplus structure may play an essential role in the development of a theory. A particularly well-known example is that of the non-symmetric sub-spaces of the Hilbert space for a collection of quantum particles generated by the action of the permutation group. It is this characteristic of these sub-spaces—namely that they are, or are representative of, surplus structure—that is crucial to Redhead and Teller's argument against 'primitive thisness' or 'transcendental individuality' in discussions of particle identity and quantum mechanics (see Redhead and Teller 1991, 1992; for a critique of their analysis, see Belousek 2000).[22]

One way of formally characterizing this notion of surplus structure, which meshes nicely with Redhead's own emphasis on embedding a structure A into a 'larger' structure B, consists in identifying the surplus structure as the extension of the structure which is isomorphic to A. In other words, in order to embed A into B, there should be a structure C, which is a substructure of B and is isomorphic to A. The *extension* of C with respect to B could then be taken as representing surplus structure. (Note that, according to this proposal, the surplus structure, strictly speaking, is obtained by *extending* the structure C; that is, by adding new elements of its domain and correspondingly new relations. From the formal point of view, it is not a matter of taking the relative complement of the structure concerned, as Redhead suggested, but rather its extension.)

The problem with this characterization is that it simply does not supply the rich framework that Redhead's notion is meant to offer. The principal idea behind this notion is that by embedding a structure into a 'larger' one, we will have a richer

---

[22] Redhead then goes on to develop his function space variant of the model-theoretic approach, as mentioned in note 21, and puts this framework to good use in discussing the difference between mathematical and physical symmetries, their heuristic power, and the relations between the symmetries of different theories.

mathematical context to work in and we can then explore the mathematical resources supplied by such a context. Moreover, given the isomorphism between A and C, we will in fact be entitled to use the extra mathematical machinery in B in order to develop A. However, and this is the difficulty faced by this characterization, the mathematical surplus required to supply this heuristic richness cannot be accommodated by the model-theoretic notion of *structure extension*, since this is nothing but the result of adding new elements to the domain of the structure, together with the corresponding new relations. There should be genuinely *new* structure, as it were, in the surplus structure, and this is not simply a matter of adding new objects and relations, at least not in the mathematical sense. What are needed are new mathematical techniques and resources, or, in other words, a whole family of other structures.

In order to accommodate this last point, we need a framework that is slightly more complex than the one just outlined. Instead of simply embedding a structure A into a larger surplus structure B, the surplus structure is actually to be found in the various further structures $C_1, \ldots, C_n$ into which B is embedded. In that case, the rich mathematical context and the structural variety required by Redhead's use of the notion of *surplus*—emphasizing in particular the heuristic role of mathematics in theory construction—can be represented in terms of the family of new structures $(C_i)_{i \in I}$. Despite its complexity, this framework is perhaps more faithful to Redhead's intentions.

Nevertheless, Redhead's account of the relationship between mathematics and physics cannot be complete, because as it stands it cannot accommodate, in a unified way, the standard ways in which the former is used to develop the latter. It certainly grasps an important aspect of the relationship: the justification of 'filling in' structures employed in physics because of the existence of an isomorphism between these structures and others found in particular domains of mathematics. It is because such structures are isomorphic that we are entitled to use the information supplied by one to articulate the other. However, not every use of mathematics in physics can be reduced to this pattern of embedding. An obvious example is given by the axiomatization of a physical theory T using, for instance, set-theoretic resources and presenting the Suppes predicate for T (for details and examples, see da Costa and Chuaqui 1988 and da Costa and Doria 1991 and 1995). In these cases, there are statements that are true in the resulting axiomatization but which cannot be derived from the relevant axioms. To axiomatize a theory in this sense is to disclose a set-theoretical formula and to consider all the structures that satisfy it. Given the decisive use of mathematics throughout this process, we see this as an important aspect of the relationship between mathematics and physics. It is clear, however, that the use of mathematics at this level cannot be reduced to the mere embedding of one structure into another, given that the aim of the axiomatization is to *determine* such a structure in the first place.[23]

---

[23] Incidentally, this example also raises a problem for Shapiro's account of the applicability of mathematics, which insists that mathematics gets applied by *instantiating* structures in reality (Shapiro 1983 and

In the next section, we shall present the approach we favour which, we believe, appropriately accommodates Redhead's useful notion of 'surplus structure' and allows us to steer a middle course between the over-emphasis on formalism displayed by the structuralists and Giere's focus on practice, at the expense of rigour. However, those of a squeamish disposition need not trouble themselves with the technical details and can skip to the end of this chapter, where we present the general account of applicability that these formal details underpin. The crucial notions we shall deploy are those of partial isomorphism and partial homomorphism. These are relations that are weaker than isomorphism and homomorphism respectively and they capture the way only some of the structure is brought across, as it were, from mathematics to physics. It is via this transfer that mathematics is applied and hence the issue of applicability can be understood.

2.5 Partial Structures

The partial structures approach (see Mikenberg et al. 1986 and da Costa and French 1989, 1990, and 2003) relies on three main notions: partial relation, partial structure, and quasi-truth. One of the main motivations for introducing this proposal comes from the need for supplying a formal framework in which the "openness" and "incompleteness" of information dealt with in scientific practice can be accommodated. This is accomplished first by extending the usual notion of structure, in order to model the partialness of information we have about a certain domain (introducing then the notion of a partial structure). Second, the Tarskian characterization of the concept of truth is generalized for such "partial" contexts, advancing the corresponding concept of quasi-truth. Here we shall focus on the former.

The first step, that paves the way to introducing partial structures, is to formulate an appropriate notion of partial relation. When investigating a certain domain of knowledge Δ (say, the physics of particles), we formulate a conceptual framework that helps us in systematizing the information we obtain about Δ. This domain is represented by a set D of objects (which includes *real* objects, such as configurations in a Wilson chamber and spectral lines, and *ideal* objects, such as quarks). D is studied by the examination of the relations holding among its elements. However, it often happens that, given a relation R defined over D, we do not know whether all the objects of D (or n-tuples thereof) are related by R. This is part and parcel of the "incompleteness" of our information about Δ, and is formally accommodated by the concept of partial relation. The latter can be characterized as follows. Let D be a non-empty set. An n-place *partial relation* R over D is a triple $\langle R_1, R_2, R_3 \rangle$, where R_1, R_2, and R_3 are mutually disjoint sets, with $R_1 \cup R_2 \cup R_3 = D^n$, and such that: R_1 is the set

---

1997). Obviously the axiomatization of a physical theory cannot be accommodated in terms of the instantiation of structures, since what has to be disclosed in this process are precisely the structures themselves.

of n-tuples that (we know that) belong to R, R_2 is the set of n-tuples that (we know that) do not belong to R, and R_3 is the set of n-tuples for which it is not known whether they belong or not to R. (Note that if R_3 is empty, R is a usual n-place relation which can be identified with R_1.)

However, in order to represent appropriately the information about the domain under consideration, we need a notion of structure. The following characterization, spelled out in terms of partial relations and based on the standard concept of structure, is meant to supply a notion that is broad enough to accommodate the partiality usually found in scientific practice.[24] A *partial structure A* is an ordered pair $\langle D, R_i \rangle_{i \in I}$, where D is a non-empty set, and $(R_i)_{i \in I}$ is a family of partial relations defined over D.[25]

A partial structure A can be extended into a full, total structure (instances of which are called A-normal structures). Let $A = \langle D, R_i \rangle_{i \in I}$ be a partial structure. We say that the structure $B = \langle D', R'_i \rangle_{i \in I}$ is an A-*normal structure* if (i) $D = D'$, (ii) every constant of the language in question is interpreted by the same object both in A and in B, and (iii) R'_i extends the corresponding relation R_i (in the sense that each R'_i, supposed of arity n, is not necessarily defined for all n-tuples of elements of D'). Notice that, although each R'_i is *defined* for all n-tuples over D', it holds for some of them (the R'_{i1}-component of R'_i), and it doesn't hold for others (the R'_{i2}-component).

As a result, given a partial structure A, there are several A-normal structures. Suppose that, for a given n-place partial relation R_i, we don't know whether $R_i a_1 \ldots a_n$ holds or not. One way of extending R_i into a full R'_i relation is to look for information to establish that it *does* hold, another way is to look for the contrary information. Both are prima facie possible ways of extending the partiality of R_i. But the same indeterminacy may be found with other objects of the domain, distinct from $a_1, \ldots, a_n$ (for instance, does $R_i b_1 \ldots b_n$ hold?), and with other relations distinct from R_i (e.g. is $R_j b_1 \ldots b_n$ the case, with $j \neq i$?). In this sense, there are *too many* possible extensions of the partial relations that constitute A. We thus need to provide constraints to restrict the acceptable extensions of A.

In order to do that, we need first to formulate a further auxiliary notion (see Mikenberg, da Costa, and Chuaqui 1986). A *pragmatic structure* is a partial structure to which a third component has been added: a set of accepted sentences P, which represents the accepted information about the structure's domain. (Depending on the interpretation of science which is adopted, different kinds of sentences are to be introduced in P: realists will typically include laws and theories, whereas empiricists will add mainly certain laws and observational statements about the domain in

---

[24] For the avoidance of confusion we should be clear that what we mean here by 'partiality' is incompleteness rather than unfair bias (thanks again to one of the readers for raising this).

[25] The partiality of partial relations and structures is due to the "incompleteness" of our knowledge about the domain under investigation—with further information, a partial relation may become total. Thus, the partialness modelled here is not "ontological", but "epistemic". (Although with suitable adjustments to the interpretation of the formalism, an ontological partiality can also be modelled if needed.)

question.) A *pragmatic structure* is then a triple $A = \langle D, R_i, P \rangle_{i \in I}$, where D is a non-empty set, $(R_i)_{i \in I}$ is a family of partial relations defined over D, and P is a set of accepted sentences. The idea is that P introduces constraints on the ways that a partial structure can be extended (the sentences of P hold in the A-normal extensions of the partial structure A).

Our problem is, given a *pragmatic* structure A, what are the necessary and sufficient conditions for the existence of A-normal structures? Here is one of these conditions (Mikenberg et al. 1986). Let $A = \langle D, R_i, P \rangle_{i \in I}$ be a pragmatic structure. For each partial relation R_i, we construct a set M_i of atomic sentences and negations of atomic sentences, such that the former correspond to the n-tuples that satisfy R_i, and the latter to those n-tuples that do not satisfy R_i. Let M be $\cup_{i \in I} M_i$. Therefore, a pragmatic structure A admits an A-normal structure if and only if the set $M \cup P$ is *consistent*.

With these concepts in hand, we can now define the notions of partial isomorphism and partial homomorphism that will be crucial for our analysis of the applicability issue.

In empirical research, we often lack complete information about the domain of inquiry. We know that certain relations definitely hold for objects in the domain, and others clearly don't; but for a number of relations we simply don't know (given our current information) whether they hold or not. Typically, we have at best partial structures to represent our information about the domain under investigation (and this includes not only empirical structures, but also theoretical ones). As a result, partial structures help us to accommodate the partiality of our information (see French 1997).

But what is the *relation* between the various partial structures articulated in a particular domain? Since we are dealing with partial structures, a second level of partiality emerges: we can only establish *partial* relationships between the (partial) structures at our disposal. This means that the usual requirement of introducing an isomorphism between theoretical and empirical structures can hardly be met. Relationships weaker than full isomorphisms, full homomorphisms, etc. have to be introduced, otherwise scientific practice—where partiality of information appears to be ubiquitous—cannot be properly accommodated (for details, see Bueno 1997, French 1997, and French and Ladyman 1997).

Following a related move made elsewhere (in addition to the aforementioned papers see also see French and Ladyman 1999 and Bueno, French, and Ladyman 2002), it is possible to characterize, in terms of the partial structures approach, appropriate notions of *partial isomorphism* and *partial homomorphism*. And because these notions are more open-ended than the standard ones, they accommodate better the partiality of structures found in science. Here they are:

Let $S = \langle D, R_i \rangle_{i \in I}$ and $S' = \langle D', R'_i \rangle_{i \in I}$ be partial structures. So, each R_i is of the form $\langle R_1, R_2, R_3 \rangle$, and each R'_i of the form $\langle R'_1, R'_2, R'_3 \rangle$.

We say that a partial function $f: D \rightarrow D'$ is a *partial isomorphism* between S and S' if (i) f is bijective, and (ii) for every x and $y \in D$, $R_1 xy \leftrightarrow R'_1 f(x) f(y)$ and $R_2 xy \leftrightarrow$

$R'_2 f(x) f(y)$. So, when R_3 and R'_3 are empty (that is, when we are considering total structures), we have the standard notion of isomorphism.

In terms of this framework, we can also represent the hierarchy of models (Suppes 1962)—models of data, of instrumentation, of experiment—that take us from the phenomena to the theoretical level (Bueno 1997, pp. 593–607):

$$A_k = \langle D_k, R_{k1}, R_{k2}, R_{k3}, \ldots R_{kn} \rangle$$

$$A_{k-1} = \langle D_{k-1}, R_{(k-1)1}, R_{(k-1)2}, R_{(k-1)3}, \ldots, R_{(k-1)n} \rangle$$

$$A_3 = \langle D_3, R_{31}, R_{32}, R_{33}, \ldots, R_{3n} \rangle$$

$$A_2 = \langle D_2, R_{21}, R_{22}, R_{23}, \ldots, R_{2n} \rangle$$

$$A_1 = \langle D_1, R_{11}, R_{12}, R_{13}, \ldots, R_{1n} \rangle$$

where each R_{ij} is a *partial relation* of the form $\langle R_{ij}^1, R_{ij}^2, R_{ij}^3 \rangle$—with R_{ij}^1 representing the n-tuples that belong to R_{ij}, R_{ij}^2, the ones that do not belong to R_{ij}, and R_{ij}^3, those for which it is not defined whether they belong or not—such that, for every i, $1 \leq i \leq k$, card(R_{ij}^3) > card($R_{(i+1)j}^3$) (Bueno 1997, pp. 600–3). The partial relations are extended as one goes up the hierarchy, in the sense that at each level, partial relations which were not defined at a lower level come to be defined, with their elements belonging to either R_1 or R_2.[26]

Moreover, we say that a partial function $f: D \rightarrow D'$ is a *partial homomorphism* from S to S' if for every x and every y in D, $R_1 xy \rightarrow R'_1 f(x) f(y)$ and $R_2 xy \rightarrow R'_2 f(x) f(y)$. Again, if R_3 and R'_3 are empty, we obtain the standard notion of homomorphism as a particular case.[27]

Using these notions, we can provide a framework for accommodating the application of mathematics to theory construction in science. The main idea, roughly speaking, is to bring structure from mathematics to an empirical set-up. In other words, mathematics is applied by 'bringing structure' from a mathematical domain (say, functional analysis) into a physical, but mathematized, domain (such as quantum mechanics). What we have, thus, is a structural perspective, which involves the establishment of relations between structures in different domains. (Note that, at the level we are considering here, the physics is already mathematized. A similar story also holds for the mathematization of a given domain in the first place, in which one would represent salient features of the domain to make it amenable to a mathematical treatment.[28]) But, given the 'surplus' structure at the mathematical

---

[26] For an additional application of this framework to the idea of partial conceptual spaces, see Bueno (2016).

[27] Note that the notions of partial isomorphism and partial homomorphism can also be formulated in second-order logic. Hence, given Boolos's (1985) plural interpretation of second-order quantifiers, they are acceptable to a nominalist.

[28] Recall Chapter 1, note 33.

level, only *some* structure is brought from mathematics to physics; in particular, those relations that help us find counterparts, at the empirical set-up, of relations that hold at the mathematical domain.[29] In this way, by 'transferring structures' from a mathematical to a physical domain, empirical problems can be better represented and examined. This assumes, of course, that we have information about the relations between the two domains, although it may well not be complete. Only the information we know to hold is used; but there are other relations at the mathematical level that may not have any counterpart at the empirical level.

It is straightforward to accommodate this situation using partial structures. The partial homomorphism represents the situation in which only some structure is brought from mathematics to physics (via the R_1- and R_2-components, which represent our current information about the relevant domain), although 'more structure' could be found at the mathematical domain (via the R_3-component, which is left 'open'). Moreover, given the partiality of information, just *part* of the mathematical structures is preserved, namely that part about which we have enough information to match the empirical domain.

The central point, then, is that by bringing structure from one domain into another, the application of mathematics has a heuristic role in theory construction: it suggests a way of searching for relations at the empirical domain, given their counterparts at the mathematical level. Obviously more can be said but this has given enough of the formal framework in which our account can be situated and, as we emphasized above, all the reader really needs to keep in mind is the idea that only some of the structure available at the mathematical level is brought across to the physical; that this can be understood in terms of establishing appropriate relations between these levels, where these relations are weaker than isomorphism; and, of course, that it is by means of this transfer of structure that mathematics is applied. The details will be spelled out shortly but let us first consider certain problems that have been raised against the above account.

2.6 Problems with Isomorphism

Let us begin by noting that both realists and anti-realists agree that scientific theories should be *empirically adequate*.[30] As we have noted, within the semantic approach, van Fraassen (1980, 1987) has characterized this notion in terms of isomorphism: in order to be empirically adequate a scientific theory should have a theoretical model

---

[29] As we noted previously (Chapter 1, note 36), there may be a question as to *which* structure is brought from the mathematics to the physics in this way and the answer, as we indicated there, is that this will depend on the context. As we say here, the issue is to focus on those relations that will have the appropriate counterparts at the empirical level, and deciding which get transferred will, crudely, be a two-step process: first of all, only certain such relations will be made available by the mathematics itself, of course; second, of those that are, broadly heuristic factors of the kind we hope to illuminate in our various case studies will be deployed to select those that get transferred. We will return to this point in later chapters.

[30] Of course, they will then disagree as to whether this should be understood as the *aim* of science, whether theories should be true *simpliciter* as well as empirically adequate, and so on.

such that all the appearances (the structures described in experimental reports) are *isomorphic* to the empirical substructures of this model (that is, to the theoretical structures which represent the observable phenomena). However, this notion may be criticized for being both too restrictive *and* too wide. Such a characterization is claimed to be too wide because given two structures of the same cardinality, there are *always* isomorphisms holding between them.[31] The problem then is to rule out those that are not interesting (see Collier 1992). On the other hand, this characterization is taken to be too restrictive since there have been cases in the history of science where scientific theories were taken to be empirically adequate despite the fact that there was no isomorphism holding between the appearances and the empirical substructures of such theories (see Suárez 1995). Both of these problems can be addressed within our framework.

In answer to the first problem, it has already been argued that the partial structures approach restricts the isomorphisms involved in a claim of empirical adequacy, by stressing the role of partial isomorphism between the appearances and the theoretical structures, and that this illuminates heuristic considerations at the level of theory pursuit (see French 1997, especially pp. 42–51; da Costa and French 2003). Here we develop one aspect of this point and spell out this suggestion that the concept of partial isomorphism, as opposed to total isomorphism, can be used to address the problem of width. Bluntly put, the idea is that some of the uninteresting isomorphisms could be ruled out on the grounds that they do not represent appropriately the relationships holding between those structures characterizing the appearances and their theoretical 'counterparts'. They map, so to say, 'inappropriately' the relations of the former onto relations of the latter. The crucial notion of 'inappropriateness' is a pragmatic concept, which depends on certain aims and expectations. For purely 'extensional' considerations there will be many isomorphisms holding between two structures, but not all of them will be appropriate or interesting. So what we need, besides the purely formal requirement that an isomorphism holds between appearances and empirical substructures, is to invoke certain pragmatic considerations to determine which of them are interesting or appropriate.[32]

It is possible to understand such pragmatic considerations in terms of *heuristic* moves employed in theory pursuit:

What we have when we compare different models as to their similarity is a time-slice of a heuristic process: the family of relations projected into the new domain forms the basis of the partial isomorphism formally holding between the relevant models. And the nature of this set—*which* relations to project—is delineated by the sorts of heuristic criteria outlined by Post.

(French 1997, p. 51; italics in the original)

---

[31] As is now well known (thanks largely to Demopoulos and Friedman 1985), this lies behind both Putnam's model-theoretic argument, and Newman's objection to Russell's structuralism.

[32] This, in a slightly different form, is one of the points of van Fraassen (1997).

As we recall, Post (1971) examined a variety of theoretical guidelines for theory construction and development, such as simplicity, the preservation of accepted symmetry principles, and, most contentiously perhaps, the General Correspondence Principle, understood, as we have noted, as the demand that a new theory must reduce to its predecessor in the domain successfully modelled by the latter (ibid., p. 228). Of course, one might substitute all or some of these criteria for others but our intention here is not to argue for one heuristic package over another. The point is simply that the partial structures framework (with the notion of partial isomorphism) supplies a formal setting in which such heuristic considerations can be accommodated, in the sense that there is room to represent the application of heuristic criteria (which of course are *not* formal!) to the (partial) models under consideration. Let us see how this works with respect to the issue of empirical adequacy, but beginning with how we understand symmetry principles within this kind of framework.

As we just noted, and as we shall see in later chapters, symmetry plays an important heuristic role in modern physics and one useful device for capturing that role can be found in van Fraassen's elegant account of how such principles can be used as problem-solving devices (see van Fraassen 1989 and 1991). Of course, for an empiricist, such as van Fraassen, that is all there is to such principles, whereas the realist will insist that those principles for which there is strong empirical support represent objective (typically unobservable) features of reality (and for the structural realist of course, these features are part of the core ontology; see French 2014). As before, we will not be taking a definite stance on either of these broad positions but, again, will suggest that our account is (for the most part) compatible with both. Our aim, then, is to examine the ways in which physicists such as Weyl, Wigner, and von Neumann used symmetry principles in their work as a crucial problem-solving tool.

Both the realist and empiricist will agree that the success of symmetry-based considerations in establishing a number of important results in modern physics is not by chance. However, it is important to recognize that the notion of symmetry has different meanings, depending on the context in which it is used, ranging from a broad, general model-theoretic sense (that can be used to systematize the different uses of the term) to more specific senses, restricted to particular domains of physics. In this section, we shall briefly review the general sense of symmetry as a transformation that leaves the relevant structure invariant. Relevance is, of course, a contextual matter, and this generates more specific forms of symmetries, which depend on what is taken to be relevant in a given context. Later we will have the opportunity to contrast the general sense of symmetry with its more specific meanings.

If we characterize the notion of symmetry in this way (as a transformation that leaves the relevant structure invariant), it becomes clear that it bears a close connection with representation. In order for one even to formulate the notion of symmetry, it is presupposed that some account of representation is adopted, since we need some such notion to talk about 'relevant structure'. In other words, it is only in a representational context that symmetry has its bite. In Chapter 3, we will present our own account of representation in general but in the current context it will be useful to recall how van Fraassen incorporates symmetry considerations into his version of the model-theoretic approach.

As we have pointed out, according to van Fraassen, to present a scientific theory is to specify a class of structures, its models (see van Fraassen 1980, 1989, and 1991). These models are used to *represent*, in particular, the states of a certain physical system. The states are characterized by physical magnitudes (observables) that pertain to the system and can take certain values (van Fraassen 1991, p. 26). Consider the example of classical mechanics: here the history of a system (its evolution in time) can be represented as a trajectory in the space of possible states of the system (its state-space). Such a trajectory is a map from time to the state-space. As an example, consider a classical mechanical system (see Varadarajan 1968). Its states can be completely specified by a $2n$-tuple $(x_1, \ldots, x_n, p_1, \ldots, p_n)$, where $(x_1, \ldots, x_n)$ represents the configuration and $(p_1, \ldots, p_n)$ the momentum vector of the system in a given time instant. Thus, the possible states of the system are represented by points of the open set nn (where n is the n-dimensional space of n-tuples of real numbers). The system's evolution is then determined by its Hamiltonian. In other words, if the state of the system at the time t is represented by $(x_1(t), \ldots, x_n(t), p_1(t), \ldots, p_n(t))$, the functions $x_i(t), p_i(t)$ (with $1 \leq i \leq n$) satisfy the equations:

$$(1) dx_i/dt = H/p_i, and\ dp_i/dt = -H/x_i (\text{with } 1 \leq i \leq n).$$

The set of all possible states of the system satisfying (1) gives us its state-space.

What is important in this representation is that it provides an immediate way of studying logical relations between statements about the system's states. Van Fraassen calls these statements, whose truth conditions are represented in the models of the system, *elementary statements* (1991, pp. 29–30). As an example, let A be the statement 'the system X has state s at time t'. Now, A is true exactly if the state of X at t is (represented by a point) in the region SA of its state-space. In other words, the state-space provides information about the physical system X. The crucial point for van Fraassen is that the family of statements about X receives some structure from the geometric character of the state-space. For instance, in quantum mechanics, if A attributes a pure state to X, the region SA it determines in the state-space is a *subspace*. In this way, by investigating the structure of the state-space, one investigates the *logic* of elementary statements (van Fraassen 1991, p. 30). Of particular interest are, of course, the following logical relations: (i) the elementary statement A *implies* the elementary statement B if SA is a subset of SB; (ii) C is the conjunction of A and B if SC is the glb (greatest lower bound), with respect to the implication relation, of SA and SB. Van Fraassen's state-spaces represent information about the structure of a physical system, by focusing, in particular, on empirical information and the study of the underlying logic of statements about these systems.

But what is the *point* of these representations? Again, both the realist and empiricist will agree that, at least in part, it is to solve problems (where they will then disagree on whether this is the whole of the point, as it were). This is quite clear in van Fraassen's case, and he stresses, in particular, the heuristic role of *symmetry* in theory construction (see van Fraassen 1989, and 1991, pp. 21–6). The idea is that in order to solve a problem we have first to devise an adequate *representation* of it: a model of the situation described by the problem. This model introduces certain variables, and what we are looking for is a rule, a function, from certain *inputs* (the data of the problem) to one, and usually only one, *output* (the problem's 'solution'). The representation of the problem brings certain symmetries, which are crucial for its solution. Now, it often happens that we may not know how to solve a problem P_1 (stated in terms of the model we initially devised), but we may well know how to solve a *related* problem P_2. One way of solving P_1 is then to show that it is 'essentially' P_2. This requires that we provide a transformation (let's call it T) from P_1 into P_2 that leaves P_1 'essentially' the same. What this means is that the structure that characterizes P_1 (the relationships between its variables) is preserved by T. The underlying heuristic move is then clear: the 'same' problems have the 'same' solutions (van Fraassen 1991, p. 25). The symmetries of the problem, being preserved by T, allow this heuristic move to get off the ground.

The crucial point that we want to emphasize is the understanding of symmetries as, in general, transformations that leave the relevant *structure* invariant. In this way, the two components of the above problem of structure meet: by specifying certain *interrelationships* between structures (appropriate symmetries), we have a powerful technique to solve problems, which in turn increases our ways of *representing* information in mathematics and in science. Thus, structure can play a decisive role in solving problems in both science and mathematics. One of the advantages of moving to a structuralist stance derives from the rich heuristic setting it provides not only for the scientist and the mathematician (since structures have a crucial role in problem-solving), but also for the philosopher (given that, moving one level up, structures also allow us to represent how problems are solved). This is a point we shall have opportunity to consider in more detail below.

Turning back now to our account and using partial isomorphisms, we can thus delineate 'the sorts of considerations that are typically drawn upon to rule out the uninteresting and unworkable isomorphisms in pursuit' (French 1997, p. 46, note 18; da Costa and French 2003). This point is made in connection with the accommodation of Hesse's notion of analogy:

One way of understanding this relationship between relationships [...] [that is, the relationship between the relations R_i in a theoretical model and those in the models of phenomena] is in terms of an isomorphism holding between some members of the R_i of one model and some members of the R_i of the other. Thus, we might say that there is a partial isomorphism between the sets of relations concerned [...], and this is another way of explicating what Hesse meant by a 'positive' analogy. The 'negative' analogy of course refers to the dissimilarities between the R_i in the two cases and the 'neutral' analogy takes in those relations which are not yet known to either hold or definitely not hold in both cases.   (French 1997, p. 47)[33]

So, the idea is that a partial isomorphism spells out the connections between the ('incomplete') information we have from the empirical level, on the one hand, and the way we represent the phenomena from our theoretical models, on the other. We recall that a partial isomorphism leaves open those relations about which we do not have enough information, suggesting what can be important lines of inquiry to be further pursued (an obvious place to apply heuristic strategies).

It might seem that since there are more partial isomorphisms than isomorphisms between any two structures, use of the former in the semantic approach makes the problem of width worse. Indeed, at least as defined in Bueno (1997), the usual notion of isomorphism is a special case of partial isomorphism. So, given two structures, we shall have more partial isomorphisms than full isomorphisms.

This again is the kind of consideration which assumes that science could be understood in terms of a purely extensional characterization of its main concepts, and underestimates, not to say simply disregards, the role of heuristic and pragmatic criteria in scientific practice. Given the interplay stressed by the partial structures approach between heuristics and partiality, it is natural to expect that heuristics plays a crucial function at this point. Indeed, one of the roles of theory construction (especially in an empiricist setting) is to articulate, in a reasonable way, the connection between theory and evidence. The partial structures formalism is advanced in order to supply a framework in which the 'openness' and 'partiality' involved in this process (in particular, with regard to our empirical information) can be accommodated. And we stress this point, for in scientific practice the construction of at least one theory that is empirically adequate (in van Fraassen's sense) is already a difficult task. So difficult that, in some instances, a theory is accepted as being empirically adequate despite the fact that no isomorphism holds between the appearances and the empirical substructures of any model of the theory.

So, let us now turn to the second problem, namely that our characterization is too restrictive, given the cases from the history of science where theories were taken to be empirically adequate despite the lack of any isomorphism holding between the appearances and the empirical substructures of such theories. One possible answer has already been given in Bueno (1997), where it was argued that a convenient, less

---

[33] Of course, French's point is that if there is a partial isomorphism between the structures under consideration—given that such an isomorphism is defined by taking into account the three components of a partial relation R, the n-tuples that satisfy R (R_1-component), the n-tuples that do not satisfy R (R_2-component), and the n-tuples for which we do not know whether they satisfy R or not (R_3-component)— the positive, negative, and neutral components of Hesse's notion of analogy can be accordingly explicated. The idea is that if a partial isomorphism holds between the structures concerned, all three components are simultaneously considered. The partial isomorphism maps partial relations onto partial relations, preserving the *incompleteness* of information represented in each structure.

strict notion of empirical adequacy can be formulated in terms of partial isomorphism, and thus we can adjust the empiricist account with the records of scientific practice. However, we shall not return to this issue here.[34]

Instead, we wish to consider a related line of argument that runs as follows. The use of isomorphism in order to characterize the notion of empirical adequacy is inappropriate, given that what actually happens is that the cardinality of the domains of the empirical substructures and of the models of phenomena is generally not the same.[35] It is, therefore, simply pointless to try to establish any isomorphism between them. Thus, any characterization of empirical adequacy which is formulated in terms of this is unacceptable. We call this argument the cardinality objection, and it can be seen as the formal counterpart to Suárez's argument against van Fraassen. However, even if we grant that this objection is sound, it does not follow that we cannot articulate a characterization of empirical adequacy having roughly the same formal features as van Fraassen's. Indeed, we can do so, through the notion of a partial homomorphism. The main idea is that, in terms of this notion, a broader version of empirical adequacy can then be formulated, and given that a partial homomorphism, as opposed to an isomorphism, does not depend upon structures whose domains have the same cardinality, the associated notion of empirical adequacy overcomes the difficulty put forward by the cardinality objection. The technical details have been given above but our central claim is that in terms of this framework we can accommodate the hierarchy of structures in science, from the high level of mathematics to the so-called phenomenological level.

With these issues resolved, we can now turn to the central theme of the book, namely the apparently surprising effectiveness of mathematics in its application to science.

2.7 The Inferential Conception

Our overall framework can now be used to underpin a general account of the applicability of mathematics that runs as follows: we begin by emphasizing the fundamental role that mathematics plays by allowing us to make appropriate inferences. It is by embedding certain features of the empirical world into a mathematical structure that it becomes possible to obtain inferences that would otherwise be extraordinarily hard (if not impossible) to obtain. Now, this doesn't mean that applied mathematics doesn't have other roles (Pincock 2012). These range from unifying disparate scientific theories through helping to make novel predictions (from suitably interpreted mathematical structures) to being used in explanations of empirical phenomena (again from certain interpretations of the mathematical formalism), examples of which we have been exploring here and will explore in later chapters.

---

[34] For a further discussion of Suárez's point, see also French (1997), p. 47, note 22.
[35] Such domains in general are finite.

All of these roles, however, are ultimately tied to the ability to establish *inferential relations* between empirical phenomena and mathematical structures, or among mathematical structures themselves. For example, when disparate scientific theories are unified, one establishes inferential relations between such theories, showing, for example, how one can derive the results of one of the theories from the other—this is, arguably, one of the points of unifying the theories in the first place. Similarly, in the case of novel predictions, by invoking suitable empirical interpretations of mathematical theories, scientists can draw inferences about the empirical world—leading, in certain cases, to novel predictions—that the original scientific theory wasn't constructed to make.

Whether novel predictions are used to support a realist reading of the theories involved or whether the anti-realist can also make sense of them, the important point for us here is that both parties should agree about the importance of *inference* in the production of novel predictions. Finally, in the case of mathematical explanations, inferences from (suitable interpretations of) the mathematical formalism to the empirical world are established, and in terms of these inferences, the explanations are formulated. These cases illustrate the crucial *inferential role* that applied mathematics can play.

To accommodate this inferential role, it's crucial to establish certain mappings in the form of partial homomorphisms (or other morphisms) between the appropriate theoretical and mathematical structures, with further partial isomorphisms (or suitable morphisms) holding between the former and structures lower down in the hierarchy, all the way down to the empirical structures representing the appearances, at the bottom (as explicated above). We then have the following three-stage scheme, called the *inferential conception* (see Bueno and Colyvan 2011, Bueno 2014, and references therein):

(a) The first step consists in establishing a mapping from the physical (empirical/ theoretical) situation to a convenient mathematical structure. This step can be called *immersion*. The point of immersion is to relate the relevant aspects of the theoretical situation with the appropriate mathematical context. The theoretical situation is taken very broadly, and it includes the whole spectrum of contexts to which mathematics is applied. (In the limit, this includes mathematical contexts as well, such as when mathematicians apply set theory to arithmetic in order to obtain results about the latter.) As we will see, several mappings can do the job here, and the choice of mapping is a contextual matter, largely dependent on the particular details of the application. Typically, such mappings will be partial, due to the presence, not least, of idealizations in the physical set-up and these can be straightforwardly accommodated, and the partial mappings represented, via the framework of partial isomorphisms and partial homomorphisms we have presented here.

(b) The second step consists in drawing consequences from the mathematical formalism, using the mathematical structure obtained in the immersion step.

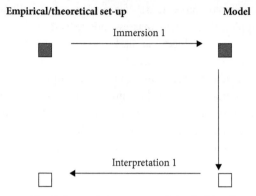

Figure 1 The Inferential Conception of Applied Mathematics

This is the *derivation* step. This is, of course, the key point of the application process, where consequences from the mathematical formalism are generated.

(c) Finally, we interpret the mathematical consequences that were obtained in the derivation step in terms of the initial theoretical set-up. This is the *interpretation* step. To establish an interpretation, a mapping from the mathematical structure to that initial theoretical set-up is needed. This mapping need not be simply the inverse of the mapping used in the immersion step—although, in some instances, this may well be the case. But, in some contexts, we may have a different mapping from the one that was used in the immersion step. As long as the mappings in question are defined for suitable domains, no problems need emerge. (This can be seen as extending Hughes' (1997) account of scientific representation to the application of mathematics while avoiding his use of denotation, which seems inappropriate in the scientific context; see French 2003a and Bueno 2006.) This sequence of steps can be diagrammed as in Figure 1.

Of course, the description above is idealized in two respects, since the immersion and the interpretation steps aren't always as clean as is suggested: (i) the mathematical formalism often comes accompanied by certain physical 'interpretations'; and (ii) the description of the physical set-up is often made already in mathematical terms. We are not assuming, in the immersion and the interpretation steps, that the physical set-up and the mathematical structures are completely distinguished components, at least not at the level of the scientific practice. The crucial point is that however the physical set-up and the mathematical structures are formulated, the application process involves establishing mappings between them. And this is the point of the immersion and the interpretation steps.

Furthermore, as we just indicated, there may be considerable choice about the mappings used in both the immersion and interpretation stages. In both cases the decision about the choice of mappings will be a matter of context, and pragmatic considerations come into play. Take the immersion stage first. If, for example, we wish to determine the combined mass of a number of objects, we should use an interpretation mapping that assigns masses—not space-time locations, not lengths, nor anything else—to each object. Similarly, at the interpretation stage we need to make a decision about how to interpret the sum obtained, because, in the un-interpreted mathematics, the sum is just a real number. Nothing forces the interpretation of this number to be the combined mass of the objects in question. Of course, in an example like this, the choices in each direction are clear. In other cases, there may be more than one way to get the desired result. And, as we've already said, the choice of interpretation will not, in general, be the inverse of the immersion map.

2.8 Conclusion

The case studies we shall present in later chapters should all be understood in terms of this schema. As we shall see, although the details become quite complicated—as one would expect from examples drawn from scientific practice—the underlying themes we shall be focusing on can all be made apparent: namely, that any apparent 'mystery' to the applicability of mathematics swiftly dissolves once we pay attention to the details of the actual practice; that the apparent indispensability of the relevant mathematics is typically only that—apparent; and that there is no need to adopt a realist attitude towards mathematical entities given the role the mathematics plays. Of course, one may still want to adopt such an attitude on other grounds but that is a different matter.

Before we move to our case studies, however, we need one further feature of our framework, namely an account of scientific representation that can accommodate the representational capacity of both scientific models and mathematics.

3

Scientific Representation and the Application of Mathematics

3.1 Introduction

We insist that any attempt to represent the relationship between mathematics and science must accommodate the points we have noted in Chapters 1 and 2. Our central claim is that such an appropriate representation can be found in the semantic or model-theoretic approach in philosophy of science, which represents theories in terms of families of mathematical models, characterized via partial structures and which captures the relationship between the mathematical and scientific via partial isomorphisms and homomorphisms, as just set out in Chapter 2.

One of the crucial features of scientific models, of course, is that they incorporate certain idealizations, as has been extensively discussed in the relevant literature (French and Ladyman 1998; McMullin 1985; Weisberg 2007—just to cite a few!). Here we will follow Frigg and Hartmann (2012) in distinguishing between 'Aristotelian' and 'Galilean' idealizations. The former involves the non-inclusion within the model of certain properties of the system, because these are deemed not to be relevant to the problem at hand. Thus one might construct a mechanical model of a system in which the component entities only have position and mass, for example, thereby effectively ignoring other properties they possess. Galilean idealizations, on the other hand, involve deliberate distortions of certain properties—the classic example being that of the frictionless plane (Frigg and Hartmann 2012). Obvious questions then arise—particularly for the realist—as to what such models might say about how the world is. An equally obvious answer is that such models can always be made more 'realistic' via some de-idealizing move (see Weisberg 2007, where this move is set out very clearly), but as we shall see in Chapter 9, it has been claimed that this is not always possible and thus that some idealizations, at least, are not dispensable in this manner.

Setting that last concern to one side for now, the more general question that arises is: How might we represent the relationship between the idealized model and the 'non-idealized' system? In terms of the partial structures approach, what we have is a relationship between parts of the structures in each case; putting it rather crudely, certain of the R_i which, strictly speaking, will typically be different for the idealized and non-idealized cases, are set the same. As we saw in Chapter 2, this relationship can be represented in terms of 'partial isomorphisms' between the relevant structures.

As we shall see in Chapter 4, the applicability of group theory to nuclear physics, for example, depended on the nucleus being regarded as an assembly of indistinguishable particles. Thus, in this case, the R_i will of course be different, because the system comprises both protons and neutrons, but treating these as the same kind then allows for the introduction of an important new physical property, namely isospin.

This framework then gives us the basis for representing the relationship between mathematical and scientific theories, viewed structurally. In applying a mathematical theory to physics, we are often 'bringing in' structure from the mathematical level to the physical. What the existence of a partial isomorphism establishes is that *some* structure is carried over in this process: namely, that involving the R_1- and R_2-components, for which enough information is available. However, there might not be a complete structural preservation when we move from one level to the other. As already noted, typically mathematics brings some 'surplus structure' with respect to physics. But a partial isomorphism, as defined in Chapter 2, requires a *bijection* between the domains of the structures under consideration. So only structures of the *same cardinality* can be partially isomorphic. As a result, it is difficult to accommodate the existence of *structural surplus* in terms of partial isomorphisms: a broader notion is required.

To motivate this we note that, as will become clear in the context of the case studies to be explored in Chapter 4, both the relevant physics and the mathematics developed in an *open-ended* manner and the applicability of the mathematics to the physics depended on certain relationships being established *within the mathematics itself*;[1] furthermore, *not all* of the mathematics is brought down to the physical level, as it were. Our claim, then, is that the appropriate way of accommodating these three features is to modify our account by introducing partial homomorphisms, as outlined in Chapter 2.

Now the three points above can be straightforwardly accommodated. First, the open-ended nature of the development of both physics and mathematics can be represented by the use of *partial* structures. In both cases, in a given instant of time, a number of issues are known to be the case, others are known not to be, and a huge amount is simply not known at all. A partial structure represents this situation, indicating, in particular, the accepted information in that context via its R_1- and R_2-components whereas the R_3-component points out where further research must be done. In terms of this latter component, the open-ended nature of both disciplines can then be accommodated.

Second, a partial homomorphism is a particular *mathematical* relationship between structures: it is a map that 'transfers' some relations from a structure into

---

[1] This adds to our comments above in Chapter 1, note 38 and Chapter 2, note 29.

another. As we shall see, the application of group theory to quantum mechanics, for example, crucially depended on the reciprocity relation between the representations of the symmetry group and the unitary group. This reciprocity provides a mathematical relation between such structures, one which allows us to 'transfer structure' from one domain into another. Partial homomorphisms capture this kind of transfer.

Third, and crucially, it's not *all* of mathematics that is used in a particular application. Some parts of the theory are crucial, but some are not.[2] And it is because of this surplus provided by the mathematics with respect to the physics that partial homomorphisms become decisive. Intuitively, there is 'more structure' at the mathematical level than at the physical. It's no wonder then that structures are 'mapped' from the former to the latter.

The representation of aspects of the physical world by mathematical structures is, of course, a particular case of the general issue of representation in science. Two questions immediately come to mind: How should the notion of representation be understood? Is there anything special about the use of *mathematical structures* in the representation of phenomena? Without any claim to completeness, we shall briefly address these issues, since they will help us to systematize a few points.

The issue of scientific representation has become very much a 'hot topic' within the philosophy of science. Formal accounts have been proposed involving isomorphisms, partial isomorphisms, and the like (French 2003; van Fraassen 2008) and have been criticized as inadequate in various respects (Suarez 2003; Frigg 2006). The account we favour is essentially based on the schema given in Chapter 2, involving the three steps of 'immersion', 'derivation', and 'interpretation', with mappings in the form of partial homomorphisms and partial isomorphisms holding between the relevant structures. We shall review and address the criticisms that have been aimed at this sort of account below. It seems to us that when considering the kinds of mathematical representations discussed here, some kind of formal account along these lines is unavoidable.

3.2 General Challenge: Reducing all Representations to the Mental

Let us begin by considering a very general criticism that insists that any such formal account should be rejected, as all forms of representation, whether in art or in science, can be reduced to one fundamental kind. Thus, Callender and Cohen insist that:

we don't need separate theories of representation to account for artistic representation, linguistic representation, scientific representation, culinary representation, and so on [and perhaps this

---

[2] It is tempting to speculate *why* some parts of group theory happen to be more crucial than others in quantum mechanics. Part of the answer arises from the structure of quantum mechanics itself, and the relations it imposes on the objects studied. As we saw, symmetry considerations are crucial for the representation of quantum particles. This is, of course, one of the most central features of group theory. The results of the latter that *don't* depend on symmetry considerations *might* not be so crucial. (We insist on the 'might', since such results may turn out to be important for other reasons.)

includes mathematical representation], but rather [they claim] that all these sorts of representation can be explained (in a unified way) as deriving from some more fundamental sorts of representations, which are typically taken to be mental states. (Callender and Cohen 2006, p. 7)

Since representation via mental states can then be described on independent grounds, the need for an account of scientific representation—such as that offered by partial structures—is eliminated.

The heart of their strategy consists in the claim that the 'vehicle' of representation represents the 'target' of the representation in virtue of the fact that the vehicle produces (in the mind of the relevant person) a mental state with the appropriate content. Which state is produced, and hence which target is represented, is ultimately, for Callender and Cohen, a matter of stipulation, although in practice some vehicles turn out to be more convenient to use than others. So, Bohr, for example, could have chosen a Manchester tart to represent the atom but instead chose the particular model he did as it was simply more convenient for creating the appropriate mental state in the minds of his intended audience or readership. Thus the choice of models, and in particular the choice of scientific models, comes down to a matter of pragmatics (where this may include empirical factors).

Now this is a bold account, particularly in the current context, but whatever the merits of such a reductionist move, one can easily see how it glosses over precisely the sorts of differences between forms of representation that we might be interested in when we consider scientific theories and models, as well as the mathematics that is brought to bear. Such differences were noted by Peirce, among others, who distinguished between three kinds of relation between representations and what they represent, corresponding to three kinds of signs: the iconic, in which the representation 'resembles', in some sense, the represented object (such as works of art, diagrams, and scientific theories); the indexical, in which the representation is *caused* by the represented object (examples include barometers, the symptoms of disease); and the symbolic, for which the relationship is stipulative (as in Western European languages). Callender and Cohen seem intent on collapsing all forms of representation into the last but, not surprisingly, that leads them into trouble, as when they present a road sign warning as a putative counterexample to the isomorphism account we defend here. Whereas the former can indeed be regarded as a stipulation determined by convention, the differences between that and, say, a work of art such as Holbein's *The Ambassadors*, or an example from science, such as Bohr's model of the atom—differences which Peirce and others express through the distinctions noted above—are so striking as to function as part of a *reductio* argument against any view which fails to appreciate them.

And, indeed, as Contessa (2007) notes, Callender and Cohen give the game away by acknowledging that in many cases our choice of a particular vehicle of representation cannot purely be a matter of stipulation as this choice is underpinned by considerations of convenience which in turn depend on an appropriate relationship holding between the vehicle and the target.

Consider the example of conventionalism with regard to geometry and space-time physics. For Poincaré, it was a matter of convention whether to use Euclidean or non-Euclidean geometry. Of course, he chose the former on grounds of simplicity, but in both cases an appropriate relationship had first to be established between the geometry and the phenomena being represented. Hence, it was not just a matter of stipulation, which, in this respect, is typical of cases of scientific representation. Callender and Cohen themselves note that it is more convenient to use an upturned right hand to represent the state of Michigan than a saltshaker because of the relevant geometric similarity. But then our preferences reveal something fundamental about the mechanism of representation—which incorporates the relevant relationship—in such cases. Callender and Cohen's reductionism effectively skips over this and leaves us none the wiser as to how either artworks or theories function as representational devices. The bottom line is that, within the category of what Peirce referred to as iconic representations, there appear to be differences between works of art and scientific theories which bear on the issue of the nature of the representational relationship and that is what we are interested in, rather than some reductionist move which is too coarse-grained to allow us to analyse such differences.

Contessa (2007) uses his criticism of Callender and Cohen as a springboard to 'disentangle' mere denotation from what he calls 'successful epistemic representation', where the latter involves 'correct surrogative reasoning' (see Swoyer 1991) by which one uses the vehicle of representation to learn about the target. Such reasoning is valid only if it is in accordance with a set of rules that supply an interpretation of the vehicle in terms of the target.[3] According to Contessa, it is only when a user adopts such an interpretation that we have a representation.

Thus, he uses as a mini-case study a comparison of Rutherford's model of the atom with Thomson's 'plum pudding' model: the latter proposed that an atom was constituted by a cloud, or 'pudding' of positive charge, in which sit the negatively charged electrons, like sub-atomic 'plums', whereas the former suggested that the positive charge—and most of the mass of the atom—were concentrated in a central core (which subsequently came to be known as the nucleus) with the electrons dispersed around it. The Thomson model can be regarded as an unsuccessful representation since it generated a false conclusion about the atom: the Rutherford model allows us to infer that the kind of scattering observed by Geiger and Marsden would occur, whereas the Thomson model does not. Thus the latter misrepresents the atom, whereas the Rutherford construct is more successful. Contessa's point is that both could be taken to denote the atom but it is only the latter that constitutes a 'successful epistemic representation' since it incorporated correct surrogative reasoning (with regard to the scattering phenomenon, for example).

We shall discuss misrepresentation below, but it is worth noting that Rutherford's model was not completely successful, of course (not least because it did not

---

[3] The rules in question need not be explicit; as is well known, rule-following is notoriously problematic.

appropriately represent the electrons as orbiting). In this regard, it is perhaps significant that Contessa does not continue the history and introduce the model that was taken to be even more successful than Rutherford's, namely Bohr's, famously regarded as inconsistent (although this has been contested; see da Costa and French 2003; Vickers 2013). It turns out that even inconsistent representations can be handled by the partial structures approach (see Bueno and French 2011), a result that Callender and Cohen seem to treat as constituting a form of *reductio* of the whole position. However, contrary to what they claim, it is not the fact of inconsistency that drives us, in (purported) desperation, to partial isomorphisms. Rather, having constructed the partial structures account, it counts in its favour that as well as accommodating more straightforward cases, it can also handle inconsistent theories.

Let us consider in particular Callender and Cohen's worry that any account of representation must distinguish between coherent and incoherent putative representations. Inconsistent representations may then be seen as examples of the latter and therefore any account that accommodates inconsistency will fail to make the necessary distinction. This is not correct, however. If we take the practice of science seriously then we need to be able to accommodate putative examples of inconsistent theories that arise in the history of science. Of course, it may turn out that in these cases the theories are not actually inconsistent, but we can accommodate such examples also—ultimately, they don't involve any inconsistency. However, if the theories are in fact inconsistent, the partial structures framework allows us to accommodate them, as we have said (see Bueno and French 2011). Furthermore, we can also distinguish inconsistent and incoherent representations. An inconsistent representation (one from which a contradiction of the form 'A and not-A' follows) can be coherent as long as it is not made trivial by the resulting inconsistency, that is, as long as not everything follows from the latter. The possibility of inconsistent but non-trivial representations emerges from the fact that the underlying logic of the partial structures account is paraconsistent (see da Costa, Bueno, and French 1998). And if the inconsistent representations in question turn out to be explanatory, empirically adequate, and formulated via principles that are logically interconnected, such representations can be considered coherent, despite their inconsistency. We shall return to consider similarly problematic representations in Chapter 7.

Thus, we see it as an *advantage* of our approach that it can handle these kinds of representations as well. Let us now move on to more specific criticisms of the kind of formal account that we favour, before detailing how we respond to them.

3.3 Specific Challenges to the Formal Account

3.3.1 Necessity

Perhaps the most well-known criticism proceeds from the argument that neither isomorphism, nor similarity in general, can be necessary for representation, since the latter may occur without any relevant similarities, or isomorphisms, being in place. If we agree that anything is similar to anything else in certain respects, then some 'criterion of relevance' (Suárez 2003, p. 235) is needed to identify the similarities that are meant to constitute representation. Relevance, of course, is not an intuitive notion. So, considering representation in art, it has been argued that in Picasso's iconic painting, *Guernica*, for example, none of the similarities that can be seen—'a bull, a crying mother holding up a baby, an enormous eye' (Suárez 2003, p. 236)—appear to be directly relevant to the targets of the representation, which can be identified with the bombing of Guernica and, more abstractly, the rising threat of fascism in Europe. The same is true, it is claimed, of an equation representing a system, where the marks on a piece of paper have no relevant similarity with the system being represented.

More specifically, it is argued, isomorphism cannot be necessary for representation since there exist examples of artworks that cannot be held to be isomorphic to *anything* (Suárez 1999). In the case of *Guernica*, again, it can be argued that it is ambiguous regarding precisely what the painting represents: on the one hand, it represents the 'concrete pain' of the inhabitants of Guernica, but on the other it also represents the 'more abstract threat' noted above. Hence it has been argued that it cannot stand in any one-to-one correspondence with (and thus be related via isomorphism to) those things that it is taken to represent (ibid., p. 78). Another example is that of certain works of Mondrian, such as *Composition No. 10*, with its famous grid-like structure, and it is suggested that to focus on representation in such cases is to miss the point of the art in that the aesthetic and emotive responses that this type of abstract painting induces do not arise in virtue of its 'representing' anything (ibid., p. 79).

3.3.2 Sufficiency

Even in cases where there is a relevant similarity, or isomorphism, it is claimed that representation may nonetheless fail to hold. Suárez urges that for A to represent B, A must lead an 'informed and competent' enquirer to consider B (2003, p. 237), so there exists an 'essential directionality of representation'. Given that similarity and isomorphism are logically symmetric, it is argued, they cannot explain how 'consideration of the target [leads] to consideration of the source' (Suárez 2003, p. 236). Hence isomorphism-based accounts are insufficient for representation.

3.3.3 Logic

Relatedly, and following Goodman's (1976) work on depiction, similarity and isomorphism are both symmetric and reflexive: 'a glass of water [for example] is similar to itself, and similar to any other glass that is similar to it' (Suárez 2003, p. 233). Representation, on the other hand, does not share these properties. Even if 'an equation represents a phenomenon, the phenomenon cannot be said to stand for the equation' (Suárez 2003, p. 232), and likewise, we would not claim that the phenomenon represented itself.[4] These concerns then lead into the next.

3.3.4 Mechanism

This derives from the fundamental question: How do models, or structures in general, represent their targets? (Frigg (2006) calls this the 'enigma of representation'.) Thus the further worry is that since isomorphisms only hold between structures, we must assume that the target 'exhibits' the relevant structure, and this will involve a descriptive element that renders the representational relation not wholly structural. Furthermore, depending on which description we choose, the target system will exhibit different non-isomorphic structures. Thus, consider the example of the shape of the methane molecule, for instance, in which four hydrogen atoms are bonded to a central carbon atom (Frigg 2006). It has been argued that depending on whether we take the vertices as objects and the edges as relations or vice versa, we get two different structures: 'The upshot of this is that methane exemplifies a certain structure only with respect to a certain description and that there is no such thing as *the* structure of methane. And this is by no means a peculiarity of this example' (Frigg 2006, p. 58; italics in the original).

3.3.5 Style

Here the concern is whether there are any constraints on the choice of style of representation in science (which Frigg calls the 'problem of style'; ibid.). It is claimed that the emphasis that proponents of structuralist accounts of representation put on isomorphism suggests that they adopt the normative view that representation must be understood in these terms. But, it is then argued, this is implausible since many representations are inaccurate in various ways (e.g. through idealizations) and also because of the possibility of misrepresentation, neither of which the isomorphism account can handle (Frigg 2006). Curiously, perhaps, it is asserted that the same holds for amended versions of the isomorphism account, such as ours, even though—given our emphasis on the partiality of information—that patently seems not to be the case. We shall return to this shortly.

3.3.6 Misrepresentation

In this case it is argued that similarity and isomorphism cannot account for the two types of misrepresentation that are encountered in practice, namely mistargeting and inaccuracy. When it comes to mistargeting, consider the following example, again from the world of art, that of Velázquez's painting of Pope Innocent X. Clearly the painting represents the Pope. Now, suppose that a friend were to disguise himself to look like the Pope (Suárez 2003). Then, upon seeing the canvas, we might mistakenly

---

[4] This seems an obvious point, given the directionality of representation. We shall consider below how one can capture this directionality within our approach.

suppose it to represent the friend. Neither similarity- nor isomorphism-based accounts can account for this misrepresentation, as there are still relations of similarity, or isomorphisms, holding between the canvas and the friend.

With regard to inaccuracy, consider the example of Newtonian mechanics used as a representation of the solar system. Here it is claimed that a similarity formulation cannot account for the fact that 'without general relativistic corrections, [Newtonian mechanics] can at best provide an approximately correct representation of the solar system' (Suárez 2003, p. 235); that is, it does not account for the quantitative differences between theory and experiment. It seems that what is being asked for here is a theory of representation which not only accounts for the respects in which the model is similar to the system it represents, but also defines the respects in which, and more importantly the extent to which, it is dissimilar. Thus, 'the interesting question is not what properties fail to obtain, but rather how far is the divergence between the predictions and the observations regarding the values of the properties that do obtain' (Suárez 2003, p. 235).

3.3.7 Ontology

This concern has to do with the issue of the ontological nature of the representational devices we are considering and it is alleged that structure-based accounts such as ours are deficient in two respects. First, it is argued that as a result of the above, and in particular of those concerns we have labelled 'mechanism' and 'style' (but to which we might also add 'necessity' and 'sufficiency'), the ontology of such devices, particularly if they are taken to be models, cannot be structural (Frigg 2006).

Second, it is argued that the variety of such representational devices that we find in practice precludes their accommodation within the kind of account we favour. Thus consider the following cases: an engineer's toy bridge representing a real bridge, a graph of a bridge again representing a real bridge, a billiard ball model representing the dynamical properties of a gas, and a quantum state diffusion equation, which represents the quantum state of a particle undergoing diffusion. It is not clear where either similarity or isomorphism enters into the account of how such a plethora of cases represents (Suárez 2003).

Let us now consider how the partial structures account might respond to these concerns in turn.

3.4 The Formal Account Defended

3.4.1 Preamble: Informational vs Functional Accounts

As Chakravartty (2009) emphasizes, the apparent dichotomy between the kind of account we favour—which he calls 'informational'—and that which is advocated by others, such as Suárez—which he designates as 'functional'—may in fact only be a confusion of means with ends, in the sense that the former is concerned with what scientific representations *are*, and the latter has the issue of what we *do* with them as its focus.[5] We do not dispute that the latter is a fundamentally significant issue and agree in fact that an important role for models in science is to allow scientists to perform the kind of 'surrogative' reasoning emphasized by Swoyer (1991), for example. Indeed, we would claim that representing the 'surrogative' nature of this reasoning effectively rides on the back of the relevant partial isomorphisms, since it is through these that we can straightforwardly capture the kinds of idealizations, abstractions, and inconsistencies that we find in scientific models.

We shall introduce our approach via Hughes' DDI account (Hughes 1997 and 2010), according to which scientific representation is a matter of denotation, demonstration, and interpretation (DDI). In particular, a model stands for, or refers to, a physical system, and elements of the model denote elements of that system. Through the 'internal dynamics' of the model we 'demonstrate' the relevant results (yielding novel predictions and the like). These results that are demonstrated within the model then have to be interpreted in terms of its subject, and we can determine whether our theoretical conclusions correspond to the phenomena, or not. In fact, although it may be true in the world of art that almost anything may stand for anything else, this is not the case in science: not anything can serve as a scientific model of a physical system since if the appropriate relationships are not in place between the relevant properties, then the 'model' will not be deemed scientific to begin with (French 2003).

Our account (see also Bueno and Colyvan 2011) overcomes this concern by replacing denotation with 'immersion', by which a mapping is established between the empirical set-up and the representational structure. In the situation we are considering here, this would be given in terms of partial isomorphisms.[6] This stage is hence explicitly 'informational' in Chakravartty's terms, given that there is information transfer from the empirical set-up to the (mathematical) model. Consequences are then drawn using the representational structure, in the 'derivation' stage, and it is here that the 'functional' role of the representation is exemplified, given that the model is used in order to derive particular results. The final step is 'interpretation', taking the results of the derivation stage back to the target system via another mapping (which may be different from the first), which is given in terms of partial isomorphisms (or some other suitable mapping). The focus here is, once again, on the 'informational' aspect, given the reverse information transfer from the (mathematical) model, suitably interpreted, to the empirical set-up. This three-stage framework thus explicitly brings the 'informational' and 'functional' approaches together, in a way that satisfies Chakravartty's suggestion. However, it crucially

---

[5] Of course, one could define what representations are in terms of what we do with them. This formulation of the 'functional' account of representation makes no room for any informational element. And given that representation does seem to have both informational and functional features, Chakravartty's distinction highlights an important point that is worth preserving.

[6] But other kinds of morphisms, such as partial homomorphism, can be—and often are—invoked, as we have already indicated (see e.g. Bueno, French, and Ladyman 2002).

depends on the relevant mappings being established. For the cases under consideration here (i.e. scientific theories and models and the systems they represent), such mappings are best articulated via partial isomorphisms (or partial homomorphisms).

The core point, as should now be clear, is that to base one's account of representation on surrogative reasoning or inferences without accepting the underlying formal aspects is, as Chakravartty notes, to engage in a confusion. Indeed, we would go further and insist that such a move makes no sense: it is only within some account of the formal relationships involved that we can understand how the relevant reasoning can be appropriately surrogative.

Furthermore, the above criticisms of informational accounts are less than decisive. Let us consider them now.

3.4.2 Necessity

First of all, and related to the above point, Chakravartty points out that it cannot be the case that similarity in general is not necessary for scientific representation, since in the absence of such, the successful facilitation of the usual practices of interpretation and inference—subsequently characterized by our account—would be nothing short of a miracle. Here it is the requirement that such facilitation be accommodated that effectively demands that there be some relation of similarity that holds between the representational device and the target system. Moreover, this relation must be such as to permit the appropriate inferences without which it would be hard to talk about *scientific* representation in the first place. The relevant mapping involved will typically be such as to map parts of the particular empirical set-up into the model and vice versa (in accordance with the inferential conception, for instance). Here and elsewhere we argue that partial isomorphism and homomorphism offer an especially appropriate way of characterizing this relation, in the scientific context (see Bueno 1997; Bueno, French, and Ladyman 2002; da Costa and French 2003). It is with regard to the above requirement that scientific representation differs from the artistic (at least insofar as it is not one of the primary aims of artistic representations to facilitate interpretation and inference) and hence putative counterexamples such as *Guernica* lose their force.

It should be noted, however, that there has to be some partial isomorphism between the marks on the canvas and specific objects in the world in order for our understanding of what *Guernica* represents to get off the ground. That understanding may vary with our historical, political, and social sensibilities, but the basic mechanism of representation involves the kinds of relations we have highlighted here. In cases such as this, these sensibilities drive the symbolic dimension of the artistic representation, but in the sense in which this dimension is understood in art, it is not as central to science. Symbolic representations may play a heuristic role in science, but when it comes to the practices of interpretation and inference touched on above, the relevant work is done by the underlying mechanism of representation (the various types of morphisms).

3.4.3 Logic and Sufficiency

With regard to the 'logical' objection, partial isomorphism is an equivalence relation if we consider only two (partial) structures, but not if we consider the domain as a whole, where other factors break the symmetry. These other factors are broadly pragmatic, having to do with the use to which we put the relevant models; even in the case where partial isomorphism is an equivalence relation, this use is crucial. This is, in effect, to repeat the point made previously and that also acts as a response to the non-sufficiency objection: namely, that we agree that partial isomorphism is not sufficient and that other factors must be appealed to.

However, as French (2003) has emphasized, the role of such factors is not straightforward and certainly not constitutive of the mechanism of representation, in the particularities (that is, although it is true that all representation must involve such factors, it is not the case that these factors are always the same). The importance of intentions is often illustrated by comparing the case of a face drawn in the sand by someone, and that of what appears to be a face, etched in the sand by the wind and waves, say. The former counts as representation—so the usual story goes—and the latter does not, because of the role of intentions in the former. But now compare a face etched in the sand by the action of the wind and waves, and the equation 'E = mc²', similarly etched. In the case of the face, the crucial issue is whether the squiggles in the sand should be regarded as a representation, or an art object in general. Here one must appeal to the relevant intentions or its causal provenance more generally, which will include the relevant causal history, the role of the artist, and so forth. (It is confusion over such provenance that has led to art installations being thrown away as rubbish.) On these grounds, the squiggles would not be taken to be representational. In the case of the equation, however, even if we had observed someone inscribing it in the sand and therefore were assured that the relevant intentions were playing an appropriate role and that the relevant provenance was in place, we would not take it as the theory, or the relevant part thereof, because of the standard objections to so regarding particular inscriptions. Whatever the intentions of the person doing the etching, we would not say that those particular squiggles *are* the Special Theory of Relativity (or the relevant part thereof) that represents the relevant phenomena. Given that the theory is something other than the inscription in the sand, the issue of the particular intentions behind that specific inscription becomes less significant when considering the representational nature of the theory.

Of course, as Chakravartty points out, we would read the squiggles as 'E = mc²' in the first place only because we possess the prior intention to use markings of that form to represent the relevant phenomena. Someone not trained or educated in the appropriate way would have no such intention and would indeed see nothing but squiggles in the sand. However, this seems a different issue, and as French (2003) went on to acknowledge, in order for us to see a painting as about a certain scene, a certain person, or whatever, we must be able to interpret the splashes of paint a certain way, and that will certainly depend on the broader context. But as he emphasized, we should not effectively incorporate a particular intention into the mechanism of representation, such that the latter becomes fixed, in the sense that Manet's *Le Dejeuner sur l'Herbe* must represent a set of bourgeois young people, or Einstein's special relativity must represent phenomena associated with rods and clocks. Rather' both artworks and theories enter into a multiplicity of representational relationships, and if the intentions of the artist or scientist are not to be privileged in a problematic way (consider the shift wrought by Minkowski with regard to the special theory of relativity), then we must allow for pragmatic or broadly contextual factors to play a role in selecting which of these relationships to focus on. 'Building' particular intentions into the representational mechanism would then be disastrous.

Returning to the example of *Guernica*, here we agree that there is a multiplicity of interpretations and hence of representational relationships, and indeed there may be no agreement about what is being represented. However, unless we have some partial mapping in place (as represented at the meta-level as it were by partial isomorphisms), then it is not even clear what is being interpreted. Without such a mapping, we would be clueless! And again, the lack of agreement can be nicely accommodated by separating off the intentions issue from that of the underlying mechanism, thus allowing Picasso's intentions to be overridden by ours.

3.4.4 Mechanism

The 'enigma' of representation derives from the point that not only must targets be described in certain ways, but different descriptions will lead to different structures and hence representational relationships. With this we can only agree and insist that we do not see it as a problem. Of course, the system must be described in such a way that one can take the relevant structure to apply. What would that description involve? Well, at the very least some minimal mathematics (and here again our account applies) and certain physical assumptions. Without those the target system could not even be said to be a candidate for scientific representation in the first place! And that different descriptions will lead to different structures is surely a fact of scientific practice that is to be welcomed since its accommodation by our account is one of the advantages we claim it has over the alternatives. Typically, at all but the most basic level, these different descriptions will involve the importation of elements and relations from other such descriptions, perhaps even from other domains, and the general framework of partial structures is clearly capable of accommodating these moves (see e.g. French and Ladyman 1997, and Bueno 2000).

3.4.5 Style

Here one must again be careful in drawing comparisons with art (Frigg 2006, p. 50). It may well be the case that artists have available a wider variety of media than scientists, and this may generate the impression that different styles need to be accommodated by the relevant accounts of representation in both cases. However, if 'style' of representation is identified with the medium of representation, we must not confuse this with the mechanism of representation. Scientists do use different media for their models—some are theoretical and 'abstract', others are physical—but one of the claims of the partial structures approach is that it can accommodate this because the underlying *relationships* have the same form and, given again appropriate descriptions of the different kinds of models so as to allow a structure to be associated with them, these relationships can be captured via partial isomorphisms (or partial homomorphisms). Again, we see it as one of the virtues of our approach that it can do this.

As for the claim that we advocate a normative stance toward partial isomorphisms but this is incompatible with their inadequacy due to the inaccuracy of scientific representations, the second part of that claim is clearly false: they are not inadequate in this sense. In effect, we are being tarred with the same brush as those who advocate isomorphisms despite our protests that our account is different in precisely this respect. With the introduction of partial isomorphism and homomorphism, no requirement is made that the structures that are used to represent other structures do so with perfect accuracy.

With regard to the normative claim, we are not demanding that scientists adopt the partial structures account, nor are we even insisting that this is the only framework for representations that philosophers of science must adopt. Certainly, we have advocated a different 'style' of representation when it comes to the relationship between mathematics and science, namely partial isomorphism or partial homomorphism, and it may well be the case that different kinds of relationships require different formal frameworks. However, the challenge then would be to show that such frameworks offer the same advantages as ours (in particular, with regard to its unitary nature) when it comes to the relationships we encounter in science which, we think, can be appropriately and straightforwardly accommodated by partial isomorphisms (or other morphisms).

3.4.6 Misrepresentation

Here again one must be careful and pay due attention to the differences between art and science. Of course, cases of mistargeting do arise in science, in the sense that scientists may think that the cause of a particular phenomenon is a certain system, and construct the appropriate representation, only to find that the cause is something else. But the difference from the example of Velázquez's painting of the Pope, as presented above, again has to do with the diachronic features of science: we discover that the representation has the wrong target and adjust accordingly. Indeed, the misrepresenting model may, and typically will, be heuristically fruitful and contain the elements that lead to its successor. The inter-theory relationships this involves are precisely the sorts of things the partial structures account was set up to capture (da Costa and French 2003). Likewise, as noted already, inaccurate or unfaithful representations are absolutely not a problem for the partial isomorphism account. And of course, with regard to the Newtonian example, the question that was posed can be answered once we have General Relativity and the respects and extent of the dissimilarity become apparent. The relationship between the two theories is then most appropriately captured via partial structures and the representational relationships between each theory and the solar system can be accommodated via partial isomorphisms.

Now those who are familiar with the debate over the nature of mental representation (see e.g. Cummins 1991) might object that the problem of misrepresentation is more basic than this, in that the mechanism of partial isomorphism (or structural similarity more generally) does not have the means to accommodate misrepresentation in its own terms. More specifically, the objection is that, since something *A* represents *B* exactly if *A* is (partially) isomorphic to *B*, our approach does not provide the conceptual resources to describe a case of misrepresentation, that is, a case in which *A* represents *B* but in the wrong way. Thus, our account of representation is unable to express the notion of misrepresentation.

However, we can respond to this along the lines already indicated. If we consider a particular partial structure, characterizing a theory or model that is proposed as representing a certain system via partial isomorphism, then that structure itself does not have the means within it, as it were, to express the possibility that it may have misrepresented that system. But that is to be expected. When it comes to scientific representations, as we have already indicated, scientists *discover* that their model is a misrepresentation (typically) through lack of empirical success, to some degree or other. They then adjust, modify, or otherwise change the model, or develop a new one entirely, to yield something that better represents the system (obviously, this is a rather crude caricature of the process!).[7] These revised or new models can of course be characterized via partial structures and their relationships to their predecessors represented by partial morphisms. Hence it is not true that because a particular partial structure does not have the means to express its own misrepresenting of the system, our account as a whole is unable to express this notion. It does so through the interrelationships between the increasingly successful models that, because of this success, can be regarded as increasingly better, more informative representations. And in the case of science, at least, this is surely as it should be.

Of course, as just noted, there is a form of misrepresentation that is accommodated within particular partial structures, namely, when the relevant model includes idealizations and such like. This is a form of deliberate misrepresentation for the purposes of simplicity, computational tractability, and so forth.

---

[7] Note that even in the case of models that are known not to represent successfully the target system, for example because they include idealizations, typically the misrepresentation involved can be explicitly identified, given that certain features of the model have no counterpart in the world.

3.4.7 Ontology

The broader concern here, to do with the ontology of models and representational devices in general, obviously involves a quite general set of issues that we do not need to resolve here. Nevertheless, some comments are in order. First of all, as da Costa and French (2003) have tried to make clear, advocates of the semantic account need not be committed to the ontological claim that models *are* structures. This presumed reification has generated considerable criticism of the approach, but it is time it should be laid to rest. Set-theoretic structures provide a useful representational (or better, perhaps, descriptive) device at the meta-level of the philosophy of science. What theories and models *are*, *qua* objects, is then a further matter.

One might think that a variety of positions on the nature of models (e.g. that they are abstract objects) could accommodate their role as representational devices. However, Hughes (1997) has argued that because models involve abstraction and idealization, they must be regarded as abstract objects, and as such cannot stand in relations of similarity with physical systems. The motivation for this last move is not clear, however. It might be grounded in a restrictive view of similarity such that the abstracting away of certain properties removes the basis for relating in this way the idealized and non-idealized models that possess them. So, to draw on a much-used example, it might be insisted that by abstracting away friction and air resistance, the (idealized) model of a simple pendulum cannot then be related by similarity to an actual pendulum. But, of course, the model and the actual system share other properties—most notably, length of string, mass of bob, and so on—via which they can still be held to enter into a similarity relationship. The possible response that an idealized length that encounters no friction at point of contact, or a mass that encounters no air resistance, cannot be held to be similar to their actual counterparts strikes us as implausible (or, at the very least, as requiring an appropriate account of property identity!).

Alternatively, Hughes' move might ultimately be grounded in the view that similarity cannot be a cross-category relation, such that it cannot be taken to hold between abstract and actual objects. However, this again seems implausible. Consider, for example, the relation between Jody and Ian such that Jody is taller than Ian. This is similar to the relevant relation between the numbers that correspond to Jody and Ian's respective heights. There seems to be no objection to maintaining the existence of such a similarity in this case, and hence we likewise see no objection to taking similarity to hold between theories, taken as abstract, and systems, taken as concrete.

With regard to the objection from the variety of models, let us consider each case in turn. The first of these examples is not particularly troubling, as it is acknowledged that 'similarity is almost always the means for concrete physical representations of concrete physical objects' (Suárez 2003, p. 231) such as this. Likewise, in the case of a graph representing a real bridge, it is recognized that 'the graph [. . .] is similar to the bridge it represents with respect to the geometric shape and proportions between the different points' (ibid.). The example of a system of billiard balls representing a gas initially appears more difficult to deal with. However, there is the similarity between the dynamical properties of the two systems (as discussed in da Costa and French 2003), and it is the dynamical properties of a gas that we attempt to explain by appeal to the billiard ball model.

The fourth case of the quantum diffusion equation may appear to be the most difficult. However, Chakravartty, again, has noted that this is no counterexample since it is the semantic content—articulated perhaps through the appropriate mathematical structure, such as configuration space—that is to be regarded as similar to the target, not the superficial representation of this content via pen and paper, or chalk and board. Indeed, we would insist that insofar as the equation is satisfied in the relevant model, it is the structure encapsulated in the latter that is truly important. Equations do not represent by themselves (or perhaps we can say that their representational capacity is 'parasitic' on that of the underlying structures); rather it is the models that function as the relevant medium. Here we bump up against broader issues to do with arguments in favour of the semantic approach to theories and models in general. Given that our account is articulated within this approach, we can turn to the accepted understanding of how it treats equations—namely, by taking them to be satisfied in the relevant models—in order to appropriately locate the underlying representational device, that is, the relevant structures.

Thus, we argue that it is possible to offer an account of scientific representation that, while emphasizing the significance of a formal framework, can still accommodate the ways representations function in scientific practice. The result is a view that, being sensitive to the partiality of scientific information, is well positioned to make sense of the way in which scientific representation is so often—despite its extraordinary success—quite incomplete. The advantages of approaching scientific representation from this perspective, and indeed of our overall framework for accommodating the applicability of mathematics, will become apparent in Chapter 4, when we consider the application of group theory to quantum mechanics.

4

Applying New Mathematics
Group Theory and Quantum Mechanics

4.1 Introduction

Let us recall where we place ourselves in the debates over the applicability of mathematics. On the one hand, many philosophers pay lip service to this issue in the context of a formal approach to the philosophy of mathematics in general but fail to provide supporting details from actual practice (see, for further examples, Shapiro 1997; Hellman 1989). On the other hand, some philosophers have presented detailed case studies but in the absence of any overarching framework (here, in particular, we have in mind, again, Steiner 1998). In the context of such studies an appropriate historical sensitivity is required if an aura of mystery surrounding applicability is not to be fabricated, as we argued in Chapter 1. Having said that, even given such historical sensitivity, there is a danger of falling prey to the kind of unwarranted optimism that we noted previously and, as a kind of prophylactic to such an attitude, the moves that are undertaken in practice by both mathematicians and scientists need to be taken into account.

In this chapter, we shall attempt to support the above claims and advance the debate in general by presenting details from the history of twentieth-century physics in the context of the Inferential Conception, underpinned by the formal framework of partial structures. The case study we have chosen concerns the introduction of group theory into quantum mechanics in the late 1920s and early 1930s (further details can be found in Bueno and French 1999; French 1999, 2000). This episode has yet to be adequately explored from both the historical and methodological perspectives. It is certainly not discussed in standard histories of quantum theory in as much detail as one might expect, although aspects of these developments feature quite prominently in the work of Steiner (1998), as we have already indicated, and Peressini (1997) for example. Here we shall be primarily concerned with the methodological issues that the episode raises. It provides a particularly useful case study since it allows us to sidestep the problem of the prior mathematization of physics (again see Stöltzner 2004): talk of the applicability of mathematics to science typically obscures the fact that the science in question is already expressed in mathematical form. What is needed is an example of the introduction of *new* mathematics.

The introduction of group theory into quantum mechanics provides a useful example, and consideration of this example sheds light on the sorts of moves, which may broadly be called heuristic, made by the participants and illustrates the kinds of idealizations and analogies that are introduced in an effort to bring mathematical and physical structures together. Group theory is a particularly interesting case as the principal participants—primarily Weyl and Wigner but also others such as Mackey and Wightman—initiated important advances on both the physical and mathematical sides.

We shall begin with a brief outline of the historical context, drawing on the work of Bonolis (2004).

4.2 The Historical Context

The history of group theory is entwined with the history of physics, although there have been periods where that engagement has been forgotten or suppressed and these have contributed to the apparent 'mystery' of applicability. As Bonolis (ibid.) notes, group-theoretical results were actually applied in the absence of a definition of the concept of 'group', which emerged as an umbrella concept covering aspects of number theory, geometry, and crystallography. But it was the problem of the solvability of algebraic equations that gave rise to this concept and it was Galois, in 1831, who understood that the algebraic solution of an equation was related to the structure of the relevant group of permutations. So, as an example, consider the following quadratic equation:

$$x^2 - 4x + 1 = 0$$

The roots of this equation are:

$$A : 2 + \sqrt{3}$$

$$B : 2 - \sqrt{3}$$

These satisfy $A + B = 4$ and $AB = 1$ and if we exchange A and B we obtain another true equation. It turns out this is the case for every algebraic equation with rational coefficients that relate these values of A and B. Thus the Galois group of the polynomial $x^2 - 4x + 1$ consists of two operations: identity and permutation, where the latter exchanges A and B (it is thus a cyclic group of order 2). The core idea here is to focus on those roots that leave a given algebraic equation satisfied under permutation.

As Bonolis observes, by setting group theory at the core of this approach, it led to a shift from computational techniques to an analysis of structure and after Galois' untimely death, it was Jordan who opened up the field with his *Traité des substitutions et des équations algèbriques* which laid down the basis of the theory of finite permutation groups. This directly influenced the work of Klein and Lie, in particular with regard to the application of group theory to geometry.[1] Here the development of non-Euclidean geometries and the introduction of large numbers of dimensions (motivated at least in parts by developments in physics, such as the theory of electromagnetism) led to concerns about capturing the core unity of geometry and the basis for classifying its different forms. These fed into Klein's 'Erlangen' programme, which placed group theory and the notion of invariance at its heart.[2] By associating each geometry with an underlying group of symmetries, the relationship between different geometries can be studied by considering the relationships between these different groups and their invariants. Take Euclidean geometry for example: the Euclidean group of symmetries keeps angles, areas, and lengths invariant. In projective geometry (rooted in the study of visual perspective), it is certain projective transformations that remain invariant but the property of parallelism (relating to parallel lines) does not, whereas it does in affine geometry. Thus, to keep things brief, classification was achieved via group theory and geometry itself was reduced to the study of invariances under the relevant group of transformations.

Lie, who, with Klein, had encountered group theory in Paris in 1870 through Jordan's book, was more interested in its application to the study of differential equations. Here the focus was on continuous transformation groups and the way they can help make partial differential equations easier to solve. Again, the relevance of developments in mathematics for physics (as well as Lie's own awareness of that) is noted:

[Lie's] research on differential equations and their symmetry group led him to investigate the structure of infinite-dimensional continuous groups, which would become fundamental in gauge theories developed by physicists in the second half of the XX century.   (Bonolis 2004, p. 8)

Indeed, Lie insisted that the principles of mechanics had a group-theoretic origin and that his ideas could be extended to optics and other areas of physics. These ideas were applied extensively to geometry and differential equations but, ironically perhaps, following Cartan's analysis of the structure and representations of the infinitesimal transformation groups, the connection with partial differential equations became obscured.

As Bonolis goes on to say:

Klein's and Lie's explicit use of groups in geometry influenced conceptually the evolution of group theory. The study of transformation groups, in extending the scope of the concept of a group, shifted the development of group theory from a preoccupation with permutation groups and Abelian groups to the study of groups of transformations.... It also provided important examples of infinite groups, and greatly extended the range of applications of the

---

[1] Bonolis (2004) also notes that Lagrange had also studied permutations the century before and implicitly employed group theory.

[2] Again, as Bonolis remarks, it is important to note the impact of work in physics on Klein's thought. Klein himself remarked on the 'intimate connection' between developments in geometry and mechanics.

group concept to include number theory, the theory of algebraic equations, geometry, the theory of differential equations (both ordinary and partial), and function theory (automorphic functions, complex functions). These developments were also instrumental in the emergence of the concept of an abstract group.   (ibid., pp. 13–14)

Further developments in the formalization of this concept occurred with the work of Cayley on algebraic invariance, regarded by Weyl as forming the birth of invariant theory (ibid.). And it was through Cayley that we obtained the abstract notion of a finite group as an arbitrary system of elements determined only by defining relations, whose structure is characterized by means of a group table. With the recognition that certain results that were thought to hold for permutation groups were actually theorems that held for finite groups in general, such formal developments had a wide and significant impact by the end of the nineteenth century and works such as Burnside's *Theory of Groups of Finite Order* played a major role in highlighting their importance.

However, one of the most significant features as far as the applicability of group theory to physics is concerned had its origin in the study of crystals. So, an obvious core concern in crystallography is to determine the structure of the crystalline material. This can be reduced to the repetition of the 'unit cell', which is the smallest group of particles (atoms, ions, or molecules) whose repetition along the principal axes forms the repeating pattern of the crystal structure. The lengths of these principal axes and the angles between them are the lattice constants that remain invariant under transformation and thus the structure of the crystal can be obtained by determining its symmetry properties. These in turn can be described by the so-called 'space groups', or symmetry groups of (usually three-dimensional) spatial configurations.[3] This latter description is quite abstract but we can always introduce a 'representation' of the group in terms of certain linear transformations (of the relevant vector space). We can thus address any problems associated with the group by studying its representation—in effect, by reducing the problem to a problem in linear algebra. As Bonolis records, it was Bravais' theoretical work on the structure of crystals that led him to study the group of transformations in three variables (for obvious reasons) which in turn influenced Jordan's work on the analytic representation of groups in general. This work was then extended by Frobenius and others, with the former establishing the theory of representations of finite groups by linear transformations. The significance of such representations has been emphasized by Mackey, who insisted that without them, group theory would only be "the shadow of itself" (Mackey 1978, p. 461).

More formally, an n-dimensional representation D of a group G is defined as follows (Bonolis 2004, p. 58): to each element $g \in G$, one associates an $n{\times}n$-matrix $D(g)$ (i.e. a linear operator on an n-dimensional vector space V) in such a way that the

---

[3] Of which there are 230.

operators $D(g)$: $g \in G$ obey essentially the same algebraic relations as the group elements themselves, and the group structure is preserved: $D(gg) = D(g)D(g)$, $D(g - 1) = [D(g)] - 1$, $D(e) = 1$ (e is the neutral element of G, 1 is the identity operator). We say in this case that the map D: $g \rightarrow D(g)$ is an n-dimensional (linear) representation of G. If the correspondence $g \rightarrow D(g)$ is one-to-one, the set of all representatives $D(g)$ forms a group which is isomorphic to the original group G.[4]

What is significant here as far as physics is concerned is that the representations in that context are the well-known transformation formulae for vectors, tensors, etc. and as Weyl and Wigner were to perceive, quantum mechanics offered fertile terrain for their deployment.

As we have just indicated, one route to this development proceeded via crystallography and it was here that, as Weyl later remarked, the most important application of group theory lay. As Bonolis (2004) again charts, the classification of crystals according to their symmetry properties had proceeded for many years in the absence of any formal group-theoretic framework. Even after Bravais's studies of crystal structures that motivated Jordan's classification of groups of motions it was some time before crystallographers started to apply groups to this area (ibid., pp. 16–18). However, here again we can discern a local entwining of the science and the mathematics, as theoreticians both applied group theory and extended the classification of the relevant symmetries through their studies.

There is considerably more historical detail covered by Bonolis' study (see also the collection of essays in Doncel et al. 1987). But for our purposes two features stand out: the motivational role of crystallography and the further development of the theory of group representations. Both famously feature in the origins of Wigner's work in applying group theory to quantum physics.

4.3 Applying Group Theory to Atoms

4.3.1 A (Very) Brief History of Quantum Statistics

The history of quantum theory began with work in quantum statistics (for details of this history, see French and Krause 2006, chapter 3).[5] Planck's classic paper of 1900 took Boltzmann's combinatorial approach to the distribution of particles over energy states from classical statistical mechanics and applied it to the distribution of energy quanta over blackbody oscillators (note the crucial 'twist' here). However, as Kuhn (1978) has emphasized, it was Einstein, five years later, who proposed the idea that

---

[4] The aim of representation theory is then to determine all the inequivalent, irreducible representations of a given abstract group to which any given representation can be reduced and thus it can be regarded as a powerful tool for studying the structure of the finite groups. With Schur's presentation of the theory based on linear algebra it also became more accessible.

[5] Summaries of this section and Section 4.4 can be found in French (2014), chapter 4, section 4.4.

these energy quanta could be conceived of as independent entities, again applying Boltzmann's framework to them, and it was Ehrenfest (Boltzmann's student) who raised the issue of the nature of the statistics involved in these applications, noting that they appeared to differ from the 'classical' form presented by Boltzmann.

Although Einstein, Ehrenfest, and others continued to work on the application of quantum theory to radiation, from about 1910 onwards attention shifted increasingly towards the problems of atomic structure. However, the publication of Bose's work in 1924 (arranged by Einstein), and the subsequent further elaboration by Einstein of the core idea, again raised issues concerning the statistical behaviour of quantum particles (for further details see Bergia 1987). Here Bose again applied the standard Boltzmannian approach but instead of applying it to the number of quanta in each of the relevant states, he considered the number of cells containing each possible number of quanta. In his own paper of 1924 Einstein then applied Bose's approach to material gas atoms (Einstein 1924), retaining Bose's counting, suitably modified to take into account the finite mass of the gas atoms and their fixed number. This paper and the two that followed it laid down the foundations of the quantum theory of the ideal gas, embodying what is now called 'Bose–Einstein Statistics'. In particular, Einstein realized that, "From a certain temperature on, the molecules 'condense' without attractive forces, that is, they accumulate at zero velocity" (quoted in Pais 1982, p. 432; see also Einstein 1925; trans. in Duck and Sudarshan 1997, pp. 91–2). We shall return to discuss this condensation phenomenon in Chapter 5 but for Einstein it was the result of a mysterious influence that could be understood through de Broglie's hypothesis of matter waves, which, as is now well known, led to Schrödinger's wave mechanics.

Bose–Einstein statistics was subsequently accommodated within a self-consistent framework by Heisenberg and Dirac, together with an alternative form, initially due to Fermi, that assumed the validity of Pauli's Exclusion Principle for the particles of the assembly concerned. This yielded what is now known as Fermi–Dirac statistics, according to which, rather than 'condensing' or 'accumulating' into the same state as Bose–Einstein particles do, the particles are constrained to occupy distinct states. Heisenberg and, independently, Dirac, then demonstrated the connection between these two forms and the symmetry characteristics of the states of relevant systems of the particles (regarded, crucially, as indistinguishable; Dirac 1926; Heisenberg 1926a, 1926b): Fermi–Dirac statistics (obeyed by electrons, for example) follows from the requirement that the wave function for the assembly be anti-symmetrical and Bose–Einstein statistics (obeyed by photons, for example) follows from the requirement that it be symmetrical. Dirac, in particular, noted that quantum mechanics itself could not determine which form of statistics was appropriate in a given situation and that extra-theoretical considerations had to be appealed to (a point we shall return to below).

These papers of Heisenberg's and Dirac's had an enormous influence on Wigner, as we shall now see.

4.3.2 The 'Wigner Programme'

Returning to the application of group theory to quantum physics, Mackey presents the relevant history in terms of two programmes (Mackey 1993): the 'Weyl programme' can be identified with a concern with the group-theoretic elucidation of the *foundations* of quantum mechanics. It was initiated by Weyl's (1927) paper that established a basis in group theory for a global form of the Heisenberg commutation relations. The 'Wigner programme', on the other hand, can be taken to focus on the *utilization* of group theory in the application of quantum mechanics itself to physical phenomena. This latter strand was concerned with the solution, or sidestepping, of dynamical problems by focusing on the underlying invariances of the situation; whereas Weyl's attempted to bring some order to the disparate models and principles of the new quantum physics by introducing group theory at the very foundations. Thus, we have here two different kinds of applicability: one to the foundations of the relevant science and the other to its representation of physical phenomena. As Mackey indicates, there is a nice inter-braiding of these threads as Wigner and Weyl actually contributed to both.[6] Wigner himself emphasized the dual role played by group theory in physics: the establishment of laws—that is, fundamental symmetry principles—which the laws of nature have to obey; and the development of 'approximate' applications which allowed physicists to obtain results that were difficult or impossible to obtain by other means (see his essay in Doncel et al. 1987, pp. 615–20 and his comments on p. 634). Here we shall consider both programmes as examples of applicability.

The Wigner programme can be traced back to Wigner's own early work on symmetry in crystals. He had been contacted by a crystallographer interested in the problem of why atoms in a crystal lattice occupy positions along and on the symmetry axes and planes, respectively, and had been urged to study group theory in order to tackle this problem. When he was introduced to the emerging 'new' quantum mechanics of Heisenberg et al. he was, in effect, already 'primed' to be sensitive to symmetry considerations. Thus he appreciated early on that anti-symmetrizing the wave function of a system (in accordance with Fermi–Dirac statistics) led to Pauli's Exclusion Principle (again, we'll come back to this shortly). More generally, he realized that the ad hoc set of rules and principles by which atomic spectroscopy was described could be related to spherical harmonics, which are certain functions defined on the surface of a sphere and used to solve partial differential equations. These functions in turn serve as the basis for the group of rotations in three dimensions, or SO(3), and Wigner immediately saw the connection with group theory. However, as he subsequently noted, his plan to study the rotation group was put on hold after Heisenberg's paper on anti-symmetrization came out

---

[6] In a footnote to his classic 1927 paper Weyl cites Wigner's work but notes that his proceeds in a different direction; we would like to thank Peter Simons for translating this passage.

(Heisenberg 1926b) and as Wigner himself puts it, he 'slithered' into the permutation group (Wigner 1963; again we'll say more on this below).

The turning point in his work came when von Neumann gave him a reprint of the famous 1906 paper of Frobenius and Schur (see Wigner in Doncel et al. 1987, p. 633). This had a double significance: first of all, it familiarized Wigner with matrices (in terms of which the symmetries of crystals were expressed), thus allowing him to immediately understand the significance of the papers by Born, Jordan, and Heisenberg on matrix mechanics and thereby to calculate atomic energy levels. Second, it pushed him to really learn group theory and, with the help of von Neumann, encouraged him to become familiar with group representations.[7] Thus, it gave Wigner a set of powerful mathematical tools that he could then bring to bear on quantum physics, following Heisenberg and Dirac's work of 1926 and 1927 on the quantum statistics of indistinguishable particles (see also French and Krause 2006, chapter 3). In particular, as noted above, this work highlighted the connection between such statistics and the symmetry characteristics of the relevant states of the particle assemblies, where such characteristics arise because of the non-classical indistinguishability of the particles; or as Weyl put it in his inimitable fashion:

the possibility that one of the identical twins Mike and Ike is in the quantum state E_1 and the other in the quantum state E_2 does not include two differentiable cases which are permuted on permuting Mike and Ike; it is impossible for either of these individuals to retain his identity so that one of them will always be able to say "I'm Mike" and the other "I'm Ike". Even in principle one cannot demand an alibi of an electron!   (Weyl 1931, p. 241)[8]

This loss of an identifying 'alibi' was associated with a fundamental new symmetry property, namely invariance under permutation, as described by the permutation group.[9] Applying this yields, in the simplest case, two possibilities: that the wave function (or eigenfunction) for the assembly of particles should be symmetric, or that it should be anti-symmetric, where, again, the former describes particles that obey Bose–Einstein statistics, and the latter describes those that obey Fermi–Dirac statistics and which are constrained by the Paul Exclusion Principle.[10]

Wigner subsequently made it clear in the preface to his 1959 book that '[t]he initial stimulus for these articles was given by the investigations of Heisenberg and Dirac on

---

[7]   See, again, Wigner's comments in Doncel et al. (1987, p. 633) and also Donini (1987, p. 110); further details of this early history are revealed in Wigner's interview with Kuhn (Wigner 1963); see also Chayut (2001).

[8]   This passage also features in the 1928 first (German) edition; we would like to thank Don Howard for confirming that.

[9]   As emphasized by Weyl in his non-technical presentation of 1929 (1968, p. 268) and also in his 1938 paper on symmetry (ibid., pp. 607–8).

[10]   As it turns out, for more than two particles, other possibilities arise, corresponding to so-called 'parastatistics' (see French and Krause 2006, chapter 3).

the quantum theory of assemblies of identical particles' (1959 [1931], p. vi).[11] In particular, as we have already noted, it was Heisenberg who explicated the connection between the above two forms of statistics and the symmetry characteristics of the relevant states of the particle systems. In his first paper (1926a), he showed that two indistinguishable systems that were weakly coupled always behaved like two oscillators for which there were two sets of non-combining states, symmetric and anti-symmetric. That these two sets of states were not connected then followed from the symmetry of the Hamiltonian of the system under a particle permutation. In his second work (1926b), Heisenberg investigated the concrete example of the two electrons in a helium atom and concluded that only those states whose eigenfunctions are anti-symmetric in their electron coordinates can arise in nature. It was this work that pushed Wigner towards group theory.

Dirac then went further by setting both Bose–Einstein and Fermi–Dirac statistics in their appropriate theoretical context (Dirac 1926). He began with what he took to be the fundamental requirement that the theory should not make statements about unobservable quantities. Consequently, he insisted, two states that differ only by the interchange of two particles must be counted as only one. Out of the set of all possible two-particle eigenfunctions only the symmetric and anti-symmetric satisfy the conditions that the eigenfunction should correspond to both of the above states and should be sufficient to give the matrix representing any symmetric function of the particles. These results were then extended to any number of non-interacting particles and with the anti-symmetric function appropriately expressed, Pauli's Exclusion Principle dropped out quite naturally. And, as we have already said, Dirac explicitly noted that the theory itself could not determine which form—symmetric or anti-symmetric—was in fact appropriate in any given situation and hence extra-theoretical considerations had to be appealed to. In particular, the solution with symmetrical eigenfunctions could not be the correct one for the case of electrons in an atom since it allows any number of electrons to be in the same orbit (ibid., p. 670; again, see French and Krause 2006).

It was precisely this analysis of quantum statistics in terms of the permutation of indistinguishable particles that heuristically motivated the construction of the 'bridge' underpinning the embedding of quantum mechanics into group theory.[12] The central problem of the 'Wigner programme' was then the following: Given a system consisting of an assembly of indistinguishable subsystems, such as the electrons in an atom, what is the effect of some (small) perturbation of the Hamiltonian of this system on the known eigenvalues of that Hamiltonian? Heisenberg had solved the problem when the number of subsystems, or electrons, is two, using elementary means.

---

[11] Wigner also acknowledges von Laue as perhaps the first of the 'older generation' of physicists to recognize the significance of group theory as 'the natural tool with which to obtain a first orientation in problems of quantum mechanics' (1959, p. v).

[12] The term 'bridge' is used in a similar context by Pickering (1995); see also French (1997).

Wigner likewise used elementary methods to solve it for three particles but for greater than three he noted that the problem could be simplified using group theory to capture the underlying invariance (see Mackey 1993, p. 247).[13] The theory of group representations as applied to the permutation group can then be used to determine the splitting of the eigenvalues of the original Hamiltonian under the effect of the perturbation (ibid., pp. 242–6).

The fundamental relationship underpinning this application is that which holds between the irreducible representations of the group, as sketched above, and the subspaces of the Hilbert space representing the states of the system.[14] In particular, if the irreducible representations are multidimensional then the appropriate Hamiltonian will have multiple eigenvalues which will split under the effect of the perturbation. Weyl speaks here of the group 'inducing' a representation in system space (see e.g. Weyl 1931, p. 185); thus, under the action of the permutation group the Hilbert space of the system decomposes into mutually orthogonal subspaces corresponding to the irreducible representations of this group. Of these representations the most well known are two already discussed, namely the symmetric and anti-symmetric, corresponding to Bose–Einstein and Fermi–Dirac statistics, respectively.[15]

However, as well as possessing permutation symmetry, an atom is also symmetric with regard to rotations about the nucleus, if inter-electronic interactions are ignored (here we have the first crucial idealization) and again as we have just mentioned, group representations can be appropriately utilized to label the relevant eigenstates.[16] It is at this point that Wigner appealed to the results established by Schur to which he had been directed by von Neumann, and also to the work of Weyl who had extended the theory of group representations from finite groups to compact Lie groups. In three classic papers of 1925 and 1926 Weyl had established the complete reducibility of linear representations of semi-simple Lie algebras. In particular, this allowed the irreducible representations of the three-dimensional pure rotation (or orthogonal) group to be deduced. Weyl's approach also gave the so-called 'double valued representations' representing spin (see Wigner 1959 [1931], pp. 157–70). The history, as represented in perhaps its most apprehensible manifestation, namely the *dates* of publication of these papers, is of fundamental significance here: not only was the relevant physics under construction at this time,

---

[13] It is precisely here that von Neumann pointed out to Wigner the relevance of group theory, as mentioned above.

[14] This was the relationship that von Neumann brought to Wigner's attention.

[15] Again, as is now well known, other representations, corresponding to 'parastatistics', are also possible (as Dirac noted). In the above terms, these would occupy other subspaces of Hilbert space, beyond the symmetric or anti-symmetric.

[16] As Wigner and von Neumann subsequently emphasized, 'This method rests on the exploitation of the elementary symmetry characteristics of every atomic system—namely, the equality of every electron and the equivalence of all directions in space' (translated and quoted in Miller 1987, p. 313); cf. Weyl 1931.

but so was the appropriate mathematics.[17] Both can thus be seen as aspects of programmes open to further development.

In a further paper of 1927, Wigner presented a systematic account of the application of the mathematics of group theory to the physics of the energy levels of an atom that covered both the permutation and rotation groups. He acknowledged, however, that his model of the atom was simplified insofar as it did not take account of the newly proposed notion of 'spin'. In a three-part paper co-authored with von Neumann and published the following year, spin was then incorporated into the analysis using Weyl's two-dimensional representations of the rotation group (Judd 1993, pp. 19–21). These results were then presented in systematic fashion in Wigner's well-known 1931 book *Group Theory and its Application to the Quantum Mechanics of Atomic Spectra*. It is to this work that elementary particle physicists returned in the 1950s when they 'rediscovered' Lie algebras and group-theoretical techniques in general.

4.3.3 The 'Weyl Programme'

Wigner's works are cited by Weyl in the latter's 1927 paper on group theory and quantum mechanics (Weyl 1927), and Weyl subsequently acknowledged Wigner's 'leadership' in this regard (Weyl 1968, vol. III, p. 679). However, as we have already noted, Weyl was also careful to emphasize the different directions taken by their work (although both programmes are represented in Weyl's classic (1931) text, *The Theory of Groups and Quantum Mechanics*). Thus, with regard to his own foundational analysis, Weyl proceeded from a different standpoint: his aim was to base Heisenberg's famous commutation relations (which yield the Uncertainty Principle) on certain fundamental symmetry principles, expressed group-theoretically (see Mackey 1993, pp. 249–51). To achieve this, he represented the kinematical structure of a physical system via an irreducible Abelian group of unitary ray rotations in Hilbert space, with the physical quantities of the system represented by the real elements of the algebra of this group (Weyl 1931, p. 275). Heisenberg's formulation then follows 'automatically' from the requirement that the group be continuous and, in particular, the requirement of irreducibility gives the relevant pairs of canonical variables. Weyl concluded that only one irreducible representation of a two-parameter continuous Abelian group exists, namely the one that leads to Schrödinger's equation. Hence, he maintained, group theory underpinned the fundamental structure of the new theory (for further discussion see Mackey 1993, pp. 249–51, 274–5). However, although perhaps 'more beautiful' than Wigner's (Wigner 1963), Weyl's book was

---

[17] Of course, the relevant physics was already articulated mathematically, although in non-group-theoretical terms. What this gave, as already hinted at, was an assemblage of models, principles, and heuristic rules (including e.g. the 'Aufbauprinzip', Heisenberg's Uncertainty Principle, Pauli's Exclusion Principle, and so forth) which, as Weyl subsequently noted, could be brought under a unifying mathematical framework via group theory. An alternative framework was, of course, provided by von Neumann's introduction of Hilbert spaces, which we shall return to in Chapter 6.

perceived as very difficult and it was the former that was widely regarded as making the methods of group theory available to physicists.[18]

Part of the reason for this perception, of course, was that the Weyl programme was informed by Weyl's own profound contributions to group theory, particularly in the three fundamental papers noted above (Weyl 1925), in which he effectively initiated the study of global Lie groups.[19] Following the work of Schur, Weyl established the complete reducibility of linear representations of semi-simple Lie algebras.[20] In the context of quantum mechanics, what this gave was a way of deducing the irreducible representations of a particularly important semi-simple Lie group, namely the three-dimensional pure rotation (or orthogonal) group mentioned above and which we shall consider in greater detail below. These representations in particular were familiar to physicists as they are the 'transformation formulae' for vectors, tensors, etc. (Wigner 1959 [1931], p. 168).[21] As Weyl emphasized, it is in the representation of groups by linear transformations that their investigation became a 'connected and complete theory' (1931, p. xxi) and 'it is exactly this mathematically most important part which is necessary for an adequate description of the quantum mechanical relations' (ibid.). Furthermore, group theory reveals '[...] the essential features which are not contingent on a special form of the dynamical laws nor on special assumptions concerning the forces involved' (ibid., p. xxi). Indeed:

All quantum numbers, with the exception of the so-called principal quantum number, are indices characterizing representations of groups. (ibid., italics in the original)

Let us now look at this relationship between quantum mechanics and group theory a little more closely:

An atom or an ion, whose nucleus is considered as a fixed center of force O, possesses two kinds of symmetry properties: (1) the laws governing it are spherically symmetric, i.e., invariant under an arbitrary rotation about O; (2) it is invariant under permutation of its f electrons. (Weyl 1968, p. 268)

The first kind of symmetry is described by the rotation group, while the second is described by the finite symmetric group of all $f!$ permutations of f things.

The nature of the relationship becomes clear when we focus on the decomposition of Hilbert space into subspaces that are irreducible and invariant with respect to a particular group (Weyl 1968, pp. 275–7). As Wigner realized, such a decomposition corresponds to the separation of the various values which are possible for a physical

---

[18] Von Neumann and Wigner apparently felt that it was 'unfair' that Weyl's book came out first (Wigner 1963).

[19] For a discussion of this work in the context of the history of the relevant mathematics, see Borel (2001, chapter III).

[20] The term 'Lie algebra' was coined by Weyl himself in 1934.

[21] To each rotation s there corresponds a linear unitary operator $U(s)$ induced in the 'system space' by that rotation. The correspondence $U: s \rightarrow U(s)$, which obeys a composition law structurally identical to the composition of rotations, is the representation of the rotation group.

quantity. Thus, the group of rotations in 'actual' space induces a group of transformations in Hilbert space under which the latter decomposes into invariant subspaces. In each such subspace, the rotation group induces a definite representation. Correspondingly, the angular momentum operator is separated into 'partial' operators (Weyl's term), each acting on one of the subspaces. The different irreducible representations of the rotation group can be distinguished by an index $j = 0, 1/2, 1, 3/2 \ldots$ and the subspace in which the representation with index j is induced has $2j + 1$ dimensions. Hence, as Weyl points out, we know the angular momentum in the subspace 'independently of the dynamical structure of the physical system under consideration' (1968, p. 276)—its components are the operators which correspond to the infinitesimal rotations in the relevant subspace and the corresponding eigenvalues belong to the relevant representations.

Physically, of course, the value of $2j + 1$ gives the 'degree of multiplicity' of an energy level.[22] To each such level of an n-electron system there corresponds a representation of the nth degree (Weyl 1968, p. 276). However, not all these representations actually occur, 'for reasons which cannot be explained without a discussion of electron spin and the Pauli principle' (ibid.). That is, group theory gives us surplus structure that we rule out for purposes of application through the invocation of further physical principles, themselves to be embedded within the mathematics (as Dirac indicated in the case of the Exclusion Principle).[23]

The power of group theory reveals itself when we consider the spectrum of an n-electron atom, where $n > 1$. In such cases, the appropriate Schrödinger equation cannot be solved exactly because of the mutual repulsion of the electrons (giving a computationally intractable expression for the potential energy). Hence approximation was viewed as the only way forward (Wigner 1959 [1931], p. 180): we ignore the mutual repulsion of the electrons, solve the resulting Schrödinger equation, and reintroduce the effects of the electrons on one another as a perturbation. The perturbation partially removes the degeneracy that results from using this initially crude approximation in which a number of eigenfunctions correspond to each of the resultant eigenvalues; in effect, the energy levels split. Nevertheless, most of the resulting levels are still degenerate and about them, 'nothing is known on a purely theoretical basis (apart from a rough estimate of their positions) except their symmetry properties' (ibid., p. 181). It is here that group theory comes in (and strictly speaking to each level there correspond three representations: one of the rotation group, one of the reflection or inversion group, and one of the permutation group to be discussed below).

How then to proceed? Ignoring the indistinguishability of the electrons and taking the rotation group as the sole symmetry group of the Schrödinger equation, the

---

[22] As Wigner noted, 'the concept of a multiplet system [...] is [...] alien to classical theory' (Wigner 1959 [1931], p. 182).

[23] Here we see the converse of the point made in note 29 of Chapter 2—heuristic factors may be deployed to rule out certain surplus mathematical structure from being transferred.

'vector addition model' was employed, based on the 'building-up principle' (or 'Aufbauprinzip'), in which the angular momenta of the single electrons are added together in order to give the total (Wigner 1959 [1931], pp. 184–93 for pitiless details). As Weyl notes:

The mathematical interpretation given this model by quantum mechanics is characterized by the two circumstances:
(1) The determination of the various numerical possibilities is to be interpreted as decomposition into invariant irreducible sub-spaces.
(2) The addition of vectors has its mathematical counterpart in the multiplicative composition of the representations induced in these sub-spaces. (1968, p. 279)

This 'interpretation' accounts for the following results:

(i) j is restricted to the values 0, 1/2, 1, 3/2 ...
(ii) The square of the absolute value of the angular momentum is $j(j+1)$ (rather than j^2 as in the classical case).
(iii) The 'inner' quantum number obtained by the 'composition' of two systems is restricted to the values $|j - j'|, |j - j'| + 1, \ldots, j + j' - 1, j + j'$. (Weyl 1968, p. 280; Wigner 1959 [1931], pp. 187–94)

Distinguishing between the spin and orbital quantum numbers one arrives 'naturally'— as Weyl put it—at Hund's empirically successful vector model of atomic spectra.[24] But, he emphasizes, this model is not to be taken literally since it can now be obtained within this new 'interpretation' according to which the above results are explained, not by ad hoc hypotheses, but on the basis of a unified viewpoint: 'This is the service rendered by the new quantum theory' (Weyl 1968, p. 281). And this was one of the results that Wigner initially thought were the most significant in his work.[25]

There are two further points to note about this. The first concerns the role of group theory in supplying a taxonomy by means of which the 'levels zoo' can be classified. This role is well known, particularly when one considers the development of isospin, SU(3), and so on, but it is worth bringing it out here. What was previously a disparate set of spectroscopic rules, explained in an ad hoc and unsatisfactory manner, comes to be embedded within a unified, coherent framework. The second point concerns the fundamental importance of group theory in producing such results given the computational intractability of the Schrödinger equation for the many-electron atom. At the beginning of his preface Wigner writes:

The actual solution of quantum mechanical equations is, in general, so difficult that one obtains by direct calculation only crude approximations to the real solutions. It is gratifying, therefore, that a large part of the relevant results can be deduced by considering the fundamental symmetry operations. (1959 [1931], p. v)

---

[24] Wigner also recalled reading Hund's book (Wigner 1963).
[25] Subsequently, he writes, he came to agree with von Laue that 'the recognition that almost all rules of spectroscopy follow from the symmetry of the problem is the most remarkable result' (1959 [1931], p. v).

What we have, then, roughly, is the following sort of scheme: idealizations are introduced, such as ignoring the indistinguishable nature of the particles, giving rise to models which can then be structurally embedded in group theory, which in turn is used to generate the appropriate results (cf. Redhead 1980, p. 156). There is a great deal more which needs to be said here, particularly with regard to the 'deduction' that Wigner highlights, but as far as we are concerned, it is important to note that the relationships between such idealized models can be appropriately captured in terms of our formal framework of partial structures and partial isomorphisms.

Let us now turn to the second kind of symmetry property, namely invariance under particle permutations. If we take two individuals, such as electrons in an atom, which are 'fully equivalent', as Weyl puts it, then, as we have already indicated, the appropriate system space (or Hilbert space) is reducible into two independent subspaces—the space of symmetric and anti-symmetric tensors of second order. Physical quantities pertaining to the system then have 'only an objective physical significance if they depend *symmetrically* on the two individuals' (1931, p. 239; italics in the original). It can then be shown, in particular, that every possible interaction between the individuals depends symmetrically on them. Hence if the system is at any time in a state in one of these subspaces, no influence whatsoever can ever take it out of it (ibid., p. 240). '[W]e expect Nature to make use of but one of these subspaces' (ibid.) but the formalism itself gives us no clue as to which one. The particular subspace in which a system is located is an initial condition, as it were, determined by the kind of particle considered, boson or fermion (exemplified by photons and electrons, respectively). As far as the representation of fermion systems is concerned, the symmetric subspaces are just so much surplus structure.

It can further be shown that every invariant subspace of the state space of f equivalent individuals, including the state space itself, can be completely reduced into irreducible invariant subspaces (1931, p. 301, Theorem 4.11). As Wigner originally recognized, this reduction implies a separation of the terms of the physical system into sets 'which no dynamical influence whatever can cause to enter into combination with each other' (Weyl 1931, p. 320). Of course, as Weyl continually insists, as far as physics is concerned it is only the symmetric and anti-symmetric subspaces that are of interest and, further, only the latter when electrons are being considered:

The various primitive sub-spaces are, so to speak, worlds which are fully isolated from one another. But such a situation is repugnant to Nature, who wishes to relate everything with everything. She has accordingly avoided this distressing situation by annihilating all these possible worlds except one—or better, she has never allowed them to come into existence! The one which she has spared is that one which is represented by anti-symmetric tensors, and this is the content of Pauli's exclusion principle.   (1968, p. 288)[26]

---

[26] See also Weyl (1931, pp. 238, 347). Cf. Huggett: 'there is no mystery at all about why non-symmetric states are never realised; they are not within the symmetrised Hilbert space that correctly represents the world, and hence do not correspond to physical possibility' (1995, p. 74).

The other, 'surplus' subspaces were subsequently shown to correspond to 'mixed-symmetry' state functions describing so-called parastatistics (as mentioned in notes 10 and 15), which, for a time in the 1960s, were thought to be obeyed by quarks (again, for details, see French and Krause 2006, chapter 3).

On this basis, one obtains not only the group-theoretic classification of line spectra of an atom consisting of an arbitrary number of electrons, taking into account the exclusion principle and spin (chapter V),[27] but also an understanding of the nature of the homopolar molecular bond (1931, pp. 341–2) and valency in general (ibid., pp. 372–7).

Thus, with regard to the construction of the so-called 'bridge' between the theoretical and mathematical structures, on the quantum mechanical side we have the reduction of the state space into irreducible subspaces and on the group-theoretical side we have the reduction of representations. In terms of our framework, it is here we have the (partial) isomorphism between (partial) structures, (weakly) embedding the theoretical structure T into the mathematical structure M. Underpinning all of this is the intimate relation between the representations of the group of all unitary transformations or the group of all homogeneous linear transformations and those of the symmetric group of permutations of f things:

the substratum of a representation of the former consists of the linear manifold of all tensors of order f which satisfy certain symmetry conditions, and the symmetry properties of a tensor are expressed by linear relations between it and the tensors obtained from it by the $f!$ permutations.   (Weyl 1931, p. 281)

Elsewhere Weyl himself refers to this correspondence between the representations as the 'bridge' and notes that since continuous groups are easier to handle than discrete, it leads from the character of the unitary group to that of the permutation group (1968, pp. 286–7).

In particular, the above reduction of the state space of equivalent particles into irreducible subspaces 'parallels' (Weyl 1931, p. 321) the complete reduction of the total group space of the symmetric permutation group into invariant subspaces. This 'reciprocity' between the symmetric permutation group and the algebra of symmetric transformations is referred to as 'the guiding principle' in Weyl's work (ibid., p. 377) and elsewhere he writes:

The theory of groups is the appropriate language for the expression of the general qualitative laws which obtain in the atomic world. In particular, the reciprocity laws between the representations of the symmetry group s_n and the unitary group G are the most characteristic feature of the development which I have here indicated; they have not previously come into their own in the physical literature, in spite of the fact that quantum physics leads very naturally to this relation.   (1968, p. 291)

---

[27] Referring to developments in spectroscopy, Weyl writes: 'The theory of groups offers the appropriate mathematical tool for the description of the order thus won' (1931, p. 245).

Interestingly, then, the construction of this bridge between quantum mechanics and group theory crucially depends on a further one within group theory itself—the bridge that Weyl identified between the representations of the symmetry and unitary groups as expressed in the reciprocity laws. So, if one is to appropriately capture the relation between group theory and quantum mechanics, it would seem that the relevant 'surplus structure' should be appropriately characterized in terms of a *family* of structures, as indicated in Chapter 2. We recall that, rather than simply embedding a structure A into a larger 'surplus structure' B, the surplus structure is actually to be found in the various further structures $C_1, \ldots, C_n$ into which B is embedded. Thus, both the mathematical context and the structural variety implied by the notion of *surplus*—with particular emphasis on the heuristic role of mathematics in theory construction—can be represented in terms of the family of new structures $(C_i)_{i \in I}$. In this way, we can represent not only the relationships between the mathematics and the physics, as it were, but also, *and crucially*, the 'internal' relationships within group theory, as manifested in these reciprocity laws. Weyl himself contributed to the elucidation of these relationships and, in the reduction of the representations of the orthogonal group, extended the group-theoretical side in such a way that spin could subsequently be accommodated. In particular, his approach yields the so-called 'double valued representations' (or, more accurately, the representations of a double 'covering' which came to be called the spinor group) that play a crucial role in our understanding and representation of spin (see Wigner 1959, pp. 157–70).

Thus the group-theoretic approach delivered on both the foundations and applications: within the Weyl programme, it gave both the Heisenberg commutation relations and Schrödinger's equation; within Wigner's, it not only provided the classification of atomic line spectra, taking into account the exclusion principle and spin (Wigner 1959, chapter V)—leading, as we just noted, to Weyl emphasising its role as the most appropriate 'mathematical tool' in this context—but also offered the basis of a theoretical understanding of the nature of the homopolar molecular bond (1931, pp. 341–2) and chemical valency in general (ibid., pp. 372–7)—prompting Heitler to famously declare, now '[w]e can [. . .] eat Chemistry with a spoon' (Gavroglu 1995, p. 54).[28]

Of course, this approach also met with some resistance. Group theory was quite widely referred to as the 'gruppenpest'[29] and many physicists like Hartree, for example, were less than enthusiastic, although he eventually admitted: 'Is it really going to be necessary for the physicist and chemist of the future to know group theory? I am beginning to think it may be' (quoted by Gavroglu 1995, p. 56). Slater

[28] Interestingly, from the formal perspective offered here, in his own discussion of the physical basis of chemical valence, Weyl presented the relationship between chemistry and quantum physics in terms of a hierarchy of structures (Gavroglu 1995, pp. 266–75).

[29] Wigner records Pauli and Schrödinger using this word (Wigner 1963). But he also recalls von Neumann dismissing them as 'old fogies' and insisting that "... in five years every student will learn group theory as a matter of course" (ibid.).

was even hailed as having 'slain the Gruppenpest' by Condon and Shortley in their 1935 work, *Theory of Atomic Spectra*. However, formally speaking, the devices appealed to by Slater, Condon, and Shortley as substitutes for group-theoretic notions, namely the linear operators of angular momenta, are nothing more than the generators of the Lie algebra of the rotation group SO(3) (Mackey 1993). Likewise, Dirac's 1929 presentation of a non-group-theoretic rewriting of Weyl's results in terms of permutation operators also foundered:

In 1928 Dirac gave a seminar, at the end of which Weyl protested that Dirac had said he would make no use of group theory but that in fact most of his arguments were applications of group theory. Dirac replied, 'I said that I would obtain the results without *previous* knowledge of group theory!' (Coleman 1997, p. 13; italics in the original)

In a later interview, Wigner noted that there was a 'certain enmity' at the time, recalling that, 'most people thought, "Oh, that's a nuisance. Why should I learn group theory? It is not physical and has nothing to do with it"' (1963, transcript 2). But then, as he continued, with regard to the notion of the quantum state:

People like to think of motions, which is not, in my opinion, and which even in that day was not, in my opinion, the right way to think about stationary states. Nothing moves, and this is what I think I digested much earlier than most people; in a stationary state nothing moves, but this is what they did not want to accept. They said, "Well, you see something going around," when actually you don't. For instance, my shells did not move, and it was evident to me that nothing moves. (ibid.)

With nothing moving, some alternative framework had to be applied, and for Wigner, of course, this was to be found in group theory.

There are two further important features of this history that we would like to emphasize at this point: first of all, behind these 'surface' relationships there lie deeper mathematical ones. One such is the reciprocity between the permutation and linear groups, which Weyl referred to as 'the guiding principle' of his work (1931, p. 377) and also as a 'bridge' within group theory leading from the character of the unitary group to that of the permutation group (Weyl 1968, pp. 286–7), as noted above: 'the substratum of a representation of the [complete linear group] consists of the linear manifold of all tensors of order f which satisfy certain symmetry conditions, and the symmetry properties of a tensor are expressed by linear relations between it and the tensors obtained from it by the $f!$ permutations' (Weyl 1931, p. 281).[30] It has a crucial practical significance since continuous groups can be more easily handled than discrete ones. Thus, the application of group theory to quantum physics was critically dependent on relationships internal to the former. Again, this leads us to conclude that the appropriate way to view the mathematics–science relationship, in such important cases, is in terms of families of structures on either

---

[30] This follows from the more general 'Reciprocity Theorem' of Frobenius (see Weyl 1931).

side. Putting it bluntly, the application of group theory to quantum physics depends on the existence of this bridge between structures within the former.

Second, it is important to acknowledge that both group theory and quantum mechanics were in a state of flux at the time they were brought into contact and both subsequently underwent further development. The relevant structures—both mathematical and physical—may therefore be regarded as significantly open in various dimensions and the partial structures programme provides the appropriate formalization of this feature. From such a perspective, both mathematical change and scientific change can be regarded as on a par. Furthermore, what we have in this case is the *partial* importation of mathematical structures into the physical realm which can then be formally accommodated by means of a partial homomorphism (Bueno, French, and Ladyman 2002). As we emphasized in Chapter 2, this is weaker than 'full' isomorphism, which would clearly be inappropriate in this context, but strong enough to underpin the applicability of the mathematics.

In general, then, from the perspective of the model-theoretic approach, what we have is a theory, quantum mechanics, which is already profoundly mathematized—in terms of matrices, wave functions, and so on—and which is then embedded in a mathematical structure, namely group theory, in the manner indicated, again, in Chapter 2. Weyl himself makes this explicit, with his talk of mathematical 'interpretation' and 'counterparts'. The isomorphisms between these counterparts in the substructure of group theory and elements of quantum mechanics are what formally constitute the embedding relation. In the case of the rotation group, the embedding is characterized by Weyl's two 'circumstances' above.

Furthermore, it is worth noting that heuristically speaking, this is obviously not a situation in which structure is imported from a well-confirmed theory modelling one domain into the domain of other phenomena, as in the case of gauge invariance for example (French 1997). Rather, what we have is the establishment of a correspondence between part of physical theory—quantum statistics—and an aspect of mathematics—group theory—which then motivates the embedding of the former into the latter. In addition, there were clearly enormous advantages to re-expressing quantum mechanics within this new mathematical structure as a disparate set of ad hoc models and broadly phenomenological principles were brought together into a coherent framework. Indeed, as far as Weyl was concerned, the result was nothing less than a 'new' quantum theory. Thus, we have a nice example of the kind of schema outlined by Redhead, as we have described in Chapter 2.

And finally, the situation was not a static one, as we have noted, on either side of the 'bridge'. Both group theory and quantum mechanics were evolving at this time[31] and both exhibited a certain structural openness that allowed for such further

---

[31] As Sarah Kattau pointed out to us, one shouldn't place too much emphasis on the fortuitous nature of these twin developments since Euler may have come close to formulating the axioms of group theory (Speiser 1987).

developments—a particular example being the incorporation of spin on the one hand and the elaboration of spinor theory on the other. Thus, the appropriate model-theoretic formulation has to be one involving openness and partiality in general or partial structures in particular[32] and the relations between the corresponding structures would consequently be those of partial isomorphism. Furthermore, each theory, group theory and quantum mechanics, is itself structured, in the way in which Post and Redhead, considering physical theories, have indicated. With this framework in hand, of partial isomorphisms holding between complex structures, we can begin to get a better grasp on one of the most important and resonant episodes in the history of modern science.

4.4 Applying Group Theory to Nuclei

As we have already noted, it is to Wigner's 1931 work, *Group Theory and Its Application to the Quantum Mechanics of Atomic Spectra*, that physicists subsequently turned when faced with the 'particle zoo' of modern elementary particle physics (see also Mackey 1993, pp. 252–4; see also the comments by Gell-Mann in Doncel et al. 1987, p. 487). However, even before completing his book, Wigner had begun to apply the theory of group representations to other problems in physics (Mackey 1993, pp. 254–78). In particular, in his well-known 1937 paper, Wigner started to tackle nuclear spectroscopy. The inspiration for this development was, again, the work of Heisenberg.

In his 1932 paper that effectively marks the beginning of nuclear physics, Heisenberg considered the forces between protons and neutrons by analogy with his earlier account of the exchange forces in the ionized hydrogen molecule (Heisenberg 1932). Exploring the nature of these new forces, Heisenberg introduced a new internal symmetry in terms of which neutrons and protons could be regarded as two different states of a single particle, the nucleon (see Miller 1987, pp. 316–20). It was precisely this idea that Wigner then exploited in order to construct an analogy between atomic and nuclear structure, by making certain idealizations that allowed him to treat these 'nucleons' in a manner similar to that of electrons.

At the very beginning of his paper, Wigner writes that 'recent investigations appear to show that the forces between all pairs of constituents of the nucleus are approximately equal. This makes it desirable to treat the protons and neutrons on an

---

[32] Suppes (1967), of course, gives group theory as an example demonstrating the advantages of employing the model-theoretic approach. The present work can be viewed as an extension of this application in line with his remark that, 'The set-theoretical definitions of the theory of mechanics, the theory of thermodynamics, the theory of learning, to give three rather disparate examples, are on all fours with the definitions of the purely mathematical theories of groups, rings, fields, etc. From the philosophical standpoint, there is no sharp distinction between pure and applied mathematics' (Suppes 1967, pp. 29–30).

equal footing' (Wigner 1937, p. 106).[33] This, of course, is to assume that the electrostatic repulsion between protons has been allowed for. Together with the approximate equality of forces, the mass of the proton is also almost equal to that of the nucleon; in other words, the two particles are approximately indistinguishable apart from their charge. It is by making these twin idealizing moves that protons and neutrons can be treated as two states of the 'nucleon', as indicated previously. Obviously, there is a difference between, on the one hand, the non-idealized model which would satisfy the 'full' Schrödinger equation that takes into account all the forces involved and the actual difference in masses, *and*, on the other hand, the model obtained by introducing these idealizations. Both can be represented in terms of partial structures, $<A, R>$, where the R are understood as partial in the sense described in Chapter 2 and the difference between them would be in the R_i in the first place.

However, these idealizations in the relations and properties of the particles then drive a difference in the As, as we move from two sets, one of protons and the other of neutrons, to one of nucleons. These kinds of idealizing moves can be represented via partial isomorphisms holding between the partial structures (French and Ladyman 1998): taking them in stages we move from protons and neutrons with non-equal forces, to a model with protons and neutrons and equal forces, to one of nucleons. At each stage, the relationship can be represented in terms of our formal framework.

Merging these moves together, the fundamental idealization is to regard the nucleus, which actually consists of two kinds of particle, protons and neutrons, as consisting of one kind, the nucleon. Thus, the nucleus can be regarded as an assembly of indistinguishable particles. By analogy with the situation in the atom this suggests the introduction of a further symmetry group and immediately after the above quote, Wigner refers to Heisenberg's work, recasting the latter's internal symmetry in terms of the new property of 'isotopic spin'.

The analogy, then, is between representations of nucleons and representations of electron spin. The decomposition of the Hilbert space for a nucleon into proton and neutron subspaces is analogous to the decomposition of the corresponding Hilbert space for the spin of an electron, say: if an 'axis' of spin is defined, the two-dimensional Hilbert space for an electron may be written as a direct sum of two one-dimensional subspaces corresponding to spin 'up' and 'down', respectively.[34]

What we have here is an isomorphism between the anti-symmetrized tensor power of the direct sum of two Hilbert spaces and a direct sum of products of anti-symmetrized tensor powers which reduces the problem of determining the interaction between n_1 protons and n_2 neutrons to that of considering $n_1 \times n_2$ 'particles' of

---

[33] The importance of experimental considerations at this point cannot be underestimated. Wigner himself subsequently commented that, in this case, these 'purely' experimental discoveries had a 'profound effect' by making it possible to introduce isospin (Doncel et al. 1987, p. 326).

[34] Indeed, the relevant groups have isomorphic Lie algebras.

the same kind, the Hilbert space of each of which is the direct sum of the proton and neutron Hilbert spaces (Mackey 1993, pp. 257–8). The analogy between atomic and nuclear structure thus reduces to that which holds between the anti-symmetrized nth power of the Hilbert space for a system of n electrons in an atom and the anti-symmetrized nth power of the Hilbert space for a system of n nucleons in a nucleus.

However, as Mackey emphasizes, the analogy is multiply incomplete (ibid., p. 259) and hence partial. First of all, the proton/neutron decomposition does not depend on choosing an 'axis'. Second, and more fundamentally, both protons and neutrons also have spin 1/2 (Wigner 1937, p. 107). Hence the representations of the rotation group in the relevant Hilbert spaces are irreducible in the electron case but the direct sum of two equivalent irreducible representations in that of the nucleons. The introduction of isospin, on the physics side, therefore requires, on the mathematical side, the use of an appropriate symmetry group which will be more complicated than in the atomic case since the corresponding Hilbert space is of higher dimension (Mackey 1993, p. 259). Thus, in section 3 of his paper Wigner attempts to 'define the analog of the multiplet system' (1937, p. 109) of spectral lines in the atom. Now, however, for every particle there are *two* spin coordinates, one for 'ordinary' spin and another for isotopic spin, giving four different sets of values. Instead of two two-valued spins, these values can be covered by one four-valued spin that, Wigner notes, 'plays the same role which the two-valued spin plays in the ordinary spin theory' (ibid.).

However, because of this four-valuedness, 'instead of the representations of the two-dimensional unitary group (or the equivalent three-dimensional rotation group), the representations of the four-dimensional unitary group will characterize the multiplet systems' (ibid.).[35] As Mackey emphasizes (1993, p. 260), the former representations were familiar to physicists from the theory of angular momentum, but the latter were not and Wigner devotes most of section 3 to presenting the mathematical results not only of Cartan and Schur, but of Weyl in particular (Wigner 1937, p. 109). Thus he shows how the theory of rotational symmetry, familiar from atomic systems, must be modified in the nuclear case when the rotation group is replaced with this higher-dimensional unitary group U(4). Indeed, what one obtains, instead of multiplets, are what came to be known as 'supermultiplets' of nuclei (see Wigner 1937, pp. 112–13).

This incomplete analogy between atomic and nuclear structure can be straightforwardly represented in terms of partial structures. In particular, here we have, in general terms, the positive analogy, represented by R_1, which holds between the atom with its electrons and central nucleus and the nucleus itself, with its nucleons and

---

[35] There is a further story to be told about the role of Young's tableaux in graphically representing this sort of situation. Introduced in 1901, they offer a useful way of capturing the representations of the symmetric group and were presented by Wigner in his 1931 book to handle the case of spin (Wigner 1959 [1931]); see also Weyl (1968, vol. III, p. 679).

centre of gravity. Going a little deeper, there is a twofold analogy (a) between the treatment of the nuclear particles as indistinguishable, that is as nucleons, and the indistinguishability of the electrons; and (b) between the spin of the electrons and the isospin of the nucleons. Mathematically this twofold analogy is covered by the permutation and unitary 'spinor' groups, respectively. Note that, with regard to (a), the indistinguishability of the nucleons is only approximate, whereas it is obviously not for electrons. Thus, we shift from a structure in which the properties of the two kinds of particle are different to one in which they are held to be the same. It is only once this idealizing move has been made that the permutation group can be applied, to the set of nucleons, and the analogy with the electron case established.

Concerning (b) the positive analogy is to be understood as holding between the direct sum decompositions into the relevant subspaces. The negative analogy, represented by our R_2, is also twofold: first, there is no 'axis' of isotopic spin in the nucleon case and more profoundly, the relevant Hilbert space is of a higher dimension since both protons and neutrons also have spin. Thus the deeper disanalogy between the two structures is that the rotation group must be replaced by the four-dimensional unitary group U(4) and a somewhat different set of group-theoretical techniques are required. Wigner's paper was doubly significant in not only giving a qualitative account of aspects of nuclear structure, such as the stability of nuclei, but also in providing a general method for classifying nuclear states. Isospin then went on to become an important feature of elementary particle physics, as the relevant structures were extended via the R_3 (see French 1997).

On the basis of this pioneering work, in 1938, Kemmer predicted the existence of the pion triplet that was subsequently discovered nine years later. More significantly perhaps, as is well known, it was the attempt to combine isospin with a further new property, namely strangeness, which led to Gell-Mann's famous classification of hadrons into the 'Eight-Fold Way'.[36] This in turn led to the quark model and, eventually, quantum chromodynamics.[37] Within this model, the symmetry associated with isospin, as captured group-theoretically by the SU(2) group, was extended to what is now called 'flavor symmetry', which has isospin as its subgroup and which can be described mathematically by the SU(3) group (also proposed independently by Ne'eman). The three quarks proposed by Gell-Mann (and Zweig, again independently) belong to the fundamental representation of this group, with two of them (the 'up' and 'down' quarks) generating the isospin symmetry (the other being the 'strange' quark). The discovery of further quark flavours ('top', 'bottom', and 'charm') generated a further expansion to the symmetry, again predicated on

---

[36] Hadrons include protons and neutrons, as well as other kinds of particle and are themselves composed of quarks, in the case of baryons, or quarks and anti-quarks when it comes to mesons.

[37] For the history of these developments see Gell-Mann (1987) and (Ne'eman 1987) and for the role of group theory in particular, see pp. 485–90 and pp. 512–14, respectively; the comments by Speiser in Doncel et al. (1987, pp. 552–3) are also useful.

'idealizing away' the relevant mass differences (which 'break' the symmetry). With these developments, isospin ceased to be regarded as 'fundamental', but then—again galloping through some further interesting history—with the development of colour and also that of the electroweak group, SU(3) too was demoted to the status of an 'accidental' symmetry (Ne'eman 1987, pp. 516–17).[38] Hence, 'isospin symmetry is just a reflection of the colour symmetry, and the fact that there happens to be two light quarks. That's a sort of accident' (comments by Fritzsch in Doncel et al. 1987, p. 633). Likewise, on the mathematical side Wigner, Weyl, Mackey, and others went on to initiate new developments, such as in the general theory of unitary group representations (see Mackey 1993, pp. 265–78; a broader but more erratic discussion is given in Coleman 1997). There is, of course, much, much more to be said here but again, from our perspective, the story is one of establishing relationships between structures that are essentially, and creatively, open or partial.

Let us conclude this section by recalling the following points:

(1) Both the physics and the mathematics were developing at the time, in an open-ended manner. And both Wigner and, in particular, Weyl were well placed to span the two fields, drawing on recently obtained results from one to illuminate the equally recently obtained results of the other.

(2) The relationship between these two fields depends on relationships 'internal' to the mathematical: in particular, the reciprocity 'laws' between the representations of the symmetry group and the unitary group, referred to by Weyl, as we have seen, as the fundamental 'bridge' and 'guiding principle'.

(3) Not all of the mathematics of group theory is brought to bear on the physics; there is a significant 'surplus' on the mathematical side (cf. Redhead 1975).

(4) Wigner's development of isospin can be seen as an *extension* by analogy with the atomic case.

(5) That analogy rode on the back of certain crucial idealizations; and

(6) the analogy was significantly incomplete, or partial.

An account of the applicability of group theory to physics that has any pretensions as to adequacy with regard to the actual practice of science must accommodate these points.

Thus, in terms of our overall framework, we can say that (parts of) group theory provide a *representation* of the subatomic world. And the use of partial structures allows us to accommodate the open-ended nature of this representation, its (intrinsic) mathematical character, and the surplus structure provided by the introduction of group-theoretic structures.

---

[38] And with the suggestion of further structure beneath the level of quarks, colour and the electroweak group may also be regarded as 'accidents' (Ne'eman 1987, p. 517).

Moreover, this account can also be used to suitably frame the isospin case. As noted above:

(i) The introduction of isospin was made via an *extension* by analogy with the atomic case.
(ii) That analogy depended on decisive idealizations.
(iii) The analogy was significantly incomplete, or partial.

With regard to (i), we can represent the extension provided by Wigner with the notion of isospin in terms of the formulation of a convenient *A-normal structure*, which *extends* the (partial) information about the atom into the nucleon itself. Concerning (ii), the idealizations involved in this process, as noted above, can be straightforwardly accommodated in terms of a partial homomorphism between the relevant structures. The R_1 and R_2 relations that are known to hold (such as the decomposition of the Hilbert space for a nucleon into proton and neutron subspaces and the decomposition of the two-dimensional Hilbert space for the spin of an electron into a direct sum of two one-dimensional subspaces) are carried over by the partial homomorphism. Finally, as for (iii), that the analogy between atomic and nuclear structure is *partial* is not surprising, for the relationship between the relevant structures is similarly partial; only a *partial* homomorphism holds between them. After all, as we noted, whereas the above decomposition for protons and neutrons doesn't depend on choosing an axis, it does in the case of electron spin. In this way, the nature of the argument used by Wigner in the introduction of the isospin (a *partial analogy* between atomic and nuclear structure) can also be accommodated.

Let us now turn our attention back to the issue of applicability, and Steiner's understanding of Wigner's point about the 'Unreasonable Effectiveness of Mathematics' in terms of the unreasonable effectiveness of mathematics in *scientific discovery* (Steiner 1998). It is precisely in the above context that Steiner sets up his central mystery, giving the example of the move from spin to isospin and then from isospin to SU(3) discussed above as a prime example of what he calls a 'formal argument' from analogy:

Suppose we have effected a successful classification of a family of 'objects' on the basis of a mathematical structure S. Then we project that this structure, or some related mathematical structure T, should be useful in classifying other families of objects, even if (a) structure S is not known to be equivalent to any physical property, and (b) the relationship between structures S and T is not known to be reducible to a physical relation.   (Steiner 1989, p. 460, 1998, p. 84)

But, he argues, 'such formal analogies appear to be irrelevant analogies and irrelevant analogies should not work at all' (Steiner 1989, p. 454). So, the success of such analogies is puzzling and 'unreasonable'.

What is meant here becomes apparent through this example of the introduction of isospin (Steiner 1998, pp. 86–8). Thus Steiner claims that Heisenberg 'conjectured boldly' that the proton and neutron could be regarded as two states of the same particle and reasoned that the nucleus was invariant under the SU(2) group. There-fore, there had to exist a new conserved quantity, *mathematically* analogous to spin, subsequently known as isospin. The emphasis here is crucial: it is the mathematics, Steiner claims, which does all the work in the analogy:

> physicists see no *physical* analogy between [...] 'spin' and 'isospin' and therefore have no explanation for the success of Heisenberg's reasoning.   (Steiner 1998, p. 90; italics in the original)

Steiner calls this form of analogy 'Pythagorean', in the sense that it is mathematical and not paraphrasable into non-mathematical language. And, in precisely this sense, he goes on to insist, it is 'anthropomorphic' rather than 'naturalist', in that it depends on certain aesthetic qualities (which we select when we construct mathematical structures).

Setting aside Steiner's reconstruction of the relevant history for the moment, one's first response might be that the success of such analogies is surely not puzzling or unreasonable, since group theory represents various symmetries, in turn understood as physical, and thus its use in discovery is empirically justifiable (Steiner 1989, pp. 464–6). However, Steiner counters, not all of the groups used in modern elementary particle physics express empirical or perceptual or geometrical symmet-ries (Steiner 1989, pp. 464–6, 1998, pp. 150–1).[39] Thus, he insists, the grounding of group theory in this regard is not straightforwardly empirical.

A second response might be to accept this point and, indeed, to take it further by noting that there was a fundamental non-perceptual and non-geometric symmetry present at the very beginning of the introduction of group theory into quantum physics, namely the permutation symmetry, which is a result of the non-classical indistinguishability of the particles. Hence, any mysterious unreasonableness cannot be a function of the move from isospin to unitary spin. What our story above illustrates is that the introduction of isospin was dependent on the empirical 'facts' of the approximate equality of forces and masses for protons and neutrons, and the resulting idealizations and (partial) analogies. That group theory worked is then no surprise or mystery at all given the fundamental idealization of regarding the nucleus in terms of an assembly of indistinguishable particles.[40]

Here again Steiner has a possible counter:

> physicists developed the SU(3) concept, for reasons unconnected to particle physics. They were attempting to classify continuous groups, for their own sake.   (1998, p. 152)

Thus, Wigner's issue is understood as an epistemological one: 'how does the mathe-matician, closer to the artist than to the explorer, by turning away from nature, arrive at its most appropriate descriptions?' (ibid., p. 154).

---

[39] For recent discussions of the issue of which symmetries can be regarded as empirical and on what basis, see Friederich (2015) and Greaves and Wallace (2014).

[40] And at this point it is worth recalling that in the context of SU(3) spontaneous symmetry breaking must be introduced to get the difference in masses.

But at this point we can hold back our historical considerations no longer! Steiner's last remark surely mis-poses the issue: *the mathematician does not 'arrive' at a description of nature*. Recall the story above: a variety of theoretical moves were made and at the highest theoretical (deepest metaphysical) level, electrons came to be regarded as (non-classically) indistinguishable—very much like mathematical points in fact! This invites a connection (represented by a homomorphism) with the permutation group and the whole family of structures making up group theory. Further developments—idealizations in particular—at the theoretical level, elaborated on the basis of the analogies then permitted, allow us to draw on further members of this family. Steiner overemphasizes the role of mathematics and creates a mystery where none exists: it is not the mathematics which does all the work in these analogies; rather it is the combination of empirical 'facts' and idealizing moves leading to indistinguishability. If we pay attention to the historical details, the details of Heisenberg's and Wigner's reasoning are readily apparent, as we have indicated above, and, furthermore, can be appropriately framed through the use of partial structures.[41]

4.5 Conclusion

Once the importance of the non-classical indistinguishability of quantum particles is grasped, group theory, particularly in the form of the permutation group, no longer seems so 'unreasonably' effective. In these cases, group theory is turned to because of mathematical intractability, within the context of the Wigner programme. As Wigner himself put it in the preface to his book:

The actual solution of quantum mechanical equations is, in general, so difficult that one obtains by direct calculation only crude approximations to the real solutions. It is gratifying, therefore, that a large part of the relevant results can be deduced by considering the fundamental symmetry operations.   (1959, p. v)

Both Heisenberg and Wigner were faced with this intractability when they considered the two-electron case by 'elementary' methods. Group theory allowed physicists to overcome the many-body problem and relate quantum mechanics to the data. This may suggest that the mathematics can be regarded as nothing more than a tool and somehow set apart from the 'real' physics. For Weyl, however, group theory was nothing less than the 'appropriate language' for quantum mechanics (see Weyl 1968, p. 291). Reading his book one has the sense that it is only through group theory that quantum mechanics approaches a coherent, unified theoretical structure. This is not simply a result of the Weyl programme, nor is it simply because it was through group theory that the tractability problem could be overcome and calculational results

---

[41] For further discussion of Steiner's approach, again see Bangu (2006), Kattau (2001), and Simons (2001).

could be obtained from quantum mechanics. It is because group theory enabled a wide range of 'ad hoc' and semi-empirical rules, such as Hund's 'vector addition model' of atomic spectra (French 1999) and Laporte's rule for dipole transitions between states of different parity (von Meyenn 1987, p. 341), to be brought under a coherent, unified framework. Here group theory illuminates the connection between symmetry and regularity (ibid.).

Thus, we recall, as far as the Weyl programme is concerned, the mathematics of group theory is applied, not to overcome a computational difficulty, but to ground the very principles of quantum theory. Weyl, in his 1927 paper, wanted to base Heisenberg's commutation relations on something deeper than an analogy with the classical Poisson brackets, namely certain fundamental symmetry principles, expressed group-theoretically (see Mackey 1993, pp. 249–51). His conclusion, given in his book but citing his 1927 paper, was that, as we noted:

> The kinematical structure of a physical system is expressed by an irreducible Abelian group of unitary ray rotations in system space. The real elements of the algebra of this group are the physical quantities of the system; the representation of the abstract group by rotations of system space associates with each such quantity a definite Hermitian form which 'represents' it.
>
> (1931, p. 275)

If the group is continuous, Heisenberg's formulation follows 'automatically' and, in particular, the relevant pairs of canonical variables then follow from the requirement of irreducibility. Weyl then showed 'by actual construction' that only one irreducible representation of a two-parameter continuous Abelian group exists, namely the one which leads to Schrödinger's equation (again, see Mackey 1993, pp. 249–51 and pp. 274–5). Here too we have representation and association and relationships between mathematical and physical structures. The motivation is different from the Wigner case, but the analysis from our perspective will be the same.

Returning to the notion of surplus structure on the mathematical side, Weyl went on to remark that 'the field of discrete groups offers many possibilities which we have not as yet been able to realize in Nature' (ibid., p. 276) and speculated that these 'holes' might be filled by applications to nuclear physics.[42] The application of U(4) to isospin is a nice example of this. However, he continues:

> It seems more probable that the scheme of quantum kinematics will share the fate of the general scheme of quantum mechanics: to be submerged in the concrete physical laws of the only existing physical structure, the actual world. (ibid.)

Ultimately, the solution to the problem of what Wigner himself referred to as 'the unreasonable effectiveness of mathematics'[43] in this case lies in the two fundamental

---

[42] Again, see our earlier comments in note 29 of Chapter 2, for example.

[43] And here the focus is more on the descriptive role of mathematics than on its role in discovery, as emphasized by Steiner, but, as we have argued, the latter is rendered mysterious by the failure to pay attention to the former and in particular to the idealizations that are introduced.

symmetries of permutation and rotational invariance. For the realist, they are aspects of 'the only existing physical structure' (see French 2014); for the empiricist, they are features of our theoretical structures, pragmatically useful for making predictions, for example. In either case, they are what drive the importation of group theory into physics. As Grattan-Guinness remarked:

> One can understand *how* the 'unreasonable effectiveness of mathematics in the natural sciences' occurs: there is no need to share the perplexity of [Wigner] on this point if one looks carefully to see what is happening. [...] The *genuine* source of perplexity that the mathematician-philosopher should consider is the *variety* of structures and of levels of content that can obtain within one mathematico-scientific context.   (1992, p. 105; italics in the original)

What we have given here is an example of the 'wrestling into agreement' of the mathematics and physics that we discussed in Chapter 1. And, as we have emphasized, the process is first of all dynamic, on both sides of the applicability divide, requiring the construction of a 'bridge' between the theoretical and mathematical structures (with, on the quantum mechanical side, the reduction of the state space into irreducible subspaces and, on the group-theoretical side, the reduction of representations); and, second, it involves crucial idealizations (such as the treatment of protons and neutrons as indistinguishable)—two aspects of the applicability problem which have not been given due attention in the literature. By paying attention to the actual details of this process within an appropriate representational framework, we can dispel the air of mystery about applicability and understand how the effectiveness of mathematics is not so unreasonable after all.

In Chapter 5, we will extend this analysis to cover not only the 'top-down' application of group theory but also the 'bottom-up' construction of models of the phenomena. Our case study will be London's explanation of the superfluid behaviour of liquid helium in terms of Bose–Einstein statistics. And our conclusion will be that the models involved and the relationships between them can again be accommodated by the partial structures approach, coupled with an appreciation of the heuristic moves involved in scientific work.

5

Representing Physical Phenomena
Top-Down and Bottom-Up

5.1 Introduction

Our aim in this chapter is to pursue our consideration of the applicability of mathematics—specifically group theory—to physics via the exploration of a series of moves that involve this form of applicability at the 'top' end, as it were, together with, at the bottom end, the applicability of theoretical structures to what van Fraassen calls the 'appearances' of the empirical domain. The example we have chosen is London's famous analysis of the superfluid behaviour of liquid helium in terms of Bose–Einstein statistics, represented via symmetrized wave functions, as described in Chapter 4. This involved not only the introduction of 'high-level' mathematics in the form of group theory but also a degree of modelling at the phenomenological level, and thus offers another nice case study illustrating the kinds of relationships we are interested in.

5.2 From the Top: The Applicability of Mathematics

Let us recall our concerns regarding various attempts to represent the relationship between mathematics and science. At one end of the spectrum are those that articulate this relationship in terms of *structure instantiation*, in which mathematical structures are taken to be instantiated at the empirical level (Shapiro 1997).[1] This sort of proposal is both rough and imprecise: it is rough because it ignores the complex hierarchy of models that exist in science and which occupy the epistemic space between data models and high-level theory; it is imprecise because, relatedly, it is not clear what it is to instantiate what might be, and often is in the case of theoretical physics, very high-level mathematics, with infinite structures.[2] At the other end of the spectrum are those accounts that represent the relationship in terms of isomorphisms holding between mathematical and physical domains (see Hellman 1989).

---

[1] As we noted in note 23 of Chapter 2.

[2] A finite physical structure underdetermines the infinity of mathematical structures that could be taken to overlap with its initial segment.

By means of such a formal relationship, the preservation of structure from the former to the latter can be accommodated. However, isomorphism is too strong (Bueno 2013a): it is typically not the case that all of a mathematical structure is imported into the physical domain, but only part, as we have repeatedly indicated.

Thus, in order to represent the relationship between mathematics and science, we need something that is less vague than instantiation, but weaker than isomorphism in order to accommodate the openness of scientific practice. Furthermore, it should accommodate Redhead's point about the heuristic fertility of mathematical 'surplus structure' (Redhead 1975), in the sense that such structure, related to the physical theory via an embedding relationship, might then be interpreted in terms of the theory and thus be used to extend it further. The structure is genuinely surplus in the sense of bringing new mathematical resources into play and, as we have suggested, the physical theory should be regarded as embedded not just into a mathematical structure, but into a whole *family* of such structures. From the perspective of the philosophy of science, then, we need something that can capture this 'surplus structure' that mathematics can bring to a physical situation and which can also accommodate the openness of these structures.[3] We have suggested here that this relationship between mathematical and physical theories can be thought of in terms of a partial mapping from one domain to the other and that this is conveniently represented by a partial homomorphism between the structures concerned, as defined in Chapter 2.

What this allows us to do is to represent, first of all, the partial importation of mathematical structures into the physical domain, and second, the interrelationships between mathematical structures themselves constituting the surplus structure. The determination of which structure, at the mathematical level, comes to be used in the construction of theories at the physical level, is an issue that again falls under the label of heuristics, as we have already indicated (see again note 29 of Chapter 2; also see French 1997). What is important for our present purposes is that our framework allows for there to be a kind of structural 'space' for the introduction of further parameters in case more structure is needed. Of course, this surplus mathematical structure cannot be represented simply in terms of more n-tuples of objects in the relevant domain. As we have indicated above, it should be represented in terms of a family of structures by means of which we can then represent how a given structure can be extended by the addition of new elements to its domain, or the addition of new relations and functions defined over these elements. The whole family of such extensions represents the surplus structure.

In terms of this framework, then, what we have is the *partial importation* of the relevant mathematical structures into the physical domain, which can be represented by a partial homomorphism holding between the structures characterizing the

---

[3] Again, as we have seen, a good example of this openness at both the mathematical and scientific levels is given by the application of group theory to quantum mechanics as detailed in Chapter 4.

mathematical surplus structure, and the structures of the physical theory under consideration. This effectively allows the carrying over of relevant structural features from the mathematical level to that of the physical theory. The heuristic fertility of the application of mathematics rests on the 'surplus' in the sense that more structure from the family can be imported if required; it is this crucial aspect that is captured by the openness of partial structures. It is important to reiterate that the standard, full, or complete notion of homomorphism cannot capture the fact that only the relevant parts of the family of mathematical structures are imported.

Of course, the above representation of the hierarchy of models leading from the data to the theory can also be further adapted and so what we have are partial homomorphisms all the way down: from the mathematical to the physical and on down through the hierarchy of structures to the data models. Let us now consider our case study from the history of superconductivity that illustrates both the moves that were made in practice and the manner in which our framework can accommodate them.

5.3 Bose–Einstein Statistics and Superfluidity

5.3.1 The Liquid Degeneracy of Helium

The example we have chosen is that of the explanation of the superfluid properties of liquid helium in terms of Bose–Einstein statistics.[4] This is a particularly interesting case study, as the phenomenon to be modelled comes to be regarded as a 'macroscopic' quantum phenomenon, on the one hand, and the explanation of it ultimately ties in to very high-level mathematical structures of a group-theoretical nature, on the other.

The history is well known (Gavroglu 1995, chapter 4; Brush 1983, pp. 172–230; Mehra 1994, chapter 17). From the late 1920s to the late 1930s a variety of experimental results indicated that below a certain temperature (the 'transition temperature'), liquid helium entered a different phase possessing extremely non-classical properties. The heat conductivity of helium II, as it was called, was many times higher than that of copper, for example. Its viscosity was extremely low and, most strangely of all, extremely small temperature differences could produce extremely large convection effects—the so-called 'fountain effect' (see https://www.scientificamerican.com/article/superfluid-can-climb-walls/). These results all indicated that liquid helium below the transition temperature could not be explained in terms of classical hydrodynamics. Furthermore, and significantly, when the specific heat was plotted against temperature a sharp maximum was noted at the transition temperature.

Fritz London was a brilliant theoretical physicist who made profound contributions to quantum chemistry, the theory of molecular forces, the theory of superconductivity,[5]

---

[4] This section provides a more detailed extension of the account presented in French (1999).

[5] For an analysis of this development from the perspective of the partial structures framework see French and Ladyman (1997) and, more recently, Bueno, French, and Ladyman (2012a, 2012b).

and the theory of superfluidity (for an excellent biography see Gavroglu 1995). In particular, he became interested in what he later called the 'mystery' of the 'liquid degeneracy' of liquid helium (ibid., pp. 147–8). The demonstration that liquid helium could not be solidified under its own pressure led to the suggestion that the liquid passed into some kind of ordered state below the transition temperature. On the urging of Mott and his brother, Heinz London, Fritz published a paper arguing that it had a diamond lattice structure. This model was subsequently abandoned in favour of a new model prompted by the 1937 Congress for the Centenary of van de Waals. Although London had no new work to present at the Congress, many of the papers took his analysis of molecular forces as the basis for further developments (ibid., p. 152). Significantly, it was at this Congress that London was led to Einstein's work on what is now called Bose–Einstein statistics, as sketched in Chapter 4 (ibid., p. 153). It is worth noting in particular that Einstein had already realized that the thermodynamical properties of an assembly of particles obeying such statistics would be significantly different from the classical, not least in that, below a certain temperature, they would 'condense' into the lowest accessible state (Einstein 1925; trans. in Duck and Sudarshan 1997, pp. 91–2).[6] As is now well documented, this results in what is in effect a new form of matter and Bose–Einstein condensates are currently the subject of intensive research (see https://www.nobelprize.org/nobel_prizes/physics/laureates/2001/popular.html).

5.3.2 The Application of Bose–Einstein Statistics

With the realization that there were now three kinds of particle statistics (classical, Bose–Einstein, and Fermi–Dirac), a number of papers were published exploring the relationship between them. Ehrenfest and Uhlenbeck, in particular, tackled the question of whether Bose–Einstein and Fermi–Dirac statistics were required by the formalism of quantum mechanics, or whether classical statistics might still be valid in certain areas (Uhlenbeck 1927). They concluded that it is the imposition of symmetry requirements on the set of all solutions of the Schrödinger equation for an assembly of particles, obtained by considering the permutations of all the particles among

---

[6] It is also worth recalling that in this paper, Einstein estimates the critical temperature for both hydrogen and helium and notes that the critical density of helium is only five times smaller than the saturation density of the ideal Bose–Einstein gas of the same temperature and molecular weight (Duck and Sudarshan 1997, pp. 96–7). He also speculates on the possibility of describing the conduction electrons in a metal as such a saturated ideal gas and thus explaining superconductivity (ibid., pp. 97–8). Intriguingly, he poses the difficulty for such an account that in order to accommodate the observed thermal and electrical conductivities compared to the very small density of the electrons, the mean free paths of the electrons in such a gas would have to be very long (of the order of 10^{-3} cm). It was a crucial element of London's theory of superconductivity that a form of long-range order is established which effectively keeps the average momentum constant over comparatively long distances. It is also interesting to note that, in a response to Einstein's work which was subsequently abandoned, Schrödinger adopted a holistic view which attributed quantum states to the body of a gas as a whole, rather than to the individual gas atoms. London was Schrödinger's assistant in 1927 (Gavroglu 1995, pp. 42–3) and, as we shall see, came to regard superfluidity as a manifestation of a kind of holistic quantum phenomenon.

themselves, that produces the symmetric and anti-symmetric combinations.[7] If no symmetry constraints are imposed, then classical, Maxwell–Boltzmann statistics results.

The issue of how to explain this non-classical condensation phenomenon subsequently came to the fore in this context of an appropriate framework for quantum statistics. Thus Schrödinger showed in 1926 that Einstein's gas theory could in fact be recovered by treating the gas as a system of standing de Broglie waves and applying Maxwell–Boltzmann statistics (attempts to retain classical statistics in the quantum context via theoretical adjustments run throughout the history of these developments; see French and Krause 2006). However, this approach ruled out the possibility of the condensation effect—an outcome that from today's perspective renders Schrödinger's approach empirically inadequate but given the uncertainty over the reality of the effect at the time was not deemed to be decisive (Uhlenbeck 1927). The issue was returned to during the van der Waals conference in 1937, mentioned above, where Born presented a paper entitled 'The Statistical Mechanics of Condensing Systems' (Born 1937). This was based on the work of Mayer, with the same title (Mayer 1937), in which the phenomenon of condensation was accepted but another attempt was made to explain it using classical statistical mechanics. Kahn and Uhlenbeck then showed that this was formally analogous to the theory of a Bose–Einstein gas (Kahn and Uhlenbeck 1938a, 1938b), having retracted the dismissal of Bose–Einstein condensation expressed in Uhlenbeck's PhD thesis. As Gavroglu has noted, it was this work of Mayer which led London to Einstein's paper (1995, p. 153).

Thus in his 1938 note to *Nature*, London writes that 'in the course of time the degeneracy of the Bose-Einstein gas has rather got the reputation of having only a purely imaginary existence' (1938a, p. 644). He began by indicating that a static spatial model of liquid helium was not possible, on energetic grounds, but that one could carry over aspects of that sort of approach, specifically the use of a Hartree self-consistent field approximation. On these grounds he claimed that it seemed 'reasonable to imagine a model in which each He atom moves in a self-consistent periodic field formed by the other atoms' (ibid., p. 643). This is similar to the way electrons were considered to move in a metal according to a theory of Bloch's, with the difference that, with helium atoms instead of electrons, 'we are obliged to apply Bose-Einstein statistics instead of Fermi statistics' (ibid., p. 644). This leads him to introduce Einstein's discussion of the 'peculiar condensation phenomenon' of the Bose–Einstein gas. Having acknowledged its reputation as a 'purely imaginary phenomenon', London then writes, 'Thus it is perhaps not generally known that this condensation phenomenon actually represents a discontinuity of the derivative of the specific heat' (ibid.) and, significantly, illustrates this discontinuity with an accompanying figure. It is significant because of the structural similarity with certain

---

[7] Again, as we have noted, and as Dirac among others realized, other combinations are also possible, corresponding to parastatistics.

features of a plot of specific heat against temperature for liquid helium (the so-called λ-point) and London continues, 'it seems difficult not to imagine a connection with the condensation phenomenon of Bose-Einstein statistics' (ibid.), even though the phase transitions are of a different 'order' (we shall return to the issue of representing such phase transitions in Chapter 9, albeit in a quite different context). Furthermore, the experimental values for the transition temperature and entropy agree quite favourably with those calculated for an ideal Bose–Einstein gas. This numerical agreement was critical in the initial stages of this investigation of the structural similarity but, as we shall shortly see, it was subsequently demoted in importance as London came to stress the qualitative explanation that Bose–Einstein statistics could give of a range of liquid helium phenomena, such as the 'fountain effect'.

These phenomena are briefly mentioned at the end of London's note, but before concluding, he also acknowledges the limitations of this model in so simplifying liquid helium as an ideal gas. In particular, according to this model the quantum states of liquid helium would have to correspond to both the states of the electrons and the vibrational states of the lattice in the theory of metals. Both the qualitative explanation and the limitations of the model are taken up in his 1938 *Physical Review* paper, where he writes:

In discussing some properties of liquid helium, I recently realized that Einstein's statement [regarding the condensation effect] has been erroneously discredited; moreover, some support could be given to the idea that the peculiar phase transition ('λ-point'), that liquid helium undergoes at 2.19°K, very probably has to be regarded as the condensation phenomenon of Bose-Einstein statistics.   (1938b, p. 947)

In the first part of the paper London presents 'a quite elementary demonstration of the condensation mechanism' (ibid.) and again presents the specific heat graph of an ideal Bose–Einstein gas with its discontinuity, highlighting the structural similarity. In §2 he considers the nature of this effect and argues that it represents a condensation in momentum space (ibid., p. 951) with a corresponding and characteristic spread of the wave packet in ordinary space.[8] Thus what we have is a 'macroscopic' quantum effect and London draws the comparison with his work on superconductivity of a few years earlier (London and London 1935; London 1935, 1937).[9] In the latter case the macroscopic phenomena can be understood in terms of a 'peculiar coupling' in momentum space, 'as if there were something like a condensed phase' in this space (London 1938b, p. 952; Gavroglu 1995, p. 158). He subsequently identified the mechanism involved as the so-called 'exchange interaction' between the electrons—something that he and Heitler also posited as underpinning the homopolar bond and hence ultimately responsible for the reduction of chemistry, leading

[8] In *Superfluids* this is expressed explicitly as a 'manifestation of quantum-mechanical complementarity' (London 1954, p. 39).

[9] For a discussion of this work see Gavroglu (1995), Cartwright, Shomar, and Suárez (1995), French and Ladyman (1997), and Bueno, French, and Ladyman (2012a, 2012b).

to Heitler's famous exclamation noted in Chapter 4. The understanding of this 'exchange interaction' in terms of anti-symmetrization of the wave function and group theory took some time to emerge clearly (see Carson 1996). And in this case, it was not only the anti-symmetrization that was crucial but also the introduction of the new quantum mechanical property of 'spin', which we shall consider in Chapter 8.

However, London went on to note, this 'assumption' of some kind of coupling could not be grounded in a specific molecular model of superconductivity, whereas it could in the case of superfluid helium. This was a serious concern for London in the former case, for which the relevant model did not emerge until the theory of Bardeen, Cooper, and Schrieffer, in 1957. London himself, like Weyl, was philosophically reflective (schooled in Husserlian phenomenology, as Gavroglu records; 1995) and had a two-state view of theory construction, whereby one first had to obtain an adequate formulation of the phenomenon before moving on to develop a theoretical model of it. The formulation itself should not be regarded as simply empirical and would typically include significant theoretical elements as in the case of superconductivity. Here, in the helium example, the first stage can be seen in the shift from the static spatial model of liquid helium to an 'entirely different conception' (French 1999). However, although certain speculative remarks were made regarding the above coupling, which signified 'an appreciable *reduction* of the mechanism which remains to be explained by the theory of electrons' (London 1937, p. 795; italics in the original; cf. also his comments in London 1948, pp. 570–1 and 1950, p. 4),[10] he was unable to obtain the kind of model he sought. In the case of superfluidity and the behaviour of liquid helium, the degenerate Bose–Einstein gas provides a good example of just such a molecular model (London 1938b, p. 952).[11]

In §3 of the paper, London re-presents, with added detail, his depiction of liquid helium as a metal in which both the lattice ions and conduction electrons are replaced with He atoms and the Fermi–Dirac statistics used in Sommerfeld's theory of metals is replaced with Bose–Einstein statistics.[12] In terms of this model the fluidity of the liquid corresponds to the conductivity of the electrons in the metal. This fluidity becomes infinite as the self-consistent field formed by the atoms tends to zero and this occurs abruptly once the temperature drops below the transition temperature (ibid., p. 953). For a fixed value of the field the 'current' is proportional to the fraction of He atoms with zero energy and velocity. Since these atoms do not

---

[10] And which served as a valuable heuristic for the later work of Bardeen, with the 'coupling' expressed via the concept of 'Cooper pairs'.

[11] Ginzburg, Feynman, and Schafroth subsequently, and independently, indicated the role of Bose–Einstein condensation in accounting for superconductivity (Gavroglu 1995, p. 246).

[12] As justification, he again draws a parallel with the case of superconductivity. There, the 'macroscopic description' is identical with that obtained by treating the superconducting material as one enormous diamagnetic atom (1938b, p. 952). Diamagnetism, of course, could only be explained by quantum mechanics but, nevertheless, one could give a classical treatment using Larmor's theorem. 'One may presume that our attempt [at explaining superfluidity] will prove as much justified as [the latter treatment]' (ibid.).

contribute to energy transport no great increase in heat conductivity should be expected below the transition temperature.

However, London goes on to note, there is another mechanism for producing heat transfer: since the van der Waals forces between the walls of the container and the atoms are much stronger than those between the helium atoms themselves, there will be a greater concentration of degenerate atoms in a layer next to the walls than in the interior.[13] The situation is analogous to that of a thermocouple, giving rise to a thermoforce which produces a 'very great' circulation of matter (ibid., p. 954). This apparent conduction of heat is strongly dependent on the temperature gradient, so that at very low temperatures, where almost all the helium atoms, in both the layer and interior, are degenerate, the thermoforce and the heat effect[14] disappear. 'All this is in qualitative agreement with the experiments', continues London (ibid.), which shows that the heat conductivity becomes 'normal' again below a very low temperature. He also notes that '[t]his mechanism of reversible transformation of heat into mechanical energy gives a very simple explanation also for the so-called "fountain phenomenon" [...] and interprets it as a pump driven by a thermoelement' (ibid.). In the same experiment in which this effect was observed, it was also noticed that in the capillary layer the flow of matter and heat would be in reverse directions as also given by London's model.

Here, then, the emphasis is on the qualitative agreement with experimental observations, rather than the numerical agreement of his earlier note. Indeed, in *Superfluids,* London suggests that this numerical agreement 'would perhaps not have deserved much attention' (1954, p. 59) had not his model offered the promise of a qualitative interpretation of the 'super' properties and 'striking peculiarities'. And the heuristic moves made by London can be straightforwardly accommodated within the partial structures approach. As we have just indicated, what was important was the structural similarity (represented graphically) between the λ-point of helium and the discontinuity in the derivative of the specific heat associated with Bose–Einstein condensation. Note that this similarity presents itself at a rather low level in the hierarchy presented in Chapter 2—at the level of models of the data in fact. However, it was on the basis of this similarity that the family of relations representing Bose–Einstein statistics at the higher theoretical level was projected into the domain of liquid helium phenomena. The analogy between a Bose–Einstein gas and liquid helium was, significantly, only partial in that the phase transitions are of different order—here we have an example of those relations known not to hold (our R_2 in the

---

[13] Van der Waals forces arise from correlations in the quantum fluctuations of nearby particle polarizations. Although weak as compared to covalent bonds, for example, they are famously responsible for the ability of geckos to walk up walls. Interestingly, London made significant contributions to the study of such forces, via his work on so-called 'London dispersion forces' between atoms that are attractive at large distances but repulsive at short distances, due to the anti-symmetry of the electron wave function under permutations (that is due to Fermi–Dirac statistics).

[14] Analogous to the 'Peltier' effect, which is the cooling of one junction and the heating of another when current is passed through a circuit consisting of two different conductors.

characterization of a partial structure). As the analogy was extended, further similarities 'emerged' from the neutral analogy, corresponding to those relations for which it is not known whether they hold or not (that is, our R_3), representing London's explanation of the fountain effect.

However, London himself was explicit about his model being 'highly idealized' (1938b, p. 953) and this aspect can also be captured within our framework. Strictly speaking, 'ordinary' collision theory cannot be applied in this case where the quantum wave packets are spatially so spread out; Ehrenfest's theorem, which allows one to apply classical mechanics to small wave packets moving in external fields, is no longer valid (1938b, pp. 952, 953). More fundamentally, intermolecular forces are not taken into account (ibid., p. 947) and one is treating a liquid as if it were an ideal gas. Hence London concludes his 1938 paper by emphasizing that:

> Though it might appear that the logical connection between the facts will not be qualitatively very different from the one we have sketched here, it is obvious that the theoretical basis given thus far is not to be considered more than a quite rough and preliminary approach to the problem of liquid helium, limited chiefly by the lack of a satisfactory molecular theory of liquids. (1938b, p. 954; see also 1938a, p. 644)

Of course, liquid helium is no ordinary liquid, as its viscosity is rather that of a gas. Thus London later writes:

> This system does *not* represent *a liquid in the ordinary sense*. There are no potential barriers as in ordinary liquids to be overcome when an external stress is applied. The zero point energy is so large that it can carry the atoms over the barriers without requiring the intervention of the thermal motion. In this respect there seems to be a greater similarity to a gas than we are used to assume in ordinary liquids. This view is supported by the extremely significant fact that liquid helium I, which on first sight appears to be quite an ordinary viscous liquid, actually has a viscosity of a type ordinarily found only in gases and not in liquids. (1954, p. 37; italics in the original)

London's model can be appropriately represented in terms of partial structures and the relationships both upwards, between the model, the theory of Bose–Einstein statistics, and further, the mathematics of group theory, and also downwards, between the model and the behaviour of liquid helium, can be captured as follows. Consider: in some respects, corresponding to those relations known to hold (our R_1), liquid helium is like a Bose–Einstein gas. Its viscosity, as noted above, is more like that of a gas than a liquid, for example. Yet in other critical respects, corresponding to those relations known not to hold (our R_2), it is clearly dissimilar; intermolecular forces of the kind that are typical for liquids are not taken into account, for example. There are crucial idealizations involved. Clearly, representing the relationships between the London model and both Bose–Einstein statistics on the one hand, and the data models on the other, in terms of a formal account involving isomorphisms would be entirely inappropriate. As we have seen, Bose–Einstein statistics, in turn, came to be understood, at the highest level, via considerations of symmetry represented group-theoretically; and as we have indicated, the framework of partial homomorphisms provides an appropriate representation of this latter relation.

Thus, in terms of the hierarchy given in Chapter 2, and putting things a little crudely, we have something like the following:

$$M_4 = \langle A_4, R_{4i}, f_{4j}, a_{4k} \rangle_{\ i \in I, j \in J, k \in K}$$
$$M_3 = \langle A_3, R_{3i}, f_{3j}, a_{3k} \rangle_{\ i \in I, j \in J, k \in K}$$
$$M_2 = \langle A_2, R_{2i}, f_{2j}, a_{2k} \rangle_{\ i \in I, j \in J, k \in K}$$
$$M_1 = \langle A_1, R_{1i}, f_{1j}, a_{1k} \rangle_{\ i \in I, j \in J, k \in K}$$

where M_4 represents the mathematical structure of group theory (where Redhead's surplus structure is represented in terms of an associated family of structures), M_3 represents the theory of Bose–Einstein statistics, M_2 represents London's model, with all its idealizations, and M_1, for simplicity, is a condensed representation of the data models, experimental models, and so forth.

Furthermore, this 'rough and preliminary' model was open to further developments—in particular with regard to the extension to a model of a Bose–Einstein liquid. Of course, this openness is delimited by the sorts of heuristic considerations mentioned previously, thus reducing the number of possible extensions, but the partial structures not only represent this openness (via the R_3), they also allow us to represent the loci or fixed points of the extensions (via those relations known to hold, namely the R_1). What is particularly significant about this case is that the point which was held fixed, as it were, had to do with the high-level symmetry considerations. Retaining the component associated with Bose–Einstein statistics, understood group-theoretically, new elements were introduced and previous idealizations could be eliminated, corresponding to a move from our R_3 and changes in the R_1 and R_2.

It was Feynman who, convinced that the liquid helium transition 'has got to do with statistics' (Mehra 1994, p. 350),[15] went on to develop the theory of a Bose–Einstein liquid and, in particular, succeeded in demonstrating that a transition would still occur analogous to that in a gas. In his first paper, Feynman employed his path-integral formulation of quantum mechanics to show that despite interparticle interactions, liquid helium did undergo a phase transition analogous to that found with an ideal gas, because of the symmetry of its states (Feynman 1953):[16]

in other words, the suggestion made by London in 1938 that the transition observed in this liquid might be a manifestation of the phenomenon of Bose-Einstein condensation was basically correct. (Mehra 1994, p. 364)

---

[15] One of the principal reasons for this belief had to do with his view that Landau's alternative approach, in constructing commutation relations for the density and velocity operators of the liquid, treated the particles of the system as 'non-identical', that is as individuals (Mehra 1994, p. 363). This, of course, violates the fundamental basis of the quantum mechanical description of an assembly of particles upon which the above symmetry considerations rest.

[16] Again, the theoretical transition is of a different order than that which the liquid helium actually undergoes. Feynman hoped that a more accurate evaluation of the path-integral would resolve the discrepancy and in 1986 this was achieved (Mehra 1994, p. 365).

In subsequent papers Feynman examined the wave function of the excited state of the system, again in terms of fundamental quantum principles and, again, in particular, in terms of symmetry considerations (ibid., pp. 367–87). The end result was a theory that, in predicting quantum vortices,[17] received conclusive empirical support.

5.4 The Autonomy of London's Model

Clearly, by emphasizing the interrelationships between different levels of our hierarchy, from the mathematical to the empirical, and in particular those that hold between London's model of liquid helium, the theory of Bose–Einstein statistics, and group theory, we find ourselves at odds with those accounts of models in science that emphasize their supposed 'autonomy' from high-level theory (Cartwright et al. 1995; Morgan and Morrison 1999; Suárez and Cartwright 2008). However, one can distinguish different senses of 'autonomy' in this context and it is not always clear from the aforementioned accounts which is being deployed (for criticisms, see da Costa and French 2000).

In its ordinary usage, 'autonomy' means something like self-government. Of course, *models* cannot be taken to govern themselves because theories are necessary to provide the theoretical concepts and laws used in them. In any case, we will consider four possible senses in which models might be said to be autonomous and shall determine which holds in the case of London's model.[18]

> *Autonomy 1*:   Model *M* is autonomous from scientific theory *T* if it is not derived, obtained from, or otherwise related to *T* by any acceptable move, where *T* is *any* theory available.

The term 'obtained' here is ambiguous between 'actually obtained', in the sense of constructed at the time the model was initially put forward, and 'subsequently obtained', that is, constructed at a later time. In either case, this is a very strong sense of autonomy and we know of no cases where a model is autonomous from theory in this sense; certainly London's model was not, as is clear from the brief history given above.

We can then introduce restricted senses of autonomy by taking 'obtained' to mean at the time of model construction and/or taking *T* to be some more fundamental theory that is known to be relevant to the context in which the model is proposed.

> *Autonomy 2a*:   *M* is autonomous from *T* if, at the time it was constructed, it could not be obtained from or related to *T*, where *T* is *any* theory available.

---

[17] Quantum vortices arise when some physical quantity, such as orbital angular momentum in the case of superfluids or magnetic flux in the case of superconductors, is quantized.

[18] These considerations are taken from Bueno, French, and Ladyman (2012b) and build on but go beyond those found in Bueno, French, and Ladyman (2002) and da Costa and French (2003).

Autonomy 2b: *M* is autonomous from *T* if, at the time it was constructed, it could not be obtained from or related to *T*, where *T* is *some relevant 'high-level' or more fundamental theory*.

Autonomy 2a might be satisfied in the case of a model that is constructed purely 'from the ground up', although again we can think of no such cases, and yet again this was not the case for the London model. Autonomy 2b, on the other hand, does seem to match some cases in the history of science, such as the liquid drop model of the nucleus (see da Costa and French 2003). Here the relevant high-level theory might be understood as a detailed quantum mechanically informed theory of the nucleus and that simply was not available at the time the model was constructed. Indeed, the liquid drop model might be seen as another case of a 'characterization' model, but where the role of analogy (in the form of the comparison with a liquid drop) was even more pronounced. In such cases, the proponents of the model may have to introduce certain crucial idealizations in order to establish the requisite connections. This was the case with London's model above, where the nature of the idealizations required to establish the connections between the quantum statistics of an ideal gas and the behaviour of a superfluid effectively forces a kind of 'autonomy' on the model. Thus, in *Superfluids* London writes that:

an understanding of a great number of the most striking peculiarities of liquid helium can be achieved, without entering into any discussion of details of molecular mechanics, merely *on the hypothesis that some of the general features of the degenerating ideal Bose-Einstein gas remain intact, at least qualitatively, for this liquid*, which has such an extremely open structure. This is an assumption which may be judged by the success of its consequences in describing the facts, pending an ultimate justification by the principles of quantum mechanics.

(1954, pp. 59–60; italics in the original)[19]

In other words, only some of the structural relationships embodied in the high-level theory of Bose–Einstein statistics (the 'general features') needed to be imported in order to account for the (low-level) qualitative aspects of the behaviour of liquid helium and this importation can be represented in terms of our framework of partial homomorphisms holding between partial structures, with mathematical structure transferred across at the appropriate level. This framework can also accommodate the manner in which the development of this model ultimately rested on a 'structural similarity' at a relatively low level, which drove a reconceptualization of superfluid phenomena at what might be called the 'phenomenological' level, *and*, further, the manner in which London's 'quite rough and preliminary approach' was open to further extensions at the high level represented by symmetry considerations.

---

[19] He acknowledges Tisza as being the first to recognize 'the possibility of evading the pitfalls of a rigorous molecular-kinetic theory by employing the qualitative properties of a degenerating Bose-Einstein gas to develop a consistent *macroscopic* theory' (ibid.; italics in the original; for further discussion of London's view of Tisza's work, see Gavroglu 1995, pp. 159–63; see also pp. 198–206 and pp. 214–17).

In this sense, then, London's model might be seen as 'autonomous', with subsequent developments reducing, but not completely eliminating, perhaps, that autonomy by indicating how the model could be obtained from a more fundamental quantum mechanical description. However, this form of autonomy seems to us unproblematic, not least because it is temporary (Bueno, French, and Ladyman 2002): as theoretical developments proceed, the autonomy is reduced.[20] What Feynman did, effectively, was to produce a new connection with high-level symmetry considerations that was not dependent on London's idealizations. Of course, it is important to acknowledge that a great deal of what goes on in science may involve such idealizations, leading to such temporarily autonomous models, but it is also important to acknowledge (a) the attitude of the proponents of the models themselves, as represented here by London's insistence that his approach should be considered only 'preliminary'; and (b) the role that is played by both high-level theoretical and mathematical considerations. In this case, we move upwards from the qualitative features of superfluid phenomena to Bose–Einstein statistics and ultimately to group theory.

Autonomy 3:   Model M is autonomous from theory T if it is not obtained from T by an appropriate set of de-idealizations applied to T.

Again, we can obtain two variants depending on whether we take the de-idealizing move to be applied as part of the construction of the model, or subsequently, as it is realized that the already constructed model can be related to some theory by such a move.

In either case, this is obviously a more restricted sense of autonomy, and certainly London's model was not obtained by de-idealizing moves from high-level theory. But, of course, models in general and London's in particular are typically tied to theory through other kinds of moves, as we have indicated. And in *this* sense, they are not autonomous from theory. Furthermore, given that London's model was characterized in the way we have outlined, one should not expect it to be obtained "by improvements legitimated by independently acceptable descriptions of the phenomena" (Suárez and Cartwright 2008, p. 68), since it was itself intended to provide, at least in part, just such an acceptable description via a new phenomenology of liquid helium.

Autonomy 4: M is autonomous from T in the sense that it acts as the locus of epistemic activities.

This is the view of autonomy that Morrison holds when she states that 'models *function* in a way that is partially independent of theory' (1999, p. 43; italics in the

---

[20] Hartmann regards such models as examples of 'preliminary physics' and has noted that this preliminary stage may last many years (Hartmann 1995). For further discussion, see Bueno, French, and Ladyman (2012b).

original). As she notes, it is the appropriate autonomy in model construction that gives rise to such functional autonomy, but even when theory does play a significant role in model construction:

it is still possible to have the kind of functional independence that renders the model an autonomous agent in knowledge production.   (Morrison 1999, p. 43)

This is a very general sense of autonomy, and again (leaving aside issues as to whether models can be regarded as 'agents') it is uncontentious. Indeed, it is surely a matter of common agreement that the roles of models in science may include acting as the locus for knowledge claims, providing the basis for further developments and so on. In this sense, London's model, and the explanatory structure that subsequently arose from it, can certainly be regarded as functionally autonomous.

Thus, we conclude that the London model of liquid helium was autonomous from theory in senses 2b, 3, and 4. However, these can be regarded as uncontentious, either because the autonomy is temporary, or because the model was never intended to be obtained via de-idealization, or because the relevant sense of autonomy relates to the way in which models become the focus of scientific practice. The model was not autonomous in the sense of either 1 or 2a, but these strike us as problematic anyway.

5.5 Conclusion

Our overall claim, then, is that, in the case of the application of Bose–Einstein statistics to superfluids like liquid helium, and moving from top to bottom, from the mathematics to what is observed in the laboratory, the models involved and the relationships between them can be accommodated by the partial structures approach, coupled with an appreciation of the heuristic moves involved in scientific work. Furthermore, we can see how this case fits with the immersion, derivation, and interpretation account of the application of mathematics. Here the phenomena associated with liquid helium are immersed in the framework of Bose–Einstein statistics and, effectively, the mathematics of symmetrization and group theory. We recall again that, as Weyl emphasized, the establishment of the bridge between group theory and quantum statistics was crucially dependent on the existence of further 'bridges' within the mathematics itself; in particular it depended on the famous reciprocity relationship between the permutation group and group of all homogeneous linear transformations. Thus, to repeat, the application of group theory to quantum mechanics depended on certain crucial relationships internal to the mathematics, as represented by this reciprocity relationship. Furthermore, as we have indicated, not all of the mathematics of group theory was brought to bear, but only those aspects relevant for the symmetries of the situation. Thus, within our framework, group theory can be conceived of as a family of mathematical structures, internally related by Weyl's 'bridges' for example, not all of which are involved in the application. What we have then is precisely the 'partial importation' of mathematical structures noted above and the appropriate representation of this relationship, as far as we are concerned as philosophers of science, is via partial homomorphisms.

This immersion then allows certain results to be derived which when appropriately interpreted describe phenomena such as the fountain effect and hence yield a qualitative understanding of the behaviour of liquid helium. However, these derivations and the subsequent understanding were limited due to the idealizations introduced, a situation that is not unusual in science, with subsequent de-idealization leading to Feynman's more developed account.

In Chapter 6 we will move on to examine a different case, one that involves the use of mathematics to bring together diverse domains, providing a unifying conception that indicates how they can be interrelated.

6

Unifying with Mathematics
Logic, Probability, and Quantum States

6.1 Introduction

In this chapter, we shall pursue a different kind of application, in which the aim is for the mathematics to unify apparently unrelated domains, such as quantum states, probability assignments, and logical inference.

In order to illustrate what is involved, we will initially compare the group-theoretic approach to quantum mechanics discussed in Chapter 4, with von Neumann's Hilbert space formalism (von Neumann 1932). In the former, group theory is, of course, the crucial feature, as we have indicated, and Weyl, for example, explored the role of symmetry to establish crucial representation theorems about quantum theory. Although group theory doesn't play a major role in the Hilbert space formalism, by the late 1930s von Neumann came to be dissatisfied with the latter (Rédei 1997). The core reason was that this approach didn't provide an adequate notion of probability for quantum systems with an infinite number of degrees of freedom. The alternative framework that he advanced was articulated in terms of his theory of operators (Murray and von Neumann 1936) and what we now call von Neumann algebras (Rédei 1998). According to von Neumann, this was the appropriate setting in terms of which we should formulate quantum mechanics (Birkhoff and von Neumann 1936). However, in order to guarantee (via a convenient representation theorem) that the notion of probability is appropriately formulated in this new framework, group-theoretic notions have to be introduced (Bub 1981). So, Weyl's emphasis on group-theoretic techniques was eventually vindicated, and the centrality of group theory in von Neumann's later approach to quantum mechanics becomes manifest. This episode therefore represents a further case study of the applicability of mathematics (specifically group theory again) but one that brings to the fore certain features and themes that have not been considered in full so far.

6.2 Group Theory, Hilbert Spaces, and Quantum Mechanics

In our brief sketch of the history of quantum physics presented in Chapter 1, we mentioned both Heisenberg's matrix mechanics and Schrödinger's wave mechanics.

As is well known, at the time, that is in 1925 and 1926, these were regarded as two entirely distinct formulations of quantum mechanics. As presented in the papers of Heisenberg, Born, Jordan, and Dirac on the one hand and in those of Schrödinger on the other, the two formulations couldn't be more different. Matrix mechanics is expressed in terms of a system of matrices defined by algebraic equations, and the underlying space is discrete. Wave mechanics is articulated in a continuous space, which is used to describe a field-like process in a configuration space governed by a single differential equation. However, despite these differences, the two theories seemed to have the same empirical consequences. For example, they gave coincident energy values for the hydrogen atom.

Schrödinger's explanation for this was to claim that the two theories were *equivalent*, and this was the main point of one of his papers of 1926. In the opening paragraph of this work (Schrödinger 1926), in which he tried to establish the equivalence, he notes:

Considering the extraordinary differences between the starting-points and the concepts of Heisenberg's quantum mechanics matrix mechanics and of the theory that has been designated "undulatory" or "physical" mechanics wave mechanics [...] it is very strange that these two new theories *agree with one another* with regard to the known facts where they differ from the old quantum theory. That is really very remarkable because starting-points, presentations, methods and in fact the whole mathematical apparatus, seem fundamentally different. Above all, however, the departure from classical mechanics in the two theories seems to occur in diametrically opposed directions. In Heisenberg's work the classical continuous variables are replaced by systems of *discrete* numerical quantities (matrices), which depend on a pair of integral indices, and are defined by *algebraic* equations. The authors themselves describe the theory as a "true theory of a discontinuum". On the other hand, wave mechanics shows just the reverse tendency; it is a step from classical point mechanics towards a *continuum-theory*. In place of a process described in terms of a finite number of dependent variables occurring in a finite number of differential equations, we have a continuous *field-like* process in configuration space, which is governed by a *single partial differential* equation, derived from a Principle of Least Action.

(Schrödinger 1926, pp. 45–6; italics in the original;
for detailed consideration, see Muller 1997, pp. 49–58)

According to Schrödinger, then, despite the conceptual and methodological differences, the two theories were equivalent because they yielded the same results. His strategy to prove the equivalence was clear enough: to establish an isomorphism between the canonical matrix- and wave-operator algebras. This was, of course, a straightforward strategy.

The problem, as Muller indicates (1997, pp. 52–3), is that he only established a mapping that assigns one matrix to each wave-operator, without establishing the converse. More surprisingly, Schrödinger himself acknowledges this point. In a footnote to his paper, he remarks:

In passing it may be noted that the converse of this theorem is also true, at least in the sense that certainly *not more* than *one* linear differential operator wave-operator can belong to a given *matrix*. [...] However, we have not proved that a linear operator wave-operator, corresponding to an arbitrary matrix, *always exists*.   (Schrödinger 1926, p. 52; italics in the original)

In other words, the equivalence between matrix and wave mechanics hasn't been proved after all. The claim that Schrödinger established the result in 1926 is therefore a "myth" (Muller 1997).

Given the importance of the two theories, and since they were thought of as having the same empirical consequences, it comes as no surprise that to establish the equivalence between them was taken as a substantial achievement. There were several attempts to do so. Dirac, for instance, provided a distinctive approach in which, as is well known, his main idea was to put each self-adjoint operator in diagonal form (see Dirac 1930). And this approach in fact succeeds in putting matrix and wave mechanics in a uniform setting. But it faces an unexpected problem: it is inconsistent! In the case of those operators that cannot be put in diagonal form, Dirac's method requires the introduction of "improper" functions with self-contradictory properties (the so-called δ-functions). However, from the viewpoint of classical mathematics, there are no such functions, since they require that a differential operator is also an integral operator; but this condition cannot be met. We shall examine how Dirac was able to dispense with this device in Chapter 7.

Given the failure of these attempts to establish the equivalence between matrix and wave mechanics, an entirely new approach was required to settle the issue. This was one of von Neumann's achievements in his 1932 book on the mathematical foundations of quantum mechanics. But von Neumann's involvement with foundational issues in quantum mechanics started earlier (we have already indicated his role in the application of group theory in Chapter 4). In a paper written in 1927 with Hilbert and Nordheim, the problem of finding an appropriate way of introducing probability into quantum mechanics had been explicitly addressed (see Hilbert, Nordheim, and von Neumann 1927). The approach was articulated in terms of the notion of the amplitude of the density for relative probability (for a discussion, see Rédei 1997). But it faced a serious technical difficulty (which was acknowledged by the authors): the assumption was made that every operator is an integral operator, and therefore Dirac's problematic function had to be assumed. As a result, an entirely distinct account was required and this led to von Neumann's 1932 work, using Hilbert spaces.

Von Neumann's now classic book is the development of three papers that he wrote in 1927 (for a discussion and references, see Rédei 1997, 1998). It is distinctive not only for its clarity, but also for the fact that there is no use of Dirac's δ-function. Probabilities are introduced via convenient trace functions, and relevant operators (projection operators) are defined on Hilbert spaces. So, von Neumann was able to claim that there is indeed a way of introducing probabilities in the quantum context without inconsistency (see von Neumann 1932).

But what should be said about the equivalence between matrix and wave mechanics? In von Neumann's hand, the problem becomes an issue of structural similarity. Bluntly put, he realized that there was a *similarity of structure* between the theory of Hilbert spaces and quantum mechanics, and that we could adequately *represent* claims about quantum systems by exploring the geometry of the former. But how can we accommodate this intuition? That is, how can we accommodate this *structural similarity* that von Neumann found between quantum mechanics and part of functional analysis?[1]

In a nutshell, it can be accommodated via an appropriate *morphism* between the structures under consideration: that is, by a transformation which preserves the relevant (features of those) structures. Of course, the strongest form of morphism is *isomorphism*—the full preservation of structure. However, the use of this notion would be inappropriate in this case, for the following reason. By 1927 quantum mechanics could be seen as a semi-coherent collection of principles and rules for applications, as again we noted in Chapter 4. And, like Dirac and Weyl, von Neumann provided a systematic approach to bring some order to this assemblage.[2] Weyl's group-theoretic approach, as presented in his 1931 book, was concerned with foundational questions, as we have discussed, although these were not exactly of the same sort as von Neumann was focused on. As Mackey points out (1993, p. 249), Weyl (1927) distinguished two such questions: (a) How does one *arrive at* the self-adjoint operators which correspond to various concrete physical observables? (b) What is the *physical significance* of these operators, that is, how are physical statements deduced from such operators? According to Weyl, (a) had not been adequately treated, and is a deeper question; whereas (b) had been settled by von Neumann's formulation of quantum mechanics in terms of Hilbert spaces. But to address (a) Weyl needed a different framework altogether; he needed group theory.

We again recall that, according to Weyl, group theory "reveals the essential features which are not contingent on a special form of the dynamical laws nor on special assumptions concerning the forces involved" (1931, p. xxi). And he continues:

Two groups, the group of rotations in 3-dimensional space and the permutation group, play here the principal role, for the laws governing the possible electronic configurations grouped about the stationary nucleus of an atom or an ion are spherically symmetric with respect to the nucleus, and since the various electrons of which the atom or ion is composed are identical, these possible configurations are invariant under a permutation of the individual electrons.

(ibid.; italics omitted)

In particular, the theory of group representation by linear transformations, highlighted previously, and regarded as the "mathematically most important part" of group theory, is exactly what is "necessary for an adequate description of the quantum mechanical relations" (ibid.). Again, we recall that the crucial connection lies in the claim that, "all quantum numbers, with the exception of the so-called

---

[1] In fact, it was due to this structural similarity that claims about quantum systems could be *represented* in terms of Hilbert spaces.

[2] Dirac's (1930) work represents a further attempt to lay out a coherent basis for the theory. However, as von Neumann perceived, neither Weyl's nor Dirac's approaches offered a mathematical framework congenial for the introduction of probability at the most fundamental level, and (initially at least) this was one of the major motivations for the introduction of Hilbert spaces.

principal quantum number, are indices characterizing representations of groups" (ibid.; italics omitted). Heisenberg's uncertainty relations and Pauli's Exclusion Principle can also be obtained via group theory and it is perhaps not surprising that Weyl concludes: "We may well expect that it is just this part of quantum physics [the one formulated group-theoretically] which is most certain of a lasting place" (ibid.).

But, as we saw, it is not only in the foundations of quantum mechanics that group theory has a decisive role; it is also crucial for the application of quantum theory, as embodied in the 'Wigner programme'. As Wigner insisted, we cannot apply Schrödinger's equation directly, but we need to introduce group-theoretic results to obtain the appropriate idealizations. In his own words:

The actual solution of quantum mechanical equations is, in general, so difficult that one obtains by direct calculation only crude approximations to the real solutions. It is gratifying, therefore, that a large part of the relevant results can be deduced by considering the fundamental symmetry operations.   (Wigner 1959 [1931], p. v)

In particular, as we discussed in Chapter 4, group theory allows the physicist to overcome the mathematical intractability of the many-body problem (involved in a system with more than two electrons) and, in this way, it allows quantum mechanics to be related to the data. Thus, as we noted previously, group theory enters both at the foundational level and at the level of application.

However, in order for the use of group theory to get off the ground, one has to adopt the prior reformulation of quantum mechanics in terms of Hilbert spaces. It is from the representation of the state of a quantum system in terms of such spaces that a group-theoretic account of symmetric and anti-symmetric states can be provided (see Weyl 1931, pp. 185–91). The group-theoretic approach also depends on the Hilbert space representation to introduce probability into quantum mechanics. Moreover, at the application level, despite the need for idealizations, Schrödinger's equation is still crucial (putting constraints on the accepted phenomenological models), and the representation of states of a quantum system in terms of Hilbert spaces has to be used. In other words, group theory is not an independent mathematical framework in which to articulate quantum mechanics; the Hilbert spaces representation is crucial. Roughly speaking, we can say that von Neumann's Hilbert spaces representation is "sandwiched" between the Weyl and Wigner programmes; that is, between the foundational use of group theory and its application. Hence, there is a close interdependence between group theory and Hilbert spaces theory in the proper formulation of quantum mechanics.[3]

However, as far as von Neumann was concerned, Weyl's approach failed to offer a mathematical framework congenial for the introduction of probability at the most fundamental level, and, as we have reiterated, this was one of the major motivations

---

[3] This may then present a challenge for the structural realist, in determining which structure she should be a realist about, although we shall not pursue this issue here.

for the introduction of Hilbert spaces. Given this, there is no way of spelling out the similarity of structure between quantum mechanics (as it was at the end of the 1920s and the beginning of the 1930s) and Hilbert spaces in terms of the existence of *full isomorphisms*. Moreover, since there is *more structure* in functional analysis than was actually used by von Neumann in his axiomatization of quantum mechanics, we can say that the relationship between those "quantum" and mathematical structures is captured by appropriate *partial homomorphisms*. The partiality involved in the formulation of quantum mechanics is then captured by the *partial* nature of the homomorphism—only those components of quantum mechanics about which we have enough information are "preserved".

An important feature of von Neumann's axiomatization was his systematic search for *analogies* between mathematical structures and between the latter and "physical" structures (i.e. structures employed in the description of physical phenomena).[4] These analogies played an important role in von Neumann's equivalence proof of matrix and wave mechanics. What von Neumann established is a mathematical relation between two systems of *functions*. And, in this case, the relation is a full isomorphism. On the one hand, we have *functions*—defined on the "discrete" space of index values $Z = (1, 2, \ldots)$—which are sequences $x_1, x_2, \ldots$, and are used in the formulation of *matrix* mechanics; on the other hand, we have *functions* defined on the continuous state-space Ω of a mechanical system (Ω is a k-dimensional space, with k being the number of classical mechanical degrees of freedom), and Ω's functions are *wave* functions $\phi(q_1, \ldots, q_k)$. Von Neumann explicitly points out that the spaces Z and Ω are quite different (1932, p. 28). It is not surprising then that alternative attempts at unification faced so many mathematical difficulties, since they assumed a *direct analogy* between Z and Ω (see von Neumann 1932, pp. 17–27). As he notes:

The method [. . .] resulted in an analogy between the "discrete" space of index values Z [. . .] and the continuous state space Ω [. . .]. That this cannot be achieved without some violence to the formalism and to mathematics is not surprising. The spaces Z and Ω are in reality very different, and every attempt to relate the two must run into great difficulties.

(von Neumann 1932, p. 28)

However, the correct analogy is not between the spaces Z and Ω, but between the *functions defined on them*. Von Neumann calls the totality of functions on Z, satisfying certain conditions, F_Z, and the totality of functions on Ω, also satisfying certain conditions, F_Ω (1932, pp. 28–9). Once this point is clearly seen, he presents his equivalence proof. But notice that what he proved is that F_Z and F_Ω are isomorphic, and therefore his theorem is restricted to the *mathematical structures* employed in matrix and in wave mechanics. Their *physical content*, as it were, is left untouched.

---

[4] The way in which the partial structures approach accommodates analogies in science has been described by da Costa and French (1990); for further details, see also French (1997) and French and Ladyman (1997). In effect, we are extending this account to mathematics.

And it was here that finding the *right analogy* paid off. For it was the existence of the mathematical equivalence (between F_Z and F_Ω) that led von Neumann to search for a more basic mathematical formulation for quantum mechanics. However, since F_Z was nothing but a Hilbert space, it was natural to adopt a slightly more abstract formulation of it—not tied to the particular features of F_Z—as the basis for shaping the structure of the theory. In von Neumann's own words:

Since the systems F_Z and F_Ω are isomorphic, and since the theories of quantum mechanics constructed on them are mathematically equivalent, it is to be expected that a unified theory, independent of the accidents of the formal framework selected at the time, and exhibiting only the really essential elements of quantum mechanics, will then be achieved, if we do this: *Investigate the intrinsic properties* (common to F_Z and F_Ω) *of these systems of functions, and choose these properties as a starting point.*

The system F_Z is generally known as "Hilbert space". Therefore, our first problem is to investigate the fundamental properties of Hilbert space, independent of the special form of F_Z or F_Ω. The mathematical structure which is described by these properties (which in any specific special case are equivalently represented by calculations within F_Z or F_Ω, but for general purposes are easier to handle directly than by such calculations), is called "abstract Hilbert space".   (von Neumann 1932, pp. 32–3; the italics are ours)

In other words, once the right analogy is found—that is, once the appropriate structural relationship is uncovered—the crucial step is taken. The rest is to explore the consequences.

However, there is a further feature worth considering in von Neumann's application of mathematics. It concerns the role of logic in this process—and as we shall see, von Neumann has both a pluralist and empiricist view about logic. According to him, there is a plurality of logics, depending on the particular context we study, and such logics should be constrained by experience. This combination of pluralism and empiricism entails that logic is motivated by experience, and strengthens the role of analogies between mathematics and physics as part of the development of physics and mathematics. In this sense, experience has a double role of generating logics and demanding further mathematical structures. In what follows, we shall elaborate on these points.

6.3 Logic and Empiricism

As we discussed in Chapter 5, various top-down and bottom-up moves were made in the context of the explanation of the behaviour of Helium III via Bose–Einstein statistics. Now, as we have just indicated, with the introduction of Hilbert spaces in quantum mechanics, von Neumann also made a top-down move from mathematical to physical theories. In this section, we shall delineate the corresponding bottom-up move in von Neumann's thought as one goes from experience through logic to highly abstract mathematical structures. Through this combination of moves, he was able to effect the kind of unification across domains that represents a further important aspect of the applicability of mathematics.

One of the outcomes of von Neumann's approach was the creation of quantum logic. He showed that we can formulate a family of propositions that describe the state of a physical system, in such a way that the geometrical structure underlying this family gives us information about the physical system. The idea is found in his 1932 book, and is developed further in the celebrated 1936 paper co-authored with Birkhoff, where "quantum logic" as such was first introduced.

We begin with the point that the study of the geometry associated with the relevant family of propositions gives us its "logic". In the case of classical mechanics, the family of propositions generate a Boolean algebra, but in the context of quantum mechanics, we have "a sort of" projective geometry. The reason for the proviso is that in the 1936 paper, Birkhoff and von Neumann only consider physical systems with a finite number of degrees of freedom (i.e. whose states can be characterized by finitely many parameters, in a state-space with a finite dimension). Now, with *this* finiteness assumption, the resulting "logic" (which is isomorphic to the projective geometry of all subspaces of a finite-dimensional Hilbert space) *is indeed* a projective geometry. But the question arises as to the logic of a physical system which has *infinitely* many degrees of freedom. Von Neumann explicitly addressed this problem in the work he produced after the 1936 paper with Birkhoff. And his solution was to provide a generalization of projective geometry, which led to the formulation of *continuous geometry* and to what is now called *von Neumann algebras* of Hilbert space operators (see von Neumann 1960, 1981; Murray and von Neumann 1936; and for a discussion, Bub 1981a, 1981b; and Rédei 1997, 1998).

In other words, it was because of von Neumann's *pluralism* that he was concerned with determining the logics adequate to each particular domain—moving from *finite* physical systems (represented by projective geometry) to *infinite* ones (represented by continuous geometry). Moreover, it was because of his *empiricism* that he undertook the search for distinct logics based on experience, that is, the various types of physical systems we have to accommodate. In other words, the case of the generalization of projective geometry (continuous geometry) is a clear example of a mathematical theory which is created from physical, *empirical* demands. We shall call this von Neumann's *first empiricist feature*. Let us consider some further features.

6.4 The 1937 Manuscript: Logics and Experience

In a manuscript, written about 1937, von Neumann discusses his view of the status of logic and alternative logical systems (see von Neumann 1937a). This work was written shortly after the publication of Birkhoff and von Neumann (1936), and von Neumann's chief task is to explore some consequences of the approach to logic left unnoticed in the 1936 paper.

The manuscript begins with a remark making it clear that it was *not* part of what was then considered work in the "foundations" of mathematics:

We propose to analyze in this note logics from a point of view which is fundamentally different from the one current in present-day "foundations" investigations. While we are not questioning at all the great importance and value of those investigations, we do believe that our present system of logics is open to criticism on other counts also. And it seems to us that these other "counts" are at least as fundamental and essential as those which form the subject-matter of the current "foundations" analysis.   (von Neumann 1937a, p. 1)

Von Neumann had several reasons for not being satisfied with the then dominant foundational research. The most important of them was the entirely aprioristic way in which that research was typically conducted. Logic was essentially taken as *classical* logic, and there was hardly any room for the development of logics that were specific and appropriate to the particular domain of science under consideration. As opposed to this, the approach favoured by von Neumann was:

much more *directly connected with and inspired by the connection of logics with the physical world*, and particularly also with probability. For this reason we even see some point in the proposition that the problems this approach generates deserve precedence over the generally considered problems of "foundations".   (von Neumann 1937a, p. 1; the italics are ours)

In his view, it is crucial to have room for the articulation of logics that are faithful to the domain in question. In fact, he goes still further, claiming that logics should be inspired by experience:

The basic idea of this approach to logic is that *the system of logics which one uses should be derived from aggregate experiences relative to the main application which one wishes to make— logics should be inspired by experience.*   (von Neumann 1937a, p. 2; the italics are ours)

In other words, it is not only the construction of mathematical theories, but also the construction of logical systems that is heuristically motivated by experience—in order for the resulting logic to be adequate to the domain in question—and a logic should also reflect the main traits of the empirical domain. Logic should be inspired by experience (this is von Neumann's *second empiricist feature*).

However, as well as being heuristically inspired by experience, a logic can also play a critical role, being open to revision on empirical grounds:

If any accepted physical theory which has a reasonable claim to "universality" permits to draw inferences concerning the structure of logics, *these inferences should be used to reform logics, and not to criticize the theory.*   (von Neumann 1937a, pp. 2–3; the italics are ours)

In other words, a logic can be revised on empirical grounds (this is the *third empiricist feature*). As a result, a logic is relative to a particular domain of inquiry. And relying on this revisability principle, von Neumann provides a further argument against the traditional approach to logic:

We hope to show on the pages which follow that an *absolutely* consequent application of this principle leads to very plausible results, and particularly to more natural ones than the usual, rigidly dogmatic, attitude in logic.   (von Neumann 1937a, p. 3; italics in the original)

All these considerations (especially the domain-dependence of logic) become completely clear with quantum logic, which, as is well known, due to the demands of quantum phenomena, lacks the distributivity of classical logic. In von Neumann's own words:

Quantum mechanics is a particularly striking example of how a physical theory may be used to modify logics—and *for this reason* it was first investigated in Birkhoff and von Neumann 1936.
(von Neumann 1937a, p. 3; the italics are ours)

However, he has something still more radical to advance:

We propose to show in this note that even classical mechanics is incompatible with the usual system of "infinite" (or "transcendental") logics that is, logics which are not related to experience. We will determine the system of logics to which classical mechanics lead. We will see that this system lends itself much better to an extension to "probability logics"—which is an absolutely indispensable one if we bear in mind the modern developments of physical theory.   (von Neumann 1937a, p. 3)

But once this is achieved, something more will be established:

We will also see that this "classical mechanical" system of logics, when combined with the "finite" "quantum mechanical" system of logics given in Birkhoff and von Neumann 1936, leads to a satisfactory, general "infinite" system of "quantum mechanical" logics, which contains also probability theory.   (von Neumann 1937a, pp. 3–4)

And *these* are the "plausible results" that he referred to above in his criticism of the "dogmatic" approach to logic.

Therefore, the picture that emerges from this paper involves, first, *pluralism* (there are as many logics as physical phenomena demand); second, *anti-apriorism* (the demand for new logics arises from experience); and third, *empirically driven change* (logic is open to change via empirical considerations). In this sense, there is more to logic than the a priori exploration of the logical consequence relation, and the main features of a logic should be obtained from, and modified by, experience (the domain of *application*).

Given the empiricist features of von Neumann's account of logic (in particular, given the idea that logic is open to change via empirical considerations), it is interesting to note that he thought that his account was comparable to the criticism of classical logic made by intuitionist and relevant logicians. As he points out in his paper with Birkhoff on the logic of quantum mechanics:

The models for propositional calculi which have been considered in the preceding sections are also interesting from the standpoint of pure logic. *Their nature is determined by quasi-physical and technical reasoning*, different from the introspective and philosophical considerations which have had to guide logicians hitherto. Hence it is interesting to compare the modifications which they introduce into Boolean algebra, with those which logicians on "intuitionist" and related grounds have tried introducing.
(Birkhoff and von Neumann 1936, p. 119; the italics are ours)

However, as Birkhoff and von Neumann note, there is a crucial distinction between the criticisms of classical logic as articulated by the quantum and the intuitionist logicians:

The main difference seems to be that whereas logicians have usually assumed that properties L71-L73 of negation namely, $(A')' = A$; $A \wedge A' = f$, $A \vee A' = t$; $A \rightarrow B$ implies $B' \rightarrow A'$ were the ones least able to withstand a critical analysis, the study of mechanics points to the *distributive identities* L6 $A \vee (B \wedge C) = (A \vee B) \wedge (A \vee C)$ and $A \wedge (B \vee C) = (A \wedge B) \vee (A \wedge C)$ as the weakest link in the algebra of logic.   (Birkhoff and von Neumann 1936, p. 119)

In their view, their own approach was closer to that of the relevant logician:

Our conclusion agrees perhaps more with those critiques of logic, which find most objectionable the assumption that $A' \vee B = t$ implies $A \rightarrow B$.   (Birkhoff and von Neumann 1936, p. 119)

All these principles (pluralism, anti-apriorism, and empiricism) are forcefully articulated in this work on quantum logic, since this logic is *required* by the quantum domain, and it is importantly different from classical logic (even if classical logic is a fragment of it). In our view, not only did von Neumann adopt these methodological and epistemological principles, but he also fruitfully explored them to develop several of his most lasting contributions. In other words, these principles informed his research, providing him with heuristic guidelines for theory construction.

Now, what we have here is a sophisticated interaction between mathematics, logic, and physics. As von Neumann articulated them, logics (such as quantum logic) are inspired by experience, but once articulated they generate new mathematical structures (such as the continuous geometry demanded by physical systems with infinitely many degrees of freedom). Such structures are in turn applied to model physical phenomena. Of course, once mathematical structures are formulated, they can also be studied in purely mathematical terms, independently of any concern with physics—as von Neumann's own mathematical work with continuous geometry beautifully illustrates (see, for instance, von Neumann 1981).

So, we have here, again, a movement from the bottom up, from experience through logic to highly abstract mathematical structures (from quantum logic to continuous geometry), in addition to the movement from the top down, from mathematical structures down to experience. The latter move was described earlier with von Neumann's use of the theory of Hilbert spaces in the formulation of quantum mechanics.

Returning now to his 1937 manuscript, we can see another aspect of his *pluralism*. In order to have a proper understanding of *classical* mechanics we need an appropriate logic, and even in *this* case, von Neumann insisted, we need logics that are non-classical ("probability logics"). This idea is explored further in another unpublished manuscript, whose subject is explicitly quantum logic (von Neumann 1937b). In this article, he outlined four possibilities concerning the interplay between the "dimension" of a physical system and its logic, each alternative leading to an appropriate logical system. These four cases are: (i) a system "which behaves in the sense of classical physics (in particular mechanics) and which possesses only a finite number of possible different states" (von Neumann 1937b, p. 11); (ii) a system similar to (i) but in which the number of states is discretely infinite; (iii) a system similar to (i) but in which the number of states is continuously infinite; and finally (iv) a system where the "finiteness of the number of states remains essentially untouched, but the 'classical' way of looking at things (as practised in (i)–(iii)) is replaced by a 'quantum mechanical' one" (von Neumann 1937b, p. 16). Moreover, von Neumann also advanced a final synthesis, combining "these two kinds of extensions", namely from the finite to the infinite, and from the "classical" to the "quantum" approach (ibid.). Unfortunately, this is an unfinished work, and he only actually discussed case (i) (see von Neumann 1937b, pp. 11–15).

But note that the three methodological and epistemological principles mentioned above play a heuristic role at this point. What led von Neumann to consider these four possibilities concerning a physical system was: (a) his logical *pluralism* (since he was searching for logics appropriate to the domain under study, whether it is classical or quantum physics); (b) his *anti-apriorism* (by refusing to adopt a logic which was not properly motivated by experience); and (c) his *empiricism* with regard to logic (which required an empirical investigation to determine the adequacy of a logic to the physical domain in question). In this sense, these principles informed and supported this programme of research.

6.5 The Status of Mathematics

It is important to notice that we are *not* saying that von Neumann was an empiricist about logic and mathematics in the naive, and untenable, sense that these disciplines *are* empirical. Once *motivated* by empirical considerations, logic and mathematics are *developed* and *articulated* as deductive disciplines in the standard way. And the adoption of this standard practice is absolutely clear in von Neumann's own mathematical writings.[5]

For instance, he criticized Dirac for the latter's introduction of the δ-function on *mathematical* grounds, along the lines we have previously indicated (see von Neumann 1932, pp. 23–7). According to von Neumann, this function lies "beyond the scope of mathematical methods generally used", and since he desired to "describe

---

[5] This empiricist approach might be interpreted in a way that clashes with our emphasis on the significance of surplus structure. Thus, Redhead associates with Hilbert and von Neumann the view that 'theoretical physics should ideally proceed by a direct axiomatization of physical concepts, with the elimination of all surplus structure' (1975, p. 88). Whatever the merits of such an approach, he notes, the development of physics itself has not proceeded in this particular manner. However, as we have just indicated, one can insist that logic and mathematics should be motivated by empirical considerations yet still acknowledge (as indeed one must!) that in developing and articulating the structures one comes up with in this manner, the possibility of embedding them in further structures will become manifest, thereby generating 'surplus' resources.

quantum mechanics with the help of these latter methods" (ibid., p. 27), he moved on to his own formulation of quantum mechanics in terms of Hilbert spaces. This surely illustrates the importance von Neumann saw in subscribing to standard methods of mathematical research. As he insists:

The method of Dirac, mentioned above (and this is overlooked today in a great part of quantum mechanical literature, because of the clarity and elegance of the theory), in no way satisfies the requirements of mathematical rigor—not even if these are reduced in a natural and proper fashion to the extent common elsewhere in theoretical physics. For example, the method adheres to the fiction that each self-adjoint operator can be put in diagonal form. In the case of those operators for which this is not actually the case, this requires the introduction of "improper" functions with self-contradictory properties. The insertion of such a mathematical "fiction" is frequently necessary in Dirac's approach, even though the problem at hand is merely one of calculating numerically the result of a clearly defined experiment.

(von Neumann 1932, pp. viii–ix)

However, he points out:

There would be no objection here if these concepts [of Dirac's "improper" functions], which cannot be incorporated into the present-day framework of analysis, were intrinsically necessary for the physical theory. Thus, as Newtonian mechanics first brought about the development of the infinitesimal calculus, which, in its original form, was undoubtedly not self-consistent, so quantum mechanics might suggest a new structure for our "analysis of infinitely many variables"—i.e., the mathematical technique would have to be changed, and not the physical theory.   (von Neumann 1932, p. ix)

This passage illustrates the extension of his *empiricism*: he would be prepared to change the standard mathematical techniques, if this were required by our physical theories. And the precedent for this in the case of Newtonian mechanics is surely well taken. Nevertheless, von Neumann continues:

But this is by no means the same case. It should rather be pointed out that the quantum mechanical "Transformation theory" can be established in a manner which is just as clear and unified, but which is also without mathematical objections. It should be emphasized that the correct structure need not consist in a mathematical refinement and explanation of the Dirac method, but rather that it requires a procedure differing from the very beginning, namely, the reliance on the Hilbert theory of operators.   (von Neumann 1932, p. ix)

That is, once we have the appropriate mathematical framework, there is no need for introducing deviant techniques in mathematics in order for us to formulate quantum mechanics.

We see then that one of the striking features of von Neumann's work is his search for an approach in which logic and the mathematical and physical theories we employ to explain the phenomena hang nicely together. This means that we should search for an account that admits different kinds of generalization, such as the following. Suppose we are considering a given type of physical system. The approach von Neumann tried to provide was such that the same framework that is used to describe a physical system with a *finite* number of degrees of freedom could be extended to accommodate a system with an *infinite* number of degrees. Moreover, the geometric structures associated with such descriptions should be such that they are also preserved and generalized when the extension to the infinite case is performed. Furthermore, the logic derived from the geometric structure should also be amenable to generalization. Finally, given the crucial role played by probability theory in physics (especially in quantum mechanics), and given the close connection that von Neumann saw between probability and logic, he searched for an approach that allowed probability to be adequately introduced in quantum mechanics, and which also made clear the relationship between logic and probability. These are broad constraints and von Neumann adopted them as heuristic devices in theory construction.

These points are clearly formulated in a paper that he presented in 1954 in the Congress of the Mathematicians in Amsterdam. Half a century after Hilbert's celebrated 1900 paper on mathematical problems, von Neumann was asked to consider further unsolved problems in mathematics. And he emphasized the importance of providing a unified approach to logic, probability, and physics. Once again, he was keen on exploring analogies and structural similarities between different domains of mathematics and physics. We shall not enter into all the details here, but again structural analogies are introduced between logic and set theory, on the one hand, and between probability and measure theory, on the other. The idea is that there are structural relations between these domains, and he explores them in order to find a unified approach. In extending this to quantum mechanics, logic and probability must go hand-in-hand in the sense that:

logics and probability theory arise simultaneously and are derived simultaneously.

(von Neumann 1954, p. 22)

There is more to say, of course.[6] But we shall end by noting, again, the significance that von Neumann placed on these explorations of the structural relations that hold between the mathematical and the physical domains.

In Chapter 7, we shall consider whether, in bringing the mathematics into application in this manner, there are grounds for taking the relevant mathematical entities to be *indispensable*. Our case study will be Dirac's use of the delta function in quantum mechanics and we shall further pursue this issue of indispensability in Chapter 8.

---

[6] It is worth noting, for instance, that von Neumann developed a form of the model-theoretic approach that bears striking similarities to both van Fraassen's state-space approach and Redhead's function space formalism.

7

Applying Problematic Mathematics, Interpreting Successful Structures
From the Delta Function to the Positron

7.1 Introduction

We have already mentioned Dirac's work in the context of the history of quantum statistics. However, among the various contributions he made to physics, one, in particular, is particularly fascinating for the plethora of philosophical issues it raises: his introduction of the delta function in 1930 (see Dirac 1958[1]). At a time when the physics community was deeply concerned about establishing the equivalence between matrix and wave mechanics as we saw in Chapter 6, Dirac made a significant and valuable attempt at proving the equivalence result. But his proposal relied, in an important way, on a problematic element: the delta function.

As is well known, and as we have already touched upon, the delta function has peculiar properties (among them, the fact that it's inconsistent!),[2] and so it's not surprising that Dirac would be suspicious about its use. (It was precisely to avoid the use of this function, as we saw, that von Neumann would later introduce the Hilbert space formalism.) Strictly speaking, the delta function is not even a function. And Dirac was, of course, aware of this. He called it an "improper function", since it doesn't have a definite value for each point in its domain (Dirac 1958, p. 58). Moreover, as he also pointed out, the delta function is ultimately dispensable. After all, it's "possible to rewrite the theory (i.e. quantum mechanics) in a form in which the improper functions appear throughout only in integrands. One could then eliminate the improper functions altogether" (Dirac 1958, p. 59).

In this chapter, we shall examine the role of the delta function in Dirac's formulation of quantum mechanics (QM), which provides a further opportunity to examine, more

---

[1] The first edition of Dirac (1958) was published in 1930.

[2] For example, the delta function entails that differential operators are integral operators, as we noted in Chapter 6, but this is never the case (for an illuminating discussion of this point, see von Neumann 1932, pp. 17–27; especially pp. 23–7).

generally, the role of mathematics in theory construction. Now, it has been argued that mathematics plays an *indispensable* role in physics, particularly in QM, given that it's not possible even to express the relevant physical principles without its use. And given that mathematical entities are indispensable to our best theories of the world, the argument goes, we ought to believe in their existence (for a thorough defence of this view, see Colyvan 2001; Baker 2009). We shall return to this argument in Chapter 8 but here we shall argue that, at least in the case of the delta function, Dirac was very clear about its *dispensability*.

After discussing the significance of the delta function in Dirac's research, and exploring the strategy that he devised to overcome its use, we examine the import- ance of this strategy for the use of mathematics in physics. We then argue that even if mathematical theories were indispensable, this wouldn't justify the commitment to the existence of mathematical entities. To illustrate this point, we examine an additional and particularly successful use of mathematics by Dirac: the one that eventually led to the discovery of antimatter. As we will see, even in this case, there's no reason to believe in the existence of the corresponding mathematical entities. Further insights on the application of mathematics thus emerge from the careful examination of Dirac's work.

7.2 Dirac and the Delta Function

7.2.1 Introducing the Delta Function

Dirac's (1958) book on the foundations of QM is another in the series of exceptional treatises produced on the subject between the late 1920s and the early 1930s, including those of Weyl (1931, originally published in German in 1928), Wigner (1931), and von Neumann (1932), discussed previously.[3] What distinguishes Dirac's book, particularly since its second edition, is its informal, concise style, and the explicit attempt to "keep the physics to the forefront", examining "the physical meaning underlying the formalism wherever possible" (Dirac 1958, p. viii). That Weyl, Wigner, and von Neumann had very definite and distinct theoretical agendas has already been made clear: group-theoretic techniques as a *foundation* for QM in the case of Weyl; group theory underpinning the *applications* of quantum theory for Wigner; and the introduction of Hilbert space to provide a mathematically coherent formulation of QM, that captured the probabilistic character of the theory, when it came to von Neumann.

Dirac tried something different. Just as in the cases of Weyl and Wigner, he explored group-theoretic techniques; just as von Neumann did, he introduced new representations. But in contrast with all three, with Dirac mathematics was never in

---

[3] There are surely interesting insights to be gained through examining the impact of these different texts on the physics community.

the forefront. He used mathematics, to be sure. But, he insisted, "the mathematics is only a tool and one should learn to hold the physical ideas in one's mind without reference to the mathematical form" (Dirac 1958, p. viii). One of our goals here is to understand what exactly this means.

In the first few chapters of *The Principles of Quantum Mechanics*, Dirac provided an algebraic framework to formulate the fundamental laws of QM (see Dirac 1958, pp. 1–52). The framework was developed in terms of the now familiar *bra vectors*, *ket vectors*, and *linear operators*.[4] Although the framework has a number of attractive features, to be able to solve certain problems, it's useful to have an additional representation in terms of numbers. As Dirac noted:

For some purposes it is more convenient to replace the abstract quantities by sets of numbers with analogous mathematical properties and to work in terms of these sets of numbers. The procedure is similar to using coordinates in geometry, and has the advantage of giving one greater mathematical power for the solving of particular problems.   (Dirac 1958, p. 53)

Using this numerical representation, he was then able to establish four main results (Dirac 1958, p. 57):

(a) The basic bras of an orthogonal representation are simultaneously eigenbras of a complete set of commuting observables.
(b) Given a complete set of commuting observables, there is an orthogonal representation in which the basic bras are simultaneously eigenbras of this complete set.
(c) Any set of commuting observables can be made into a complete commuting set by adding certain observables to it.
(d) A convenient way of labelling the basic bras of an orthogonal representation is by means of the eigenvalues of the complete set of commuting observables of which the basic bras are simultaneously eigenbras.

For the purposes of this chapter, the details of this representation are not important. What *is* important is to note, as Dirac does, the representation's power and generality.

However, despite these features, the representation is still limited. How could one characterize the *lengths* of the basic vectors invoked throughout the discussion? As is well known, if we were considering only orthogonal representations, it would be natural to normalize the basic vectors. The trouble, however, is that vectors can be normalized only if the parameters that label them have *discrete* values. If the

---

[4] The introduction of bra and ket vectors is actually an innovation of the second edition of *The Principles of Quantum Mechanics* (see Dirac 1958). This innovation was preserved in all subsequent editions. In many ways, the most substantial revisions in the *Principles* were made in the second edition, which adopted a less formal style than the one developed in the first edition—a feature that many people praised, including Heisenberg.

parameters have *continuous* values, it's impossible to normalize them. As a result, in this case the lengths of the vectors cannot be defined. Dirac makes the point very clearly:

We have not yet considered the lengths of the basic vectors. With an orthogonal representation, the natural thing to do is to normalize the basic vectors, rather than leave their lengths arbitrary [...]. However, it is possible to normalize them only if the parameters which label them all take on discrete values. If any of these parameters are continuous variables that can take on all values in a range, the basic vectors are eigenvectors of some observable belonging to eigenvalues in a range and are of infinite length [...].   (Dirac 1958, pp. 57–8)

Given this difficulty, an entirely new approach is required. Dirac is explicit about what is demanded:

Some other procedure is then needed to fix the numerical factors by which the basic vectors may be multiplied. To get a *convenient method* of handling this question, a new *mathematical notation* is required [...].   (Dirac 1958, p. 58; italics added)

This "convenient method" and "new mathematical notation" is the *delta function*.

Before examining the details of the function, note the way in which he describes the latter. It is characterized as a "convenient method", which highlights the function's *pragmatic* character. Moreover, the function is only thought of as a "mathematical notation", which downplays any *ontological significance* the function may have. As we will see shortly, Dirac had a justification for this way of describing the delta function—or, at least, he tried to provide such a justification, which has to do with the strategy he devised to dispense with the function altogether.

But what exactly *is* the delta function? As Dirac pointed out, it is a quantity $\delta(x)$ that depends on a parameter x and satisfies the following conditions:

$$\int_{-\infty}^{+\infty} \delta(x)dx = 1$$

$$\delta(x) = 0 \text{ for } x \neq 0$$

In order to "visualize" the behaviour of the delta function, consider a function that is identical with 0 at every point except 0, and in the immediate neighbourhood of 0, although the value of the function is not exactly defined, it is so large that its integral adds up to 1. A function behaving in this way would be a delta function. In Dirac's own words:

To get a picture of $\delta(x)$, take a function of the real variable x which vanishes everywhere except inside a small domain, of length ε say, surrounding the origin $x = 0$, and which is so large inside this domain that its integral over this domain is unity.   (Dirac 1958, p. 58)

Now, it's perfectly natural to ask whether any object satisfies the conditions above. And Dirac immediately acknowledges that there is something peculiar about the delta function. First, it is *not* exactly a function, given that at $x = 0$, it doesn't have a sharply defined value—only its *integral* has a value:

$\delta(x)$ is not a function of x according to the usual mathematical definition of a function, which requires a function to have a definite value for each point in its domain, but is something more general, which we may call an 'improper function' to show up its difference from a function defined by the usual definition. (Dirac 1958, p. 58)

Second, if at $x = 0$ the delta "function" lacks a sharply defined value, in what sense is its integral defined at all? Although Dirac did not explicitly address this second question, clearly there is something delicate going on with the "function". As a result, the use of the latter has to be restricted. He was very clear about this point:

$\delta(x)$ is not a quantity which can be generally used in mathematical analysis like an ordinary function, but its use must be confined to certain simple types of expression for which it is obvious that no inconsistency can arise. (Dirac 1958, p. 58)

In other words, a strategy to delimit the function's appropriate scope needs to be found. And by determining the function's scope, it's then possible to dispense with it altogether.

7.2.2 Dispensing with the Delta Function

Given the worries mentioned above about the delta function, it comes as no surprise that Dirac provided a strategy to dispense with it. Although, as we will argue below, there's more going on here than simply a worry about the peculiar behaviour of the function, it's worth being clear about this dispensability strategy.

The first step consists in identifying two significant properties of the delta function. These properties provide algebraic rules to manipulate the latter, and they motivate the function's dispensability. The first property crucially depends on the fact that the integral of the delta function is 1 in the neighbourhood of 0. As Dirac pointed out:

The most important property of $\delta(x)$ is exemplified by the following equation

$$\int_{-\infty}^{+\infty} f(x)\delta(x)dx = f(0) \tag{3}$$

where $f(x)$ is any continuous function of x. We can easily see the validity of this equation from the above picture of $\delta(x)$. The left-hand side of (3) can depend only on the values of $f(x)$ very close to the origin, so that we may replace $f(x)$ by its value at the origin, $f(0)$, without essential error. (Dirac 1958, p. 59)

In other words, if f is a well-behaved, continuous function, with definite values for each argument in f's domain, the value of the integral in the left-hand side of (3) ultimately only depends on the values of f close to 0. After all, in this context the integral of the delta function is 1.

The second property of the delta function also highlights an additional algebraic feature. Dirac formulated it in the following way:

By making a change of origin in (3), we can deduce the formula

$$\int_{-\infty}^{+\infty} f(x)\delta(x-a)dx = f(a) \tag{4}$$

where a is any real number. Thus *the process of multiplying a function of x by δ(x - a) and integrating over all x is equivalent to the process of substituting a for x.*
                                      (Dirac 1958, p. 59; italics in the original)

The same move discussed above in connection with the first property of the function is also found in the second property. Again, the crucial feature of the function—the fact that its integral is 1, when values close to 0 are considered—is used to obtain the second property.

Given these two properties, the strategy to dispense with the delta function is not hard to figure out. As long as the delta function occurs in an integral, it can be eliminated. As Dirac noted:

Although an improper function does not itself have a well-defined value, when it occurs as a factor in an integrand the integral has a well-defined value. In quantum theory, whenever an improper function appears, it will be something which is to be used ultimately in an integrand. Therefore *it should be possible to rewrite the theory in a form in which the improper functions appear all through only in integrands. One could then eliminate the improper functions altogether.*   (Dirac 1958, p. 59; italics added)

In other words, as long as the use of the delta function is restricted to a factor in an integrand, it becomes clear that the function is dispensable. After all, due to the two properties discussed above, if the delta function occurs in an integral, it can be completely eliminated. The conclusion is then clear:

The use of improper functions thus does not involve any lack of rigor in the theory, but is merely a *convenient notation*, enabling us to *express in a concise form* certain relations which *we could, if necessary, rewrite in a form not involving improper functions*, but only in a *cumbersome way* which would tend to *obscure the argument.*
                                      (Dirac 1958, p. 59; italics added)

With these remarks, Dirac highlights the *pragmatic* character of the delta function. It is a "convenient notation", useful to "express in a concise form" certain relations. And it's only in a "cumbersome way", which "would tend to obscure the argument", that we could express—without invoking the function—the relations we want to express. Although ultimately dispensable, the function is no doubt useful. But its usefulness is a pragmatic matter, which confers no ontological significance on it.

7.2.3 The Status of the Delta Function

It might be argued that the dispensability of the delta function paves the way for the function's replacement by something else. In fact, the argument goes, with hindsight someone can indeed maintain that the delta function is not really a function. Not in the sense acknowledged by Dirac (that is, the delta "function" doesn't have a definite value for each point in its domain), but in a deeper sense. Given the "reformulation" of the delta "function" in Schwartz's theory of distributions, the "function" is actually a *distribution* (see Colyvan 2001, pp. 103–4, note 20).[5] This also explains why the delta function is dispensable: it is not the relevant sort of object. What is needed to consistently achieve what Dirac was aiming at is not a function, but a distribution.

Is this response justified? Well, not quite. The suggestion that the delta function is ultimately a distribution misses a significant point about Dirac's strategy: namely, the *pragmatic* role that the function plays in describing—in an elegant and simple way—the relevant relations. In fact, the introduction of the theory of distributions increases hugely the complexity of quantum theory as well as the size of the function space for QM. Not surprisingly, introducing distributions ends up "obscuring the argument", in pretty much the way Dirac warned about. There's something to be said for his strategy after all.[6]

Moreover, any identification of the delta "function" with a distribution cannot be right. Even an improper function is *not* a distribution. To stipulate that they are the same objects based on the role they play in the respective theories (Dirac's on the one hand, Schwartz's on the other) is to assume that the objects in question behave in the same way. But they don't—and they can't. If Dirac's formulation is simpler and inconsistent, and Schwartz's is consistent but much more complex, there obviously is a significant difference between the two theories and their corresponding objects. No literal identification of these objects can be made.

Of course, this is not an argument for adopting Dirac's formulation. It is simply an argument that highlights the difficulty of absorbing Dirac's theory within Schwartz's. The relation between the two theories is much more complex than a simple identification of their ontologies may suggest.[7] Indeed, we would speculate—although we shall not go into details here—that our framework of partial homomorphisms could

---

[5] With his theory of distributions, Schwartz created a calculus that extends the class of ordinary functions to a new class, the distributions, but which still preserves several of the basic operations of analysis, such as addition, multiplication by C^∞ functions, and differentiation. As there defined, the space of distributions consists of continuous linear functionals on C^∞, that is, the dual space of the space of the so-called testing functions, equipped with a given topology that involves the convergence of derivatives of all orders. This entails that each distribution can be represented locally in terms of a finite sum of derivatives (in the sense of the distribution theory) of continuous functions. (For details about the theory of distributions, see Halperin and Schwartz 1952; see also Brezis and Browder 1998, pp. 97–9.)

[6] Chris Mortensen provides an interesting approach to the delta function, preserving its properties in the context of paraconsistent mathematics (see Mortensen 1995, pp. 67–72).

[7] And we would speculate that something similar could be said about other such 'identifications' in the history of mathematics.

quite naturally capture both the open-ended nature of Dirac's theory and the manner in which it can be related to Schwartz's. However, our point here is straightforward: the indispensability of the delta function cannot be argued for on the basis of a 'post-hoc' identification with Schwartz's distribution.[8]

Nevertheless, Dirac did have a point when he insisted that, although ultimately dispensable, the delta function helps to express the relevant relations "in a concise form". And even though we could rewrite QM without it, this would be achieved "only in a cumbersome way which would tend to obscure the argument" (Dirac 1958, p. 59).

Now, if failing to use the delta function tends "to obscure the argument", does it mean that the function has an *explanatory* role? It appears not. The function does allow us to define the length of the basic vectors even when we can't normalize them (because the parameters that label the vectors take on continuous values). But it's not clear that the function *explains* why this definition is possible. As Dirac noted, the delta function only provides a useful notation that allows the definition to be carried out—a notation that is ultimately dispensable, and so it's unclear whether it plays any explanatory role here.[9]

Two additional worries can be raised at this point: (a) given that the delta "function" is admittedly not a function in the strict sense, how could Dirac have known that using it within an integral yields something meaningful? We have good reason to believe that *functions* are well behaved in the context of integrals. But clearly, we don't have any such reasons with regard to *improper functions*. He could only *hope* that improper functions behave as functions in the context of an integral. But do they really behave in this way? (b) Moreover, the really puzzling feature about the delta function is not just that it lacks a sharply defined value at some point in its domain and yet has an integral over the whole domain. The problem, rather, is that the non-zero integral seems to arise from the undefined point. How can the integral of any function be non-zero over a single point?[10]

These are reasonable concerns, and in an attempt to alleviate them, at least in part, Dirac indicated how the delta function could be dispensed with altogether. As a result, the function ultimately wouldn't be needed. Of course, this doesn't completely dispel the worries, given that, with regard to (a), only *functions* and *numbers* can be meaningfully used in an integral. If the delta function is neither of them, how can it be so used? In Dirac's view, despite not being a function in its complete generality, the delta "function" behaves like a function under most circumstances—that's the reason why he highlighted the two properties of the function discussed above (see equations (3) and (4)). These properties yield the intended results "without

---

[8] Again we are grateful to one of OUP's readers for urging us to make this clear.

[9] Similarly, the delta function does not play an explanatory role when it's used to establish the equivalence between matrix and wave mechanics. Ultimately, the function only provides a notational device to relate differential and integral operators. And as von Neumann showed, the use of the function in this context is clearly dispensable (see von Neumann 1932).

[10] We thank Mark Colyvan (personal correspondence) for pressing this point.

essential error" (Dirac 1958, p. 59)—again, this is a significant *pragmatic* move that he made. There certainly is the possibility of error—and Dirac openly admitted that. But given equations (3) and (4), that possibility is substantially restricted, allowing the results to be obtained without significant trouble.

Similarly, with regard to (b), Dirac's idea was that the delta function "vanishes everywhere except inside a small domain, of length ε say, surrounding the origin $x = 0$" (Dirac 1958, p. 58). However, the value of the function "is so large inside this domain that its integral over this domain is unity" (ibid.). So, strictly speaking, the non-zero integral is not arising from the undefined point, but from the *small domain in the immediate neighbourhood* of that point. Of course, Dirac was perfectly aware of the issue as to whether there is a function satisfying this condition. That's why he ultimately insisted on the function's dispensability.

However, was he justified in claiming that the delta function is dispensable? To answer this question, recall, first, that Dirac's elimination strategy involves identifying two properties of the delta function (expressed in equations (3) and (4), above). These properties not only allow one to operate algebraically with the delta function, but also to eliminate it eventually; after all, on the right-hand side of equations (3) and (4), the delta function doesn't figure any more. So, as long as the delta function is only used as a factor in an integrand—which is the only context in which Dirac needed to use the function—and as long as equations (3) and (4) are invoked to operate with the function—which is something he had to do in any case—his strategy works. In other words, Dirac's dispensability strategy was crucially based on two mathematical properties of the delta function—those expressed by equations (3) and (4).[11] Moreover, by using the delta function as a factor in an integrand, Dirac had no difficulty in using it to define the length of the basic vectors of observables with continuous spectra. And by invoking later these two mathematical properties of the delta function, he was then able to dispense with the function altogether.

7.3 The Pragmatic and Heuristic Role of Mathematics in Physics

The discussion above indicates that Dirac was clear about his attitude with regard to the role of mathematics in physics and the nature of applied mathematics. Mathematics plays a *pragmatic* role: it is *useful* in expressing relations among physical quantities, and in establishing (and shortening) derivations. The expressive and inferential usefulness of mathematics doesn't justify a commitment to the *truth* of the relevant mathematical theories. As Dirac emphasized, mathematics is *only* a tool:

---

[11] So, Dirac's dispensability strategy is *not* based on the idea that mathematical theories are ultimately tools of derivation in science. Rather, the strategy explores mathematical properties of the delta function that enabled it to be dispensed with.

Mathematics is the tool specially suited for dealing with abstract concepts of any kind and there is no limit to its power in this field. For this reason a book on the new physics [i.e. QM], if not purely descriptive of experimental work, must be essentially mathematical. All the same *the mathematics is only a tool* and one should learn to *hold the physical ideas in one's mind without reference to the mathematical form.* (Dirac 1958, p. viii; italics added)

Dirac's attitude toward mathematics is, of course, not uncommon among physicists. After all, what matters is not the mathematical content of the empirical theories under consideration, but the *physical* content. To highlight this instrumental role of mathematics, Dirac also emphasized the fact that he tried (at least in his 1958) to "keep the physics to the forefront", and, in addition, that he examined "the physical meaning underlying the formalism wherever possible" (Dirac 1958, p. viii). This suggests that there is a significant demarcation between the mathematics and the physics, in that the physics leads the way while the "physical meaning" of the mathematical formalism is explored.

None of this is particularly new, of course. But it's worth highlighting the point, given that it paves the way for Dirac's own attitude with regard to the delta function. The function clearly has a crucial role in his approach, since it guarantees that every self-adjoint operator can be put in diagonal form.[12] Moreover, as we saw, the delta function is also used in the case in which the eigenvalues of an observable are not continuous, and so the basic vectors cannot be normalized (see Dirac 1958, p. 62). In this case, "to fix the numerical factors by which the basic vectors may be multiplied" (Dirac 1958, p. 58), the delta function was introduced as a "convenient method of handling this question" (ibid.).

Throughout this discussion, as we saw, Dirac stressed the *pragmatic* character of the delta function. He also considered the function simply as a "new mathematical notation" (Dirac 1958, p. 58). The crucial question then arises: Was his attitude towards the delta function—in particular his attempt to establish the function's dispensability—only the result of its peculiar properties (its inconsistency and unusual character), or was there something more general going on? That is, should Dirac's pragmatic attitude towards the delta function, and his strategy of dispensing with it, be extended to other mathematical entities used in physics? Or is the pragmatic attitude only warranted in the case of ill-behaved mathematical entities (such as inconsistent objects or objects defined in an unusual way)?

We would suggest that there was something more general going on, and that Dirac's attitude should be extended to any mathematical entity referred to in physics. Whether this attitude was adopted or not in distinct cases of practice is a matter for further historical analysis. Our point here is that it can be extended to any mathematical entity, and thus *should* be so extended, undercutting the supposed indispensability on which realism towards such entities is grounded. In fact, Dirac's

---

[12] This is the property that Dirac needed in order to establish the equivalence between wave and matrix mechanics. For an insightful discussion of this point, see von Neumann (1932), pp. 3–33.

emphasis (a) on the physical content of his mathematical formulation of QM, and (b) on the pragmatic use of mathematics already indicates that he would resist the reifying moves of those who believe in the existence—and indispensability—of mathematical objects. The fact that the delta function is inconsistent and unusually defined is certainly a source of worry. But, interestingly enough, Dirac *doesn't* raise the issue of the function's inconsistency (or the way in which it has been defined) as a motivation for its dispensability. The inconsistency only motivates a *restricted use* of the delta function. As he pointed out: "$\delta(x)$ is not a quantity which can be generally used in mathematical analysis like an ordinary function, but *its use must be confined to certain simple types of expression for which it is obvious that no inconsistency can arise*" (Dirac 1958, p. 58; italics added). In other words, even though the delta function's inconsistency *might have been* a reason for its dispensability, it *actually* wasn't. Not, at any rate, according to Dirac.

Rather, Dirac's attitude towards the delta function could easily have been the same as that advanced by von Neumann. If our physical theories *require* inconsistent concepts, we should incorporate the latter concepts into our current mathematical theories in order to preserve the physics. In fact, it wouldn't be the first time in which the scientific community changed standards and theories in mathematics due to a demand from physics: the introduction of the calculus in the context of Newtonian mechanics did exactly that. We recall von Neumann's insistence that:

There would be no objection here [to the introduction of inconsistent mathematical concepts, such as the delta function] if these concepts, which cannot be incorporated into the present day framework of analysis, were intrinsically necessary for the physical theory. Thus, as Newtonian mechanics first brought about the development of the infinitesimal calculus, which, in its original form, was undoubtedly not self-consistent, so quantum mechanics might suggest a new structure for our "analysis of infinitely many variables"—i.e., the mathematical technique would have to be changed, and not the physical theory.   (von Neumann 1932, p. ix)

Of course, we are not suggesting that von Neumann or Dirac would have introduced inconsistent mathematical concepts without hesitation. The point is that they would have introduced such concepts, albeit reluctantly perhaps, if the latter turned out to be *indispensable* to our physical theories. So, the issue has less to do with the *status* of the concepts in question (e.g. whether they are inconsistent or not), and has more to do with the *role* that these concepts play in physics (i.e. whether they are indispensable or not).

As it turns out, the delta function is *doubly dispensable*: Dirac devised one strategy (namely, to use the function only in restricted conditions) and, as we saw, von Neumann devised another strategy (namely, to bypass the use of the delta function altogether, by formulating QM in terms of Hilbert spaces). One might, of course, wonder why they adopted such different strategies and the answer will have something to do with not only their different mathematical abilities but also their different points of view: Dirac was focused on the 'local' and specific behaviour of the delta function and the context in which it could be reliably used, whereas von Neumann was considering the broader landscape of the functions defined on the relevant mathematical spaces of each of the two formulations of QM. However, this does not, we think, undermine our suggestion above that Dirac's pragmatic attitude can be extended to other mathematical entities, since the point is that even in cases where scientists are unable to do what von Neumann did and reformulate the very foundations of the theory, they will typically adopt a pragmatic attitude towards the relevant mathematical entities.

To return to this particular case, as we said, the delta function is not indispensable after all, and so there is no reason to postulate its existence. We shall return to consider this issue of dispensability and claims for the existence of mathematical entities in Chapter 8.

7.4 The Discovery of Antimatter

Before we do, let us consider a further interesting case study from Dirac's work. This concerns his famous discovery of antimatter (Dirac 1928a, 1928b, 1930, and 1931) where it appears that the mathematical formalism helped guide Dirac to his prediction of the existence of (what later would be called) the *positron*. In this case, it might seem, the mathematics *does* play an indispensable role.

Colyvan has presented this point as follows:

In classical physics one occasionally comes across solutions to equations that are discarded because they are taken to be "non-physical". Examples include negative energy solutions to dynamical systems. This situation arose for Paul Dirac in 1928 when he was studying the solutions of the equation of relativistic quantum mechanics that now bears his name. This equation describes the behaviour of electrons and hydrogen atoms, but was found to also describe particles with negative energies. It must have been tempting for Dirac to simply dismiss such solutions as "non-physical"; however, strange things are known to occur in quantum mechanics, and intuitions about what is "non-physical" are not so clear. So Dirac investigated the possibility of negative energy solutions and, in particular, to give an account of why a particle cannot make a transition from a positive energy state to a negative one.

(Colyvan 2001, p. 84)

This investigation would turn out to be extremely fruitful. After all:

Dirac realised that the Pauli exclusion principle would prevent electrons from dropping back to negative energy states if such states were already occupied by negative energy electrons so widespread as to be undetectable. Furthermore, if a negative energy electron was raised to a positive energy state, it would leave behind an unoccupied negative energy state. This unoccupied negative energy state would act like a positively charged electron or a "positron". Thus, Dirac, *by his faith in the mathematical part of relativistic quantum mechanics and his reluctance to discard what looked like non-physical solutions, predicted the positron.*

(Colyvan 2001, p. 84; italics added)

Colyvan then concludes:

Dirac's equation play[s] a significant role in predicting a novel entity despite the relevant solutions seeming non-physical [...]. It is hard to see how a nominalised version of Dirac's theory would have had the same predictive success.   (Colyvan 2001, p. 84)[13]

However, it turns out that the situation was more complicated, and even in the case of the positron, the use of mathematics was at best pragmatic. First of all, here we have a nice case illustrating the significance of surplus structure: the mathematics yields a set of solutions to the dynamical equations that are initially deemed to be 'non-physical'. At this point this extra structure is clearly 'surplus' and like other such 'non-physical' solutions, would typically be dismissed as having no ontological—that is, physical—referent. However, Dirac, famously, went on to exploit this structure, assigned it a referent—via his 'sea' of energy states and, subsequently, in the form of the positron—and, to cut a quite complex story short, predicted the existence of antimatter.[14] Note, however: there is no doubt that via the presentation of this surplus structure, the mathematics played a *heuristic* role, but this is not the same as an *indispensable* role. Ultimately, the work is done by *interpreting* the mathematical formalism—by assigning a physical meaning to the formalism—rather than by the formalism alone. But, as we will see, the *mathematical formalism doesn't uniquely determine its physical interpretations*, given that it is compatible with radically different physical scenarios. And so, ultimately, the mathematical formalism *doesn't* establish what is physically significant. As a result, the formalism cannot have a "significant role in predicting a novel entity" (Colyvan 2001, p. 84), since it is compatible with fundamentally different physical entities. It is no surprise that the acceptance of the positron only happened after *independent evidence* was found for it— that is, only after certain *criteria of existence* had been satisfied.

It is, of course, no trivial task to specify such criteria. Here we simply suggest the following conditions (we shall return to them in Chapter 8): we have *observed* a given object (a particle say);[15] we have *interacted* with it; we have *tracked* it; we have developed mechanisms of *instrumental access* to this object (mechanisms that can be *refined*). Although any of these conditions can be defeated, taken together they provide significant (independent and sufficient) grounds to be committed to the existence of the objects in question.[16] And in the case of the positron the satisfaction

---

[13] For additional discussion of this episode, see Schweber (1994), pp. 56–72, Pais (1986), pp. 346–52 (see also pp. 290–2 and 362–4), Jacob (1998), pp. 46–61, and Pais (1998), pp. 11–19. We draw on these sources in the discussion that follows.

[14] For further, if brief, discussion of such shifts in ontological reference see Redhead (1975, p. 88). In effect this history sets out the kinds of broadly heuristic factors previously touched on as providing the basis for selecting which of the mathematical structures made available get transferred to the empirical level.

[15] Obviously realists and empiricists may understand 'observe' here in a wider or narrower sense.

[16] Azzouni provides (tentatively) one set of *criteria of existence* in terms of what he calls "thick epistemic access" (see Azzouni 1997). In his view, we have a *thick* form of epistemic access to an object if the

of such criteria was only established after much controversy about what the experiments established.[17] Given that the mathematical entities quantified over throughout this episode *don't* meet these criteria of existence, nothing requires their existence—and as we will see, Dirac didn't claim they did (and again we shall return to this distinction between mathematical and physical entities in Chapter 8).

It is important to realize that Dirac himself was initially unsure about how to interpret this surplus structure, namely the existence of negative energy solutions to his equation. In his first 1928 paper, Dirac explicitly acknowledged the problem. The familiar relation between energy, momentum, and mass, $E^2 = c^2p^2 + m^2c^4$, has two roots:

$$E = \pm C\sqrt{(p^2 + m^2c^2)}.$$

But it's not clear what should be done with the negative solutions. As Dirac pointed out:

One gets over the difficulty on the classical theory by arbitrarily excluding those solutions that have a negative [E]. *One cannot do this in the quantum theory, since in general a perturbation will cause transitions from states with* [E] *positive to states with* [E] *negative.* Such a transition would appear experimentally as the electron suddenly changes its charge from -e to e, a phenomenon which has not been observed. The true relativity wave equation should thus be such that its solutions split up into two non-combining sets referring respectively to the charge -e and to the charge e.   (Dirac 1928a, p. 612; italics added; see also Schweber 1994, p. 61, and Pais 1986, pp. 347–8)

Three points should be noted here. First, by 'arbitrarily excluding' the negative energy solutions, classical physics treats them as corresponding to genuinely surplus mathematical structure.

Second, this is because there are no grounds in this theoretical context for taking that structure to have any heuristic significance. But in quantum physics, we do have such grounds. Thus, Dirac emphasizes the existence of a *physical reason* why in quantum theory negative energy solutions to the energy-momentum-mass equation *cannot* be excluded. After all, a perturbation could cause transitions from positive

epistemic access to this object (i) is *robust*, (ii) can be *refined*, (iii) enables us to *track* the object, and is such that (iv) certain properties of the object itself play a role in *how we come to know* other properties of the object. Observation provides the paramount example of thick epistemic access. Azzouni then distinguishes this thick form of epistemic access from a *thin* form of access. According to the latter, our access to an object is *through a theory* that has five virtues: (i) simplicity, (ii) familiarity, (iii) scope, (iv) fecundity, and (v) success under testing. (These are the familiar Quinean theoretical virtues; see Quine 1976, p. 247.) The point can be presented in the following way: as opposed to thick epistemic access, thin epistemic access provides *no* reason to *believe* in the existence of the objects in question—at best, it provides a reason to *accept* the theory in which these objects are studied. (Azzouni doesn't use the distinction between acceptance and belief to make his case, but, as will become clear, the distinction is useful in the discussion that follows.)

[17] Interestingly enough, the empirical detection of the positron was largely undertaken independently of Dirac's theory, as we shall see.

energy states to negative energy states. If such transitions are not observed, it's crucial to identify the *physical* reason for that. So, the issue isn't whether our intuitions about what is "non-physical" in QM are clear or not (Colyvan 2001, p. 84). The issue is to identify the *physical process* that explains why we *don't* observe transitions in which an electron changes its charge from $-e$ to e. In other words, as Dirac conceptualized the problem, the issue is physical, *not* mathematical.

Third, however, at this point in 1928, because Dirac wasn't clear about the appropriate physical interpretation of the equation, he argued that eventually the negative energy solutions should, in fact, be excluded. As he insisted, in the end, "half of the solutions must be rejected as referring to the charge $+e$ of the electron" (Dirac 1928a; see also Pais 1986, p. 348). In other words, Dirac effectively reverted to regarding the corresponding mathematical structure as genuinely 'surplus'; in the absence of further grounds, the heuristic potential of this structure was not realized (at least not immediately).

It was only *after* Dirac developed a physical interpretation of the equation—an account that identified the physical process described by the negative energy solutions—that he took this structure, corresponding to the latter solutions, seriously. So, Dirac was not blindly following the mathematics wherever it took him. It was *not* a matter of having "faith in the mathematical part of relativistic quantum mechanics" and being reluctant "to discard what looked like non-physical solutions" (Colyvan 2001, p. 84). The surplus structure in and of itself has no heuristic force—it is only when there are the relevant grounds, or reasons, in play that the structure has significance.[18] Thus, Dirac first identified what he considered to be physically significant features of the situation, and only then took seriously the negative solutions. In the absence of these physically significant features, he would simply rule out the negative energy solutions to his equations as, again, just so much mathematical surplus. In the end, these physically significant features were only identified in 1930 (and so two years after the original formulation of the problem). And the features were developed in terms of Dirac's "hole" theory (see Dirac 1930). As Dirac claimed, in a now famous passage:

> Let us assume that there are so many electrons in the world, *that all the states of negative energy are occupied except perhaps a few of small velocity.* [...] *Only the small departure from exact uniformity, brought about by some of the negative-energy states being unoccupied, can we hope to observe.* [...] We are therefore led to the assumption that the holes in the distribution of negative electrons are the protons.
>
> (Dirac 1930, p. 362; italics in the original; see Schweber 1994, p. 62)

In other words, the issue here is of a physical, not mathematical nature. And even though the mathematics does play a heuristic role here, Dirac didn't consider the

---

[18] Again, this is to reinforce the point already made about the role of contextual factors here; see also note 29 of Chapter 2.

negative energy solutions as significant until a *physical process* was clearly put forward. This is, of course, an additional example of his strategy of keeping "the physics to the forefront" (Dirac 1958, p. viii)—a strategy that informed so much of his approach.

The important point—and Dirac himself was explicit about this—is that what was guiding the physical construction here, and thus underpinning this heuristic signifi-cance, was *not* the mathematics, but actually, it was *chemistry*, in particular, the chemical theory of valency. Commenting on Dirac's strategy, Schweber puts the point very clearly:

> The idea of the hole theory was suggested to Dirac by the chemical theory of valency in which one is used to the idea of electrons in an atom forming closed shells which do not contribute at all to the valency. One gets a contribution from an electron outside closed shells and also a possible contribution coming from an incomplete shell or hole in a closed shell.
>
> (Schweber 1994, p. 62)

In fact, Dirac insisted:

> One could apply the same idea to the negative energy states and assume that normally all the negative energy states are filled up with electrons, in the same way in which closed shells in the chemical atom are filled up.   (Dirac 1978, p. 50)

But, alas, Dirac's interpretation, in terms of "holes", was far from adequate! Heisenberg, Pauli, Oppenheimer, and Weyl, among others, raised several objections. For example, in Dirac's theory, the electron and the proton get the same mass, and the theory entails a much too high rate of annihilation of protons and electrons into γ-rays to be empirically adequate.[19] Because of these criticisms, in 1931 Dirac eventually gave up his "hole" theory:

> It thus appears that we must abandon the identification of the holes with protons and *must find some other interpretation for them.*[20] Following Oppenheimer, we can assume that in the world as we know it, all, and not nearly all, of the negative energy states are filled. *A hole, if there were one, would be a new kind of particle, unknown to experimental physics, having the same mass and opposite charge to an electron.* We should not expect to find any of them in nature, on account of their rapid rate of recombination with electrons, but if they could be produced in high vacuum they would be quite stable and amenable to observation.
>
> (Dirac 1931, p. 61; italics added; see Schweber 1994, p. 67)

This prediction, as is well known, was ultimately vindicated (even though the data didn't support the existence of "holes" as Dirac initially conceived of them). In fact, in 1932, Anderson reported the detection of the positron (Anderson 1932a, 1932b).

---

[19] See Schweber (1994, pp. 65–7), for a discussion of and reference to the arguments put forward by the critics of Dirac's hole theory.

[20] Note that the crucial issue, as Dirac emphasized, is to identify a physical interpretation of the formalism—that's where the work is done.

And in the following year, other experimentalists confirmed Anderson's result (see Blackett and Occhialini 1933; Blackett 1933). But the positron was certainly not the "hole" initially postulated by Dirac![21]

Given all the vicissitudes of the process of theory construction that led to the positron's detection, it is no wonder that Anderson thought that the discovery of the positron was completely accidental. When asked about the influence of Dirac's work on his own ideas, Anderson was very clear:

Yes, I knew about the Dirac theory. [...] But I was not familiar in detail with Dirac's work. I was too busy operating this piece of equipment to have much time to read his papers.

(Anderson 1966; see Pais 1986, p. 352)

Anderson then added:

[The] highly esoteric character [of Dirac's papers] was apparently not in tune with most of the scientific thinking of the day. [...] The discovery of the positron was wholly accidental.

(Anderson and Anderson 1983; see Pais 1986, p. 352)

Of course, the considerations above don't diminish at all the importance and significance of Dirac's discovery. They only highlight that the work was not done by the mathematics alone. Of course, it has to 'provide', as it were, this surplus structure. But the significant work, as one would expect, was done by the physics—by developing a fruitful *physical interpretation* of Dirac's equation and granting that surplus structure its heuristic significance.[22] This interpretation identified the *type of entity* and the *physical properties* that this entity should have for the theory to be empirically adequate. As Dirac pointed out, what needed to be found was a "new kind of particle", one that had "the same mass and opposite charge to an electron" (Dirac 1931). But these features are *not* determined by the mathematical formalism. The latter is compatible with both (i) Dirac's initial identification of his "holes" with protons, and (ii) Dirac's later identification of the "holes" with a new kind of particle. And so, clearly, the interpretations are *underdetermined* by the formalism. Given the underdetermination, the mathematics doesn't determine what is physically relevant, and given that the mathematical objects referred to in the formalism fail to satisfy the relevant criteria of existence, no reification of the mathematics is justified either. Thus, just as with the delta function, the use of mathematics in the case of the positron was *heuristic* and *pragmatic* at best. There's no need to reify mathematics to accommodate this episode.

---

[21] As Pauli wrote in a letter to Dirac in 1933: "I do not believe in your perception of 'holes', even if the existence of the 'anti-electron' is proved" (quoted in Schweber 1994, p. 68). Talking about Anderson's discovery and the confirmation by Blackett and Occhialini, Bohr added: "Even if all this turns out to be true, of one thing I am certain: that it has nothing to do with Dirac's theory of holes!" (also quoted in Schweber 1994, p. 68).

[22] We shall briefly return to this point in Chapter 10, in the context of Bangu's alternative account of Dirac's prediction (Bangu 2008, 2012).

In other words, (a) in the case above, the crucial work was done by *interpreting* the mathematical formalism (without a physical interpretation, no empirical predictions could ever be obtained from the latter). But (b) the mathematical formalism *under-determines* its physical interpretations. So, it's ultimately silent about what is *physically significant*—as one would expect from a mathematical theory. Thus it would be misleading to say that the mathematics played an *indispensable* role (particularly if this is used to justify a commitment to the existence of mathematical entities). Mathematics certainly played a *heuristic* role. But this is a very different issue.

Still, one might insist that there is more to it than that. One could argue that granted that the mathematical formalism alone did not determine the properties of the positron, granted also that a physical interpretation and argument were needed, nevertheless it was not the case that the mathematics itself was silent about what was physically significant or relevant. Indeed, it may be argued, the mathematics played a significant role in effectively pointing to the holes that needed to be filled in the first place. Dirac's reasoning depended on there being holes (granted, again, that they were part of the surplus mathematics), on the possibility that they corresponded to physical energy states (supported by the relevant physical reasons), and on the need to fill them by means of some appropriate interpretation. He could only have had confidence in the (admittedly tentative) trustworthiness of his reasoning if he had had reason to believe the first mathematical step, that there were holes to begin with. The question then is how he could have had such confidence if the mathematics was acting purely as a pragmatic device. Thus the underlying concern here is that even if some device is 'merely' heuristic or pragmatic, that does not mean that it lacks *objective* content.[23]

This is an interesting point, suggesting, as it does, an intermediate position between indispensability and heuristic usefulness. However, we do not think that such a position is viable. Although in order for mathematics to be applied certain moves need to be made on both 'sides'—for example, certain idealizing moves when it comes to physics and certain 'bridge-building' moves when it comes to the mathematics, as we described in Chapter 4—such moves will still typically leave 'surplus' mathematical structure. The mathematics itself does not 'point' to this structure—it is just 'there', as it were, as an inevitable concomitant or 'by-product' of the structural fecundity of the mathematics itself, even when constrained by the aforementioned moves. The issue then is whether this surplus structure is exploitable and only the physical context can determine that. In the case of Dirac's 'holes', we have perhaps quite a simple form of surplus structure, namely negative energy solutions, akin to the $\sqrt{4}$ being $+2$ or -2. In the classical context, as we have already indicated, this 'surplus' would be simply dismissed as unphysical; it is only in the relevant quantum context that it becomes exploitable and open to the possibility of a

---

[23] Again, we owe this argument to one of the anonymous readers of an early draft of the book.

physical interpretation. Thus there is no 'pointing' on the part of the mathematics—it simply provides or makes available the structure that, in a sense, it already 'contains'.

As for Dirac then having confidence in the trustworthiness of his reasoning, it would be a mistake to suppose, as the argument above does, that when he had reason to believe the first mathematical step, this can be 'cashed out' as his having reason to believe there were holes to begin with. He certainly had grounds to trust the mathematical step precisely because it already followed from the mathematics he already had to hand, as it were—just as the possibility of the $\sqrt{4}$ being -2 follows from the mathematics. But he had *no* reason to believe there were holes *qua* physical entities—this only follows once a physical interpretation has been provided. To suggest otherwise would be to beg the question at issue! Thus any objectivity that accrues to the mathematics is associated with its deductive nature and certainly cannot be said to amount to some form of 'objective content' about the physical world that could then allow it to occupy a third position between being heuristic and being indispensable.

So how, then, do we distinguish between the indispensable and the heuristic roles of mathematics? A mathematical theory plays an *indispensable* role if to achieve this role (whatever it turns out to be) the mathematical theory *cannot be eliminated*. A mathematical theory plays a *heuristic* role if the mathematical theory is fruitful in helping to generate new ideas (particularly, but not exclusively, if those turn out to be well supported) by, for example, providing surplus structure that, with the appropriate grounds, can then be physically interpreted. As we shall emphasize in Chapter 8, that the mathematics plays *this* role provides no support for claims about the existence of mathematical entities. In the case of the positron, the mathematics at best *suggested* that there might be some object yet to be discovered. But it didn't determine which object that was, nor did it specify its physical properties. As we saw, the mathematics was compatible with radically different physical interpretations. So, the role of mathematics clearly was heuristic.

7.5 Conclusion

The picture of the application of mathematics that emerges from Dirac's view is one in which the crucial work is carried out by suitable physical interpretations of the mathematical formalism, including relevant surplus structure, without the commitment to (i) the truth of the relevant mathematical theories or (ii) the existence of the corresponding mathematical objects. With regard to (i), as argued above, in order to play a heuristic role, mathematical theories need not be true;[24] they only need be consistent with the empirical data, given suitable physical interpretations of the

---

[24] Hartry Field provided a very interesting programme to justify the claim that good mathematical theories need not be true (Field 1980, 1989; for some discussion see Bueno 1999, 2003). But as opposed to the approach suggested in the present chapter, Field's programme is not based on mathematical or

mathematical formalism. And even if mathematical theories were to play an indispensable role, that wouldn't entail that we ought to be ontologically committed to the corresponding objects either, since quantification over mathematical entities alone doesn't require commitment to their existence.[25] After all, moving now to (ii), the commitment to the existence of certain objects—including mathematical entities—requires the satisfaction of certain criteria of existence. If mathematical objects fail to meet such criteria, nothing requires us to be ontologically committed to them. Thus, even if mathematics is taken to be true, this would not commit one to Platonism—more is needed, namely the satisfaction of the above criteria.

We would like to emphasize again two crucial points, in this regard. First, as we have already indicated, Dirac himself had a double-edged view of mathematics, setting great store in what he regarded as the mathematical 'beauty' of certain physical theories, yet also taking mathematics to be no more than a 'tool' (perhaps reflecting his engineering background). Kragh makes the further point that in his cosmological work in particular, Dirac also adopted something akin to the 'Pythagorean Principle' 'according to which numerical coincidences and regularities in nature are not fortuitous but are manifestations of the order of the laws of nature' (Kragh 1990, p. 283). This led him to emphasize the role of the natural numbers, which are bound up with the theory of functions of complex variables, which in turn, Dirac held, had 'a good chance of forming the basis of the physics of the future' (quoted by Kragh in ibid.). However, there is nothing here to suggest that Dirac adopted a form of Platonism according to which numbers exist above and beyond the regularities in nature. Instead the idea seems to be that the presence of the natural numbers in our fundamental physics is suggestive of a certain order in the relevant laws.[26] Furthermore, and relatedly, the quote from Dirac himself supports the claim that the role of mathematics in this case is primarily heuristic.

---

scientific practice (it's not meant to be), and it yields a much more complex strategy to dispense with mathematical entities.

[25] Thus one could adopt the view that certain mathematical theories are true, on the grounds that they play an indispensable role, in some sense, but that one is not committed to the existence of the entities apparently referred to or implied by these theories. If one takes the sense of indispensability here to be stronger than mere heuristic usefulness and the notion of truth to be understood in the standard realist correspondence sense, then this might strike some as a rather odd position. After all, if we take a scientific theory to be true in that sense, we typically take the relevant theoretical terms to refer. One might try to understand the view being suggested in terms of mathematical structuralism for example, in that this position might be interpreted as taking mathematics to be true but without commitment to the relevant entities, such as numbers for example. (This suggestion was offered by one of the anonymous readers of this work.) However, we can simply broaden the reach of what we mean by 'entity' to include structures and the sense of oddness returns. Nevertheless, our point is that one *could* insist on the truth of mathematics but still refuse to be committed to the relevant entities (objects or structures) on the grounds that they do not satisfy the relevant criteria of existence. (We shall elaborate further on this point in Chapter 8.)

[26] Eddington, of course, also famously set significant store by such numerological coincidences in his work in cosmology.

Second, even granted the Pythagorean point, whatever Dirac may have thought about the natural numbers in this particular context, his own practice does not motivate or require a Pythagorean stance. As we have repeatedly emphasized, whatever work was done was not done by the mathematics alone—interpretation was required. Indeed, as we have noted, the Dirac equation, as it stands, is compatible with three quite different empirical set-ups.

Furthermore, even though we acknowledge that here we are examining just one particular case study from the history of science which obviously cannot support a general claim as to the dispensability of mathematics, it is worth noting that this episode has played a significant role in supporting certain realist claims (by Colyvan, for example). The point of our analysis is to suggest that such claims may not be as well-grounded as might be thought—not least because the mathematics is being used in a more complex manner than can be summarized via its purported role in predicting new physical entities. Indeed, this case study is illustrative of a situation that recurs throughout the history of science, whereby complex mathematics is introduced together with certain concomitant requirements that constrain the use of that very mathematics.[27]

In Chapter 8, we return to the purported indispensability of certain forms of mathematics and its putative explanatory role which has been taken to support the claim that certain entities have a *hybrid* mathematico-physical nature. We shall emphasize, again, that mathematics plays no explanatory role in such cases but acts only in a representational capacity and that claims as to the hybridity of certain entities fail to fully grasp the details of the interrelationships between mathematical and physical structures in general and the distinction between the mathematical formalism and its interpretation in particular.

---

[27] Again we are grateful to one of the readers for pressing us on these issues.

8

Explaining with Mathematics?
From Cicadas to Symmetry

8.1 Introduction

In this chapter and the next, we shall focus on the following question: What are the philosophical implications of the role of mathematics in physical explanations? In particular, here, we shall distinguish between two kinds of claims that have been made: the first (Strong) is that this role implies the existence of mathematical objects; the second (Weak) is that it implies that physical quantities have a 'hybrid nature'. The Strong claim has been much discussed in the recent literature particularly with regard to the so-called 'indispensability arguments'. We shall begin by considering these in the context of our considerations of applicability set out in Chapters 4–7. In particular, we recall that in the case of the positron, the significance of supplying an interpretation was crucial for displaying the dispensability of the mathematics and that this fits nicely into the framework of applicability we presented in Chapter 2. We shall then move on to more recent forms of this kind of argument that emphasize the explanatory role of mathematics and we shall show that a certain counter-response to criticism of them can in fact be closed off.

We shall turn to the Weak claim in the second half of the chapter and argue that it too can be blocked. In Chapter 9 we then consider the concern that these discussions all focus on mathematical *objects*, but ignore the role of certain *operations*, that, it is argued, force a re-evaluation of the nature of explanation and applicability. On the contrary, we shall maintain, this role can also be straightforwardly accommodated within our framework.

8.2 The Strong Claim and Indispensability

Our discussions in previous chapters raise the broader issue of the conditions under which we would be justified in believing in the existence of mathematical entities (referred to in our physical theories). The anti-realist answer is that we are never justified in believing in these entities. Nothing in the practice of physics actually *requires* the commitment to these entities. And as we saw, physicists such as Dirac have emphasized that we should "keep the physics to the forefront", articulating "the physical meaning underlying the formalism wherever possible" (Dirac 1958, p. viii). This obviously suggests that it is possible to distinguish the mathematical and the physical content of a physical theory, a possibility that is disputed by those who endorse the 'Weak' claim noted above. However, if our commitment is to the physical rather than mathematical content, then this needs to be articulated appropriately even though physical theories can only be expressed in terms of mathematical notions. So, the idea is that although mathematics is used in theory construction, this does not imply any claim about the existence of the corresponding mathematical entities. The running motif throughout this book is that the framework presented in Chapter 2 offers just such an appropriate articulation and hence that it does not follow from the role (or roles) played by mathematics in science that one must adopt a realist stance towards mathematical entities.

However, the following worry then arises: Are we, and physicists such as Dirac, then committed to a double standard, denying the existence of those very entities that we use to formulate the best physical theories? A Quinean would say so (see Quine 1953, 1960): we ought to be ontologically committed to all (and only) the entities that are indispensable to our best theories of the world. This is, of course, the first premise in Quine's famous indispensability argument, which is meant to provide the reason why we should believe in the existence of mathematical entities. The second premise of the argument states that mathematical entities *are* indispensable to our best theories of the world. From which it then follows that we ought to be ontologically committed to mathematical entities. (See Colyvan 2001, 2015; Maddy 1997; Azzouni 2004; and references quoted therein.)

However, the conclusion is not warranted. First, we often quantify over entities whose existence we have no reason to believe. Things like "The average star has 2.4 planets" and fictional characters (such as Sherlock Holmes or Macbeth) provide obvious examples. There is no way of dispensing with 'the average star' since we cannot, of course, count all the stars (see Melia 1995)![1] There is also no way of understanding the Holmes stories without asserting things like "There is a detective who lives in Baker Street". Our best theories of fiction require us to assert such existential claims. But this doesn't mean that we are committed to the existence of Sherlock Holmes! Similarly, there may not be any way of formulating quantum mechanics without quantifying over abstract entities (whether they are vectors in a

---

[1] Of course, Quine insisted that prior to considering which is our 'best' theory in terms of the usual theoretical virtues, such as simplicity, scope, explanatory power, etc., the relevant candidates must be put in an appropriate canonical form, such that one can then assert that 'to be is to be the value of a variable' falling within the scope of the relevant (first-order) quantifier. Such a reformulation would then be expected to paraphrase away such examples as 'The average star...'. However, following Melia, we are dubious about whether this can be achieved (a point on which one of the anonymous readers also insisted). Furthermore, Quinean canonical reformulation has generally fallen out of favour within recent philosophy of science, not least because it comes 'pre-loaded' with certain metaphysical baggage. For instance, a reformulation in terms of classical first-order logic brings with it a certain understanding of what is taken to be an individual that some find problematic (see French and Krause 2006).

Hilbert space or complex numbers). But this doesn't mean—and physicists typically don't mean—that we are committed to the existence of the corresponding entities. But if mathematics is being *used*, what are the moves that allow us to justify our *lack of commitment* to mathematical entities?

There are two moves here. The first is to indicate that the relevant mathematical notions are not indispensable after all. They are, at best, convenient notational devices that help the derivation of the intended results. In principle, the latter results could be established with no recourse to mathematical notions. As we saw, this is precisely Dirac's strategy in the case of the delta function—"a new mathematical notation", in his own words (Dirac 1958, p. 58). Of course, it's not enough simply to *say* that the offending notions are only useful notations, one needs to *establish* that they can actually be dispensed with. And, as we also saw, that's exactly what Dirac did as well, providing a mechanism in which the "improper functions" would only appear in an integrand. As a result, he concluded, "one could then eliminate the improper functions altogether" (ibid., p. 59).

Of course, this provides at best a piecemeal dispensability strategy, which ultimately would need to be applied to each individual mathematical notion used in physics—obviously, a virtually endless task.[2] So, a more systematic approach needs to be developed, an approach that doesn't depend so much on the particular features of the mathematical notions involved, but could be applied across the board. To articulate this approach, the second move is made.

This is to introduce ontologically neutral quantifiers: we begin by distinguishing between *quantifier commitment* and *ontological commitment* (see Azzouni 1998, 2004; Bueno 2005). A quantifier commitment is simply a commitment we incur when we quantify over certain entities; that is, we are not committed to their existence but only to the recognition that they are in the scope of what we quantify over. An ontological commitment, on the other hand, *is* a commitment to the existence of certain entities. Note that the former doesn't entail the latter, since we may quantify over entities whose existence we are not committed to. (The cases mentioned above of fictional entities and average stars clearly illustrate that.) Quine conflated these two types of commitment in the case of entities that are indispensable to our best theories of the world: ontological commitment is determined by the quantification over those entities that are indispensable to our best theories (as canonically formulated). After all, if an entity is indispensable to our best theories, we can't rewrite such theories without quantifying over the entities in question. And if the theories are the best we have, we can't do without them either. And so, if we

---

[2] Moreover, as Mark Colyvan pointed out (in correspondence), Dirac's strategy may not be conclusive either. After all, the delta function is being replaced with *additional mathematics*. And for the anti-realist at least, the additional mathematics would still need to be dispensed with. But note, first of all, the general point that problematic 'pieces' of mathematics such as the delta function tend to be replaced, albeit at some cost, such as simplicity or ease of use (so there is a trade-off between rigour and convenience). Second, the anti-realists, or agnostics, can avail themselves of the second move outlined below.

quantify over X, and X is indispensable to our best theories, we are ontologically committed to X, and X exists. So, in Quine's picture, in the case of indispensable entities, quantifier commitment and ontological commitment go hand in hand.

However, there is no need to follow Quine in identifying these two types of commitment—even if we grant the indispensability of the entities in question. As noted above, we often quantify over (and thus have a *quantifier commitment* to) fictional characters. But we don't take this quantification to provide any reason to believe in the *existence* of these fictions (even if we grant that quantification over these entities is indispensable). After all, to claim that something *exists*—to be committed to the existence of the objects in question—we require an additional condition: a *criterion of existence* needs to be satisfied.[3]

As we noted in Chapter 7, it is, of course, no trivial task to specify such a criterion. And for our present purposes, there's no need to. It's enough to provide sufficient conditions for someone to believe in the existence of certain objects. This is crucial, since it allows us to avoid counter-claims of question begging. Here we recall some such conditions previously suggested: we have *observed* a given object;[4] we have *interacted* with it; we have *tracked* it; or we have developed mechanisms of *instrumental access* to this object (mechanisms that can be *refined*). Again, although any of these conditions can be defeated, taken as a whole they provide significant grounds to be committed to the existence of the objects in question. If such conditions are *not* met, it doesn't entail that the objects in question don't exist (the conditions are only meant to be sufficient, after all). But, in that case, the scientific community will typically be suspicious about the existence of the corresponding entities, and doubts will be raised.

Now, let us contrast the conditions mentioned above with something much weaker. Suppose that the criteria of existence (or the conditions just mentioned) are *not* satisfied, but we quantify over objects that play a *theoretical role* in the description of the phenomena. Suppose that quantification over these objects helps to *simplify* the description of the phenomena, yields a theory that is *familiar, fecund*, and has a large *scope*. Do these theoretical virtues provide reasons to believe in the existence of the corresponding objects—even though the criteria of existence are *not* met? It doesn't appear so. The point is, of course, contentious, and typically, realists and anti-realists assess the issue very differently, with the latter insisting that unless it can be shown what simplicity, familiarity, fecundity, and scope have to do with truth, then they should be taken to be *pragmatic* reasons (see van Fraassen 1980). Consequently, if a theory satisfies the above theoretical virtues, we have at best reasons to

---

[3] Thus, even if Quine's suggestion were adopted, that by our 'best' theories we mean, minimally, those that are appropriately canonically formulated, we would maintain, following Azzouni, that quantifier commitment does not imply ontological commitment.

[4] This term can be understood in a general sense, and not in any way that should be taken to conflict with a structuralist ontology, for instance!

accept the objects in question, but, as far as the anti-realist is concerned, we *don't* have reasons for *belief*.[5] It may be *useful*, at best, to postulate such objects. What about the realist? Some will certainly admit that if only the theoretical virtues are satisfied, and the criteria of existence are not, then we don't have reason to believe in the existence of the objects in question.[6] Others may put less emphasis on instrumental access,[7] for example, and focus more on the explanatory role played by the relevant terms (see e.g. Saatsi 2008). But note, first, that this role is taken to be played within our 'best'—that is, most (empirically) successful[8]—theories; and second, as we shall argue in Chapter 9, mathematics does not play such an explanatory role and hence cannot be the repository for realist beliefs on these grounds.

Where does this leave us with regard to the existence of mathematical entities? In a nutshell, we have, at best, *pragmatic* reasons to *accept* such entities. Mathematical entities are simply not the kind of thing we are able to track, interact with, or devise mechanisms of instrumental access to.[9] Our only access to these entities is through our mathematical theories. These may be simple, fecund, familiar to us, etc. But this gives us at best reasons to *accept* and work with these theories. It doesn't give us reason to *believe* in the existence of the corresponding objects.[10]

In this way, by clearly distinguishing quantifier commitment and ontological commitment, it's possible to resist the indispensability argument. Its first premise is denied: the fact that we quantify over certain entities—even indispensably so—doesn't entail that we ought to be ontologically committed to these entities. As a result, since we have no reason to accept the first premise of the argument, we are no longer obliged to accept its conclusion. In the end, the issue is left open as to whether mathematical objects do exist or not. The point is that simply by quantifying over

---

[5] The distinction between acceptance and belief is articulated in detail by van Fraassen (see van Fraassen 1980, 1985). Realists may want to avail themselves of the analysis of this distinction given in da Costa and French (2003).

[6] Maddy provides a very interesting case study about belief in the existence of atoms that beautifully illustrates this claim (see Maddy 1997). Theories invoking the existence of atoms have been around for a long time, and these theories even satisfied the theoretical virtues discussed above. But the scientific community only stopped being sceptical about the existence of atoms when there was independent evidence for these objects. In other words, it was only when the criteria of existence were satisfied that the community incorporated atoms in the ontology of science. (Although Maddy doesn't present the point in these terms, this is a plausible reading of her case study.) For further details, see Psillos (2011).

[7] So, one might worry about what sort of instrumental access we have to black holes, for example (see Hacking 1989).

[8] Some may even drop this requirement, as has been suggested in the case of string theory for example (see Dawid 2013).

[9] Of course, Quine also incorporated a pragmatic attitude into his philosophical views, but our point is that when it comes to considerations of indispensability, mere quantifier commitment is not enough—we must also have the relevant instrumental access (even if indirect). For Quine, in addition to quantifier commitment, reference to the relevant objects needs to be indispensable. We do not think, however, that indispensability is sufficient for existence.

[10] Using Azzouni's terminology as given in note 16 of Chapter 7, we have at best *thin* epistemic access to mathematical entities; we never have *thick* epistemic access to them.

these objects one need not be committed to their existence—nothing in scientific practice requires that.

For these reasons, it's not surprising that Dirac adopted the attitude he did towards the delta function. As we saw, he always emphasized the pragmatic character of the function, and never thought that the function's usefulness provides any reason to believe in its existence. Moreover, the second move—highlighting the distinction between quantifier commitment and ontological commitment—also explains why Dirac only considered mathematics as a *tool*, as something that provides convenient notational devices to express relations among objects, but nothing beyond that. As long as mathematical objects fail to satisfy the criterion of existence, we will always be justified in avoiding being committed to their existence—as Dirac certainly was.

However, it might be objected that even if we grant the importance of distinguishing between quantifier commitment and ontological commitment, the particular criteria of existence suggested above (or the conditions that motivate such criteria) are inadequate. Recall that these involve being able to observe a given object, interact with it, track it, etc. But the trouble here, the objection goes, is that, even granted their acceptance by certain realists, these criteria seem to favour a broadly empiricist epistemology (or something like an empiricist epistemology, one that even a realist would accept). After all, the criteria emphasize one form or another of *empirical* access to the objects we claim to exist—and obviously mathematical objects fail this. As a result, the criteria beg the question against the existence of mathematical entities. (As we shall see shortly, similar question-begging claims have been made with regard to a further form of the indispensability argument that emphasizes the explanatory role of mathematics.)

In response, recall that the criteria suggested here provide only sufficient conditions for belief in the existence of certain objects; they do not provide necessary conditions as well. So, the fact that the criteria are not met by mathematical objects doesn't entail that such objects don't exist. Moreover, note also that the criteria suggested here seem to inform much of the theoretical practice of physicists. Dirac's and von Neumann's own attitudes towards the role of the delta function in theory construction, and more generally their attitude towards the role of mathematics in physics, can be better understood as the result of adopting these criteria. And so, at least to this extent, the criteria seem to be reflected in the physicist's practice. Furthermore, if there is something about the way in which mathematics is used in science that doesn't allow (or, at least, doesn't require) the users to believe in the existence of mathematical entities, this is not a prejudice against mathematical entities. It's simply a reflection on the way in which the practice goes. Of course, the point is that the practice doesn't *require* us to be realists about mathematical entities. Thus, to the extent that the criteria above seem to be informed by the practice, and so are motivated independently of the locus where philosophical disputes are typically conducted, no questions should be begged here. Again, we are not suggesting that scientific practice alone settles the issue surrounding such metaphysical controversies. Our point is that, even though the practice may not uniquely determine the relevant criteria, it seems to suggest that considerations such as those discussed above *do* play a role in the practice of physics. Given that the practice is typically independent of this sort of controversy, the proposal advanced here doesn't beg the question.

In the case of the delta function, its dispensability is more or less apparent. But what about the case of the 'prediction' of the positron? Was the role of the mathematics here also indispensable? That is, *could* Dirac have predicted the existence of antimatter without referring to mathematical entities? We think he could, because in the end, that's what he actually did. But it's a controversial issue, how to assess a counterfactual condition such as the above. The realist about mathematical entities will insist, of course, that reference to such entities cannot be eliminated, and so we ought to believe in their existence. But is this really the case? The considerations above indicate that, when the details of the process of theory construction are examined, there is no need to assign any ontological significance to the indispensable role of mathematics. After all, the predictive work was ultimately done by the physical interpretation of the formalism. In other words, even if mathematics played an indispensable role, it doesn't follow that we are justified in believing in the existence of mathematical objects. To believe in the latter objects, the criteria of existence need to be satisfied. Given that the criteria are not met, we are not required to believe in the existence of these objects. So, even if the use of mathematics turned out to be indispensable, this is not sufficient to draw ontological conclusions about mathematics. In the end, we emphasize again, the mathematics plays *at best* a heuristic role.

However, there is a further objection that could be made. Drawing on Dirac's work again, at the beginning of the *Principles*, he highlights two methods of presenting the "mathematical form" of quantum mechanics (Dirac 1958, p. viii). The first is the "symbolic method", which "deals directly in an abstract way with the quantities of fundamental importance (the invariants, etc., of the transformations)" (ibid.). The second is the "method of coordinates or representations", which deals with "sets of numbers corresponding to these quantities" (ibid.). As Dirac points out, although the method of coordinates is more familiar (at least it was in the 1930s!), the symbolic method "seems to go more deeply into the nature of things" (ibid.). The latter method also has advantages in expressive power, given that it "enables one to express the physical laws in a neat and concise way" (ibid.), and it is the method that Dirac eventually uses in his book. Now, and here comes the objection, it's not possible to make sense of the idea that the symbolic method goes "more deeply into the nature of things" except by reifying mathematics. That is, except by maintaining that mathematical theories are true, and mathematical entities exist. Thus, in the end, Dirac needs to be committed to the existence of mathematical entities to make sense of his own practice!

However, note that, once again, Dirac highlights the *pragmatic* character of mathematics—at least of the mathematics used in physics. As he points out, the symbolic method "enables one to express the physical laws in a *neat* and *concise* way" (Dirac 1958, p. viii; italics added). This ability to express physical laws neatly and concisely is a *pragmatic* feature. It indicates that mathematics is useful to us. But, as pointed out above, this usefulness provides no reason to believe that the mathematical theories in question are true. Thus, when Dirac asserts that the symbolic method goes "more deeply into the nature of things" (ibid.), this is a reflection on the symbolic method's expressive power, not a reification of mathematics. After all, using this method, physical laws can indeed be formulated in an elegant and concise form. But, to achieve this result, as we saw, there's no need to reify mathematics—and Dirac didn't.

Of course, one can always try to reify mathematics—mathematical realism is not incoherent after all! For example, the realist may understand Dirac's claim that the symbolic method "goes more deeply into the nature of things" (ibid.) as an expression that the method yields valuable explanations that couldn't be obtained otherwise. But, as suggested above with the case of the delta function, nothing in the practice of mathematics or physics requires this reification.

Now, how does the indispensability argument look in the light of our more formal—that is to say, structural—considerations regarding applicability? Peressini has approached this issue by first making a clear distinction between what he calls 'pure' mathematics and its 'formal analog' in a physical theory in the context of the example of group theory that we covered in Chapter 4:

> What is the relationship between pure group theory and its 'formal analog' at work in the physical theory? It is clear enough that pure mathematical group theory is not the *same* theory as the analog present in the theory of the spin of quantum particles, for they are about different things. The groups/members that appear in quantum theory are taken to be specific groups/members that are interpreted as the physical properties (spin) of physical objects (particles). The propositions of pure group theory, on the other hand, lack any such physical interpretation. This physical interpretation, which is at the heart of the difference between pure theory and physical application, is far from trivial as a look at the details of mathematical application will reveal.   (Peressini 1997, p. 213; italics in the original)

Furthermore, he insists, the physical application of pure group theory requires 'empirical bridge principles' to underwrite the physical interpretation:

> These principles are what distinguish pure mathematics from mathematized physical theory and enable claims about the physical world to be deduced from the latter.   (Peressini 1997, p. 214)[11]

Clearly, Peressini's picture is very different from the model-theoretic one advocated here which effectively blurs the pure/applied distinction. However, 'bridge principles'

---

[11] In linking pure mathematical vocabulary to the physical object/property vocabulary, these bridge principles supply a semantics (generating a further applicability problem according to Steiner 1998, who adopts a Fregean approach).

ignore the location of the 'object vocabulary', in this case at a very high level in the hierarchy of models, idealizations, etc., and presuppose that, on one side of the bridge, we have a clear grasp of the physical object, when in fact, we may not. In particular, they presuppose a mathematically independent grasp but achieving such may not be so clear-cut.[12] In particular, the question arises: If the fundamental basis of what grasp we have of the fermionic nature of electrons lies with group theory, how can we separate the 'pure' mathematical vocabulary from the physical object vocabulary? It may not be clear in this case that the physics and the mathematics are indeed 'about different things', an issue to which we shall return below in the case of spin.

Nevertheless, the point is well taken, of course, namely that certain indispensability arguments, at least, assume a form of confirmational holism that encourages 'global' realism about the objects referred to by the theory.[13] However, on the science side, Glymour (1980) has long urged resistance to such holism, arguing that his 'bootstrap' theory of confirmation allows us to adopt a piecemeal approach. For the realist, this has the obvious advantage of selectivity in what has to be accepted into her ontology. For the anti-realist, this selectivity is similarly welcome, since it allows her to avoid ontological commitment to (at least some) unwanted entities.

On the mathematical side, in the kinds of applications that we have sketched in previous chapters, typically only some of the mathematical structure is 'carried over' as we have repeatedly stated; thus, not all will be confirmed. Furthermore, as we have also emphasized, the applicability of mathematics in certain cases may depend crucially on appropriate idealizations being made on the physical side. Any attempt to run a form of indispensability argument in these cases is first going to have to come up with an account of confirmation that can accommodate such idealizing moves. In effect, the question has to be answered: Why should we be realists about those mathematical objects or structures which feature in idealized models of the physical phenomena? Given that scientific realists are typically cautious about the existence of idealized entities, such as mass points, rigid rods and, indeed, nucleons, so the mathematical realist should adopt a similar attitude. And for obvious reasons, these considerations are also welcomed by the mathematical anti-realist.

8.3 The Enhanced Indispensability Argument and Explanation

More recently, however, the focus of debate has shifted to the so-called 'Enhanced Indispensability Argument' (EIA) that is taken as the principal support for the claim that the role of mathematics in explanation implies the existence of mathematical

---

[12] If one adopts the 'hybrid' view that we shall be considering below, in which the objects of quantum physics are understood to be more mathematical than material, then no such grasp may be achievable.

[13] Not all—see Putnam (1979). For further discussion on the differences between Putnam and Quine on the indispensability argument, see Bueno (2013b).

objects (see Baker 2009; Daly and Langford 2009; Saatsi 2011). The argument runs as follows:

(P₁)   We ought rationally to believe in the existence of any entity that plays an indispensable explanatory role in our best scientific theories.

(P₂)   Mathematical entities play an indispensable explanatory role in science.

Hence, we ought rationally to believe in the existence of mathematical entities.[14]

Two of the most well-known examples that have been given of the apparently indispensable explanatory role of mathematical entities in science are:

(i) *Honeycombs*:   Here the relevant *explanandum* is the hexagonal shape of bee honeycomb cells, and the relevant mathematical *explanans* states that the optimal shape for tiling the Euclidean plane (in the sense of the minimum total side length for arbitrarily large areas) is a hexagon (Hales 2001).

Due to concerns over appropriately drawing the distinction between physical and mathematical geometry, there has also been considerable focus on:

(ii) *Cicadas*:   Here the *explanandum* is the prime periodic life cycles of cicadas in the US (13 and 17 years depending on the species) and the *explanans* has to do with the mathematical fact that prime periods minimize the relevant intersection together with the biological claim that a life cycle period that minimizes intersection (with other periods) is evolutionarily advantageous (see e.g. Cox and Carlton 2003).

Now, both of these examples can be contested. Thus, Wakil and Justus (2017) have pointed out that, in the case of the honeycombs, biologists going back to Darwin have suggested alternative hypotheses that claim that the hexagonal shape is just a by-product of hive construction; and when it comes to the cicadas, the explanation of the prime life cycle may have to do with an original nine-year life cycle being delayed due to 'nymphal crowding'.[15] In other words, there are biological explanations of the phenomena concerned that do not appeal to the explanatory role of the mathematics used. Now, of course, it is not our intention to enter into discussion about which of these various explanations is the 'best'; nor do we claim that when biologists refer to

[14] This is different from the original 'Putnam–Quine' Indispensability Argument in emphasizing the indispensable *explanatory* role of the entities concerned. Thus, the former can be summarized as follows: (P₁) We ought to have ontological commitment to all and only the entities that are indispensable to our best scientific theories; (P₂) Mathematical entities are indispensable to our best scientific theories; hence we ought to have ontological commitment to mathematical entities (see Colyvan 2015, which also contains extensive references to the original literature). The shift reflects the 'explanationist' tendency that has arisen within realism in general—to the effect that it is to those entities referred to by terms that play the appropriate roles in our best theoretical *explanations* that the realist should be committed.

[15] Whereas Goles, Schulz, and Markus (2001) argue that prey with prime numbered life cycles will avoid encounters with predators more than those with non-prime numbered life cycles. Thanks to Manuel Barrantes for pointing this out to us.

the 'primeness' of the relevant life cycle they are invoking mathematics in an explanatory role. It is only when the relevant account is reconstructed philosophically that one can allege that the above 'mathematical fact' is playing an *ineliminable* explanatory role.[16] More importantly, perhaps, given the range of rival accounts of these phenomena, the onus is on the mathematical realist to show that the explanation that takes the above 'mathematical fact' as part of the *explanans* is, in some sense, the best. This is a point to which we shall return throughout this discussion.

Furthermore, even if we were to grant that last point, one might wonder what it is one is expected to rationally believe in that case: With regard to the cicada explanation, is it simply the numbers 13 and 17, or all prime numbers or all numbers?[17] Consider for example the sequence of explanations involved in the discovery of the electron and particularly Thomson's work showing that the negatively charged particles that explained Crooke's experiments on 'cathode rays' were also produced by heated, illuminated, and radioactive materials (although see Falconer 1987 for a rather more nuanced account). In such cases the *explanans* typically involves the existence of a kind of thing, here electrons, the positing of which is extended to cover other *explananda*. In the cicada case, although we can envisage the initial positing being likewise extended to numerous other cases, it is just not clear what we are being invited to accept. By comparison with the electron case, we might take it to be the kinds '13' and '17' but withholding belief in other kinds, although appropriate in the electron case, does not seem so here: once we have established the existence of the number 13, it seems we can quickly infer that of 12, 14, and the rest of the reals (since, of course, one of the fundamental properties that numbers possess is being related by succession). What the explanation thus gives us grounds to believe in the mathematical case is, at best, a certain number *structure*. This suggests that the advocate of the EIA should be, at most, a structural realist when it comes to mathematics.

However, again leaving aside the above issue of the precise nature of the phenomena that is acting as the *explanandum*, the EIA can also be resisted by emphasizing the kinds of issues to do with representation and explanation that we have been tracking here.

8.3.1 Indexing and Representing

One well-known set of responses to the EIA insists that mathematics has only an 'indexing' (Melia 2000) or representational role (Saatsi 2011), rather than an explanatory one. Melia writes:

[Although we may express] the fact that a is 7/11 metres from b by using a three-place predicate relating a and b to the number 7/11, nobody thinks that this fact holds in virtue of

---

[16] Again, we are grateful to Manuel Barrantes for encouraging us to make this clear.

[17] Cf.: 'the point is that the explanandum of the biological theory is only that the periods are 13 or 17, not that the period is some n, where n is prime' (Saatsi, quoted in Baker 2009, p. 616).

some three place relation connecting *a*, *b* and the number 7/11. Rather, the various numbers are used merely to index different distance relations.   (Melia 2000, p. 473)

Similarly, Saatsi argues:

Mathematics can help us learn about the world by virtue of playing a representational (and, derivatively, inferential and justificatory) role in science. The fact that mathematics can play such a role is something that calls for an explanation. But *this* issue, and any argument for Platonism that turns on this issue, is independent from EIA. The latter concerns not the representational capacity of mathematics in general, but more specifically the capacity and role that mathematics per se has in explaining and yielding understanding about the concrete world.   (Saatsi 2011, p. 145; italics in the original)[18]

Indeed, it is argued that we can effectively dispense with the relevant mathematics by, for example, using physical rods of a certain length to represent cicada lifetimes (ibid., p. 150). Here we take a number of rods or sticks of lengths 14, 15, 16, 17, and 18 cm (take twenty such rods of each length, say). Place on a table sticks of each kind, one after another, to find the least common multiple of each pair. It can be discovered very straightforwardly that the least common multiple is almost always longer for pairs in which one rod is 17 cm long. If we take the linearity of length to be a faithful representation of the linearity of time—which seems uncontentious in this context—then we have an account of our *explanandum* that does not invoke ineliminable mathematics.[19]

Thus, even when biologists refer to the prime nature of the relevant life cycle in their explanation, we are not compelled, in our philosophical reconstruction, to take that mathematical feature itself as playing an ineliminable explanatory role; rather we can understand it as *representational*. Indeed, insofar as we can use the resources of first-order logic to dispense with numbers, so we can deploy these resources to dispense with the specific prime numbers invoked in the explanation of cicada life cycles. So, the relevant prime numbers, 13 and 17, can be dispensed with along the following lines: consider the claim to the effect that there are two objects, for example. This can be written as:

$$\forall x \forall y \left(\left((ox \wedge oy) \wedge x \neq y \right) \wedge \forall z \left(oz \rightarrow (z = x \vee z = y) \right) \right)$$

and we can straightforwardly, if lengthily, extend this to the numbers 3 and 17.

One option for the advocate of EIA would be to resist this move by arguing that the mathematics plays a crucial (and, again, ineliminable) role in establishing the

---

[18] Saatsi in particular includes details of how explanations can be given of the above examples that avoid the mathematical premises.

[19] Of course, one might insist that merely by invoking 'length', one is committed to mathematics playing a role in the *explanans*. But note: the mathematics describes the measure of length—it is the length itself, as a *physical* quantity, that is playing the relevant explanatory role here. For further discussion, see Saatsi (2011).

generality of these explanations, in the sense that "...any periodical organism with periodical predators is likely to evolve a life-cycle period that is prime" (Baker 2012, p. 257). Here the idea seems to be that the explanation that appeals to the mathematics is more explanatory by virtue of being more general: it captures that which is common to both cases (life cycles of 13 and 17 years) and thereby allows us to project the explanation to further possible cases. However, as Saatsi again notes (2011), there is another mathematics-free alternative that can also do this unificatory work.

We begin with the biological claim that having a life-cycle period that minimizes the intersection with other relevant periods is evolutionarily advantageous. We also take the physical fact that there is a unique intersection-minimizing period T_{min} for periods within a given range. We then add the ecological constraint that cicadas within a certain ecosystem type are limited, biologically, to that given range of periods. We thus conclude that cicadas in that ecosystem type are likely to evolve with T_{min} life-cycle periods (ibid.). Here we have an argument pattern that is generalizable and can be used to make predictions, but again the mathematics is reduced to playing only a representational role, at best.

A further option would be to argue that this claim that mathematics is playing only a representational or indexing role is at odds with scientific practice—thus, in at least some cases, biologists invoke the primeness of the life cycles in their explanations (see again Cox and Carlton 2003). However, as we have just noted, that biologists refer to the prime nature of the life cycles should not be taken to straightforwardly imply that this 'primeness', *qua* mathematical feature, is itself playing the appropriate explanatory role. As we have just seen, Saatsi, for example, argues that it should be understood as merely representational. Furthermore, we are generally sceptical of such invocations of scientific practice in this kind of context; again, it is through the relevant *philosophical* reconstruction that the ontological implications become apparent. We shall return to this point in Chapter 9.

The advocate of indispensability could also argue that the mathematics plays more than a representational role by virtue of the purported fact that it provides the relevant explanation with a certain modal strength, in the sense that the *explanandum* is *necessary*; that is, it is independent of any contingent ontological structure (see Lange 2013). Consider, for example, the physical impossibility of dividing twenty-three strawberries among three people (ibid.)—the *explanans* here is the mathematical fact that 23 cannot be evenly divided by 3. Lange argues that the explanatory power is grounded in the necessity provided by the mathematics in virtue of which we can understand that the *explanandum* had to occur, irrespective of what the world is like ontologically. On this view, then, the role of the mathematics goes beyond that of simply representing or indexing the relevant phenomena.

However, as Saatsi (2016) notes, even if we grant this sense of explanatory power, it is unclear why it should compel us to regard the mathematics in realist terms. Even if we accept that such mathematical explanations explain by showing how the *explanandum* is independent of any physical ontology, it is a non sequitur to conclude that the mathematics that is deployed this way must be interpreted realistically. What we need is an *argument* that takes us from this modal feature to the relevant ontological commitment and the onus is on the mathematical realist to provide such an argument. Thus, as Saatsi (ibid.) points out, even if we grant the indispensability of mathematics in this sense, there is still no prima facie connection with ontological commitment.

Note, furthermore, that the necessity of the mathematics is 'internal' to it, in the sense that it follows from the relevant mathematical principles and the underlying logic. If either of those were to change, the necessity would change accordingly. Of course, if the mathematics is applied to a physical system, any modality that emerges would be a feature of the properties of the system in terms of which the mathematics would be interpreted. The mathematics allows for more generality in accounting for the phenomena, insofar as it represents the phenomena in a certain way. When it comes to strawberries, the impossibility of dividing twenty-three evenly among three people is a *physical* fact—in effect, one runs out of strawberries before everyone has been evenly served!

The upshot, then, is that even if we were to agree that these features of generality and modality are explanatorily significant and are lost in the reconstructed versions of purported explanations such as that of the cicada life cycles, there would be no clear implication with regard to ontological commitment when it comes to the mathematics. Of course, more needs to be said about the sense of explanation involved here (see, again, Saatsi 2016) and we shall return to that in Chapter 9. But the point we wish to insist upon is that we do not see anything in such features that should give us pause.[20]

Another option would be to point to more complicated examples, where the mathematics is not so straightforwardly dispensable. Consider for example the case studies presented in the previous chapters, where certain mathematical structures played an apparently indispensable role in the explanation of a variety of phenomena in both atomic and nuclear physics. In such cases, it may not be so clear what physical entities one could invoke to do the same job. Indeed, some have argued that in at least some highly significant cases in physics (such as those that involve the behaviour of systems near phase transitions; see Batterman 2010) no such separation can be achieved and the mathematics must be accorded an explanatory role. However, as we shall insist below, and in Chapter 9, even in the more complex cases, the indispensability argument can be resisted.

It may also be suggested that such a non-mathematical representation is not adequate for all purposes, in which case a kind of 'weaseling' manoeuvre has been proposed, according to which we posit certain mathematical entities in order to

---

[20] Once again, we are grateful to Manuel Barrantes for urging us to consider these moves.

construct the appropriate representation, only to subsequently deny the existence of these entities. So, using an example from mereology, Melia shows that a theory T can have nominalistic (that is, anti-realist) consequences that its nominalistically rewritten version T' does not. By adding new predicates to T and new entities to our ontology, it turns out that we can guarantee the existence of a certain kind of concrete entity whose existence we are unable to guarantee in T'. We shall not enter into the details here but the point is that the relevant non-mathematical representation may not in fact be capable of doing the job required of it:

> If the nominalist simply takes his theory to be the set of nominalistically acceptable sentences entailed by some platonist theory, he has no guarantee that his theory actually has the same nominalist content as the platonist theory.   (Melia 2000, p. 461)

Melia's 'weaseling' response is to insist that we can assert the theory T, with its mathematical representation, but deny the existence of mathematical entities. After all, scientists often introduce certain terms, in order to construct their theories, yield the relevant equations, make certain predictions, and so on, where these terms are understood as idealizations and hence reference to the putative associated entities is denied. Consider the case of rigid rods in Special Relativity. Einstein can be thought of as engaging in a form of weaselling here: introducing such rods in order to elaborate his theory but then (effectively) denying that there can be such rods, on the basis of Special Relativity itself. The mathematics can then be seen as '...the necessary scaffolding upon which the bridge must be built. But once the bridge has been built, the scaffolding can be removed' (Melia 2000, p. 469).

However, obvious concerns arise with this manoeuvre (for criticism, see Colyvan 2010; and for a response, see Liggins 2012). First of all, idealization terms do not have the same status as mathematical ones. In the case of the former, an anti-realist stance is typically associated with them right from the start, on the basis of some justification for taking them as idealizations. When it comes to mathematical terms, this is not the case (or at least it is not without begging the question). As for the case of rigid rods, here the idealization terms were effectively replaced as the theory came to be seen as 'about' something other than clocks and rods, namely Minkowski space-time. Again, this does not appear to be the case here where the issue is precisely that we may not be able to replace the mathematical entities, at least not straightforwardly. We shall return to this below, noting that although, again, in the cases we have considered, the mathematical scaffolding appears to be an integral part of the edifice, still the EIA can be resisted. Finally, given what we've just said, it may not be clear when we should 'weasel' and when not. This can be contrasted with the separation of quantifier commitment from ontological commitment deployed above; in this case there are no such concerns as we first scan the theory to see what putative entities we should provisionally accept (tracking the relevant quantified commitment) and then determine which we have the relevant access to in order to decide which we should believe in.

Returning to the EIA, Baker insists that if one is going to maintain that the mathematics functions in these cases as a non-explanatory component of a larger explanation, then a principled distinction needs to be drawn between the non-explanatory and explanatory aspects. Here he presents what he calls the 'Steiner test':

The difference between mathematical and physical explanations of *physical* phenomena is now amenable to analysis. In the former, as in the latter, physical and mathematical truths operate. But only in mathematical explanation is [the following] the case: when we remove the physics, we remain with a mathematical explanation—of a mathematical truth!

(Steiner 1978, p. 19; italics in the original)

However, Baker sees this test as fatally flawed, since:

the evidence from scientific practice indicates that the internal explanatory basis of a piece of mathematics is largely irrelevant to its potential explanatory role in science.   (Baker 2009, p. 623)

Thus, in the case of the honeycomb explanation, the attribution by the biologists to Hales' proof (of the Honeycomb Conjecture) of a key role in the explanation of the shape of bee honeycombs appears to have little or nothing to do with the details of the proof, or whether it counted as an explanation in mathematical terms. Baker's conclusion, then, is that:

mathematical explanations of physical phenomena do not map onto mathematical explanations of mathematical results in the neat way that Steiner claims.   (ibid., p. 624)

Here we are inclined to agree with Baker that there is no 'ready made' test for the explanatoriness of a given piece of mathematics. However, when he asks who must carry the burden of proof—the realist about mathematical entities or the anti-realist—our response is that given that the practice of science—as exemplified in the case studies presented here—cannot supply or underpin such 'proof', and given that the anti-realist stance admits of fewer ontological commitments, it is up to the realist to provide further arguments.

Although mathematics is assigned only a representational role in the response to the EIA, surprisingly little has been said about the framework in which this representational role can be understood. Here we can move beyond Melia and Saatsi in spelling out the kind of framework that would do the job. As we have noted, in this context the usual arguments against isomorphism-based accounts of representation have diminished force (see van Fraassen 2008). In particular, characterizing, at the (meta-)level of the philosophy of science, the representational capacity of mathematics within science by means of the framework outlined in Chapter 2 allows us to capture the sense in which mathematics provides 'surplus structure' in this context, as well as the associated heuristic fertility and the applicability of mathematics more generally. In particular, as we have said, it allows the carrying over of relevant structural features from the mathematical level to that of the physical theory, as in the examples we have presented previously.

We recall that this representational framework is articulated in terms of a three-stage scheme, involving immersion, deduction, and interpretation, where the mappings involved in the first stage will typically be those we have emphasized throughout, namely partial isomorphisms and homomorphisms. Our presentation of Dirac's 'prediction' of the positron in Chapter 7 provides a useful exemplification of this three-stage process: the initial immersion yields the relativistic wave equation; consequences are drawn from that piece of mathematics, involving both positive and, crucially, negative energy solutions; but finally, and crucially, these latter are interpreted in a certain way. In the case of the application of group theory, as we saw, the immersion has a 'fine-grained' nature as the relevant mappings must effectively track what Weyl called the 'bridges' within the mathematics, and must also accommodate the crucial role of the relevant representations. And in the case of the application of Bose–Einstein statistics to liquid helium phenomena, the immersion must be similarly nuanced on the physical side. In all these cases, of course, the mathematics is playing representational and heuristic roles, where these do not compel us to adopt a realist attitude towards the mathematical entities invoked.

8.3.2 Explaining

But of course, from the representational standpoint, it is hard to put one's finger on what it is about mathematics that makes it representational and *not* explanatory; one could certainly use the partial structures approach as a framework for an account of explanation as well.[21] So, with the representational 'force' of mathematics characterized in the same manner as that of scientific theories, the focus must turn to the context of scientific explanation (Saatsi 2016). Here the defender of the EIA can (not surprisingly) expand the remit of the Deductive-Nomological approach to include mathematical laws and principles (Baker 2005), but this still does not explicate how such laws and principles can themselves have explanatory power (see Daly and Langford 2009). Although the D-N approach also suffers from well-known deficiencies, alternative causal accounts of explanation obviously invite accusations of question begging in this context.

Strevens' (2008) 'kairetic' approach, however, may provide the defender of the EIA with an entry point. This places the focus on *difference-making*, in virtue of which one state of affairs bears causal-explanatory relevance to another. Using an

---

[21] Manuel Barrantes has pointed out that in terms of our approach, Steiner's 'test' above, regarding putative mathematical explanations of physical phenomena, can be understood as a requirement that what we call the derivation step should consist of, or include, an explanatory proof of the relevant mathematical structures invoked in the overall explanation. Of course, as we have just indicated, we reject the very basis for this test, and hence the subsequent requirement, with the onus placed on the mathematical realist to provide arguments in favour of it. Nevertheless, as Barrantes suggests, we might still insist that not all moves in that step are permissible insofar as they may be deemed to fail in supporting the explanation as a whole. Thus, for example, trivial mathematical derivations from inconsistent premises should be excluded (assuming the adoption of classical logic; these inferences are immediately blocked in a paraconsistent setting). We shall return to this useful point in Chapter 9.

optimizing procedure to identify these difference makers, Strevens is able to show how irrelevant influences can be eliminated and a stand-alone explanation obtained in which one state of affairs bears causal-explanatory relevance to another. Mathematics then qualifies as an 'explanatory tool' because it is through grasping mathematical dependences and independences that we are able to grasp *causal* dependences and independences (ibid., p. 331; see also Pincock 2012).

Now, an immediate objection here is that functioning as an explanatory tool is not quite the same as playing the sort of explanatory role that supports the kinds of existence claims associated with the Indispensability Argument. Indeed, allowing us to grasp causal dependences is precisely one of the things that representational devices do. Furthermore, when it comes to the core claim of Strevens' account, it might be insisted that mathematics *makes no difference* when it comes to physical states of affairs and hence cannot bear the kind of explanatory relevance that Strevens has in mind.[22]

However, Baker has rejected as fatally flawed the arguments that have been given in support of the claim that mathematics makes no difference and argued that indispensability considerations are in fact crucial to the evaluation of this claim (Baker 2003). In particular, the intuitions that typically lie behind it rely on an implicit analogy between abstract objects and remote, concrete ones. The latter, when posited by a theory, get dismissed as dispensable, precisely because their failure to make a difference is indicative of their lack of causal role. A similar line of reasoning runs the risk of question begging when it comes to mathematical objects and Baker concludes that if the mathematical objects concerned are taken to be indispensable, then there is no determinate answer to the question whether their existence makes a difference.

The advocate of the EIA might see a way out of this impasse in Strevens' comment that the central difference-making criterion "takes as its raw material any dependence relation of the 'making it so' variety, including but not limited to causal influence" (Strevens 2008, p. 179). The idea is that once we have established the relevant dependence relation between some state of affairs and some set of 'entities', the criterion will tell us what facts regarding those entities underpin the relation's 'making it so' (ibid.). If this is not limited to causal influences, could the defender of the EIA adopt this approach and insist that there is a sense in which the mathematics 'makes it so'?

A useful example, given in response to the standard causal account, is that of the explanation of the halting of the gravitational collapse of white dwarf stars by the Pauli Exclusion Principle (Colyvan 1999; here we recall our discussion of the Principle in Chapter 4). Here the *explanandum* is the halting of the gravitational collapse of white dwarf stars: stars of a certain size and at a certain point in their life

---

[22] We are grateful to Juha Saatsi for raising this point.

cycle begin to collapse as they are unable to sustain certain fusion reactions sufficient to create enough thermal pressure to resist the gravitational attraction; but then at a certain point, this collapse stops. The *explanans* is the Exclusion Principle and the core of the explanation usually given is that according to the Principle, no two electrons can occupy the same state, and so as the energy levels become filled, the electrons are effectively forced to occupy higher and higher levels, creating an effective 'Pauli pressure' that balances the gravitational attraction.

The claim, then, is that insofar as Pauli's Principle cannot be regarded as a causal law, this represents an example of an acausal explanation of the behaviour of physical systems that, in turn, opens the door to the explanatory role of mathematics. Indeed, the defender of the EIA can argue that it is the Exclusion Principle that 'makes it so' with regard to the halting of stellar collapse and that, given an appropriate construal of this principle, this exemplifies the relevant indispensable explanatory role of mathematics.

The strength of such a claim obviously depends on how we understand the Exclusion Principle itself (see Massimi 2005). Here philosophy has not served us well in providing a useful analysis. Lewis, for example, talks of the Principle as representing 'negative information':

A star has been collapsing, but the collapse stops. Why? Because it's gone as far as it can go. Any more collapsed state would violate the Pauli Exclusion Principle. It's not that anything caused it to stop—there was no countervailing pressure, or anything like that. There was nothing to keep it out of a more collapsed state. Rather, there just was no such state for it to get into. The state-space of physical possibilities gave out. [...] [I]nformation about the causal history of the stopping has been provided, but it was information of an unexpectedly negative sort. It was the information that the stopping had no causes at all, except for all the causes of the collapse which were a precondition of the stopping. Negative information is still information. (Lewis 1986, pp. 222–3)

Although this could be too easily dismissed as an ultimately futile attempt to shoehorn the Principle into the causal account and as suggesting that the lack of causal information is still indicative of causal relevance (Colyvan 1999), the initial suggestion regarding restrictions on the state-space is at least on the right track, as we shall shortly see.

Strevens also considers the Principle within his kairetic approach:

What relation holds between the law [PEP] and the arrest, then, in virtue of which the one explains the other? Let me give a partial answer: the relation is, like causal influence, some kind of metaphysical dependence relation. I no more have an account of this relation than I have an account of the influence relation, but I suggest that it is the sort of relation that we say 'makes things happen'. (Strevens 2008, p. 178)

The lack of an account of the metaphysical dependence relation could be taken to provide appropriate room for the defender of the EIA: if the Pauli Exclusion Principle can 'make things happen' via some kind of 'metaphysical dependence relation', then, one might ask, why can't mathematical objects? In particular, why should grasping mathematical dependences and independences be indicative of our grasp on causal dependences and independences only? Why couldn't this grasp reveal a dependence relation holding between the physical state of affairs and the relevant mathematical entities themselves, as represented via the framework of partial homomorphisms advocated here? (As we shall see, a similar thought to that which lies behind these questions also plays a role in the suggestion that certain properties are a hybrid of the mathematical and the physical.)

One way of responding to these questions is to offer a causal *explanans*, along the following lines: what is actually going on in the case of a collapsing white dwarf star is that the gravitational attraction is actually being resisted, and eventually is counterbalanced, by what physicists call a 'degeneracy pressure'; that is, a pressure that (supposedly) results from the existence of states that correspond to the same energy, but which can be occupied by electrons with different spin, in accordance with Pauli's Principle (Skow 2014). Thus there is no need to search for alternative explanatory frameworks in this case—the standard causal account will do just fine—and certainly no need to open the door to an explanatory role for mathematics by virtue of allowing Pauli's Principle to play such a role.

An immediate and obvious counter-response would be to insist that this is in effect a throwback to earlier and subsequently discarded attempts to understand the Principle in terms of 'exchange forces' and the like (Carson 1996). As the nature of the Principle came to be better understood, and in particular its grounding in permutation symmetry, as discussed in Chapter 4 (and we'll return to this shortly), such attempts fell by the wayside, at least as far as foundational studies were concerned, although such terms as 'degeneracy pressure' continue to be deployed in a mostly pedagogic context. But of course, the advocate of a causal account might insist that the notion of 'pressure' is here being extended beyond its origins in some force-based conception, in a manner that physical concepts often display as the relevant science develops. Thus it has been argued that 'pressure' in this kind of case can be understood simply in terms of the disposition of a system to transfer energy and thus we are perfectly entitled to refer to the behaviour of a white dwarf star in these terms (Skow 2014, pp. 458–9).

However, the very term 'degeneracy' gives the game away: it is standardly accepted that the physical origin of any such degeneracy in a (quantum) system's states is typically the presence of some symmetry in that system. And in this case, as we have already said, that symmetry would be that associated with the permutation group, discussed in Chapter 4 as one of the crucial new features—indeed, for Weyl, one of the very foundations—of the emerging new quantum physics. Given that the basis of the Pauli Exclusion Principle is the requirement that the relevant wave function of the system be anti-symmetric, and given that that requirement 'drops out' of the imposition of permutation symmetry (in effect as one of the various possible representations of the permutation group—two (bosonic and fermionic) for two-particle systems, or many more (bosonic, fermionic, and parastatistical) for systems with three or more particles; see French and Rickles 2003)—it is that symmetry that is the ultimate *explanans* in this case.

Now that conclusion might restore some measure of hope to the mathematical realist, not simply by virtue of opening the door to indispensability claims for mathematical entities as Colyvan suggests, but because of the way this symmetry is represented in group-theoretical terms. Indeed, one might be tempted to argue that such symmetries are, in some sense, 'nothing but' the mathematics, and so what we are appealing to here as *explanans* is mathematical structure (see e.g. Cao 2003; for a response see French and Ladyman 2003). Indeed, the question might be pressed: Why should that particular group structure (that of the permutation group) be applicable in this case? Answers to this question can be placed along a spectrum: at one extreme we have the view that it is a priori. Weyl (1952, p. 126), for example, insisted that "all a priori statements in physics have their origin in symmetry". Not surprisingly, empiricists such as van Fraassen have resisted this line (van Fraassen 1989), as we noted in Chapter 2, and, at the other end of the spectrum, offer a broadly pragmatic answer. From this perspective, permutation symmetry comes to be seen as nothing more than a problem-solving device (see Bueno 2016).

Occupying the middle ground between these extremes we have the following alternative answers to our question:

(1) It is just a brute fact. Here permutation symmetry might be seen as simply an 'initial condition' imposed on the situation (see French 1989). This of course might appeal to the advocate of the EIA, who will argue that given its mathematical nature, and the indispensable role of this symmetry in explanations involving quantum statistics (such as those presented in previous chapters), one may be compelled to adopt a realist attitude towards it, understood as exemplifying mathematical structure. Of course, this depends on taking permutation symmetry as brutish, and there are alternatives to doing so.

(2) It is to be understood as reflecting the metaphysically peculiar nature of the particles themselves. However, given that the particles can also be described in a metaphysically straightforward way—as individuals—this option is always going to require some further principle whose status may be less well grounded than that of permutation symmetry itself (see French and Krause 2006).

(3) It is to be understood as reflecting a structural aspect of the world (French 2014). From this perspective, that the permutation group is applicable is neither a simple brute fact nor metaphysically derivative, in the sense of mathematically representing the nature of the particles, but rather it represents something profoundly structural about the world.

(4) It is to be understood as a law-like feature of the world.[23] Just as one wouldn't argue that because Newton's Laws are presented in terms of the differential calculus, say, these laws have to be understood as purely 'mathematical structure' in some (non-physical) sense, so one shouldn't, or, at least, doesn't have to, take symmetry principles as mathematical just because they are presented via group theory, say. Indeed, Wigner famously suggested that such principles should be regarded as 'meta-laws' that constrain the more familiar laws of physics, so one might adapt the Deductive-Nomological account to incorporate them. Alternatively, one might incorporate them within an extension of the Woodwardian account, where, in the case of permutation symmetry, the counterfactual feature of this account is manifested via the different possible statistics that result (French and Saatsi 2018).[24]

Whatever answer we choose, as a result of the action of permutation symmetry, the relevant Hilbert space can be thought of as divided up into subspaces, corresponding to the different representations (as outlined in Chapter 4) and hence different statistics; and the symmetric nature of the appropriate Hamiltonian is such that once 'in' such a subspace, particles cannot get out, as it were. It is in this sense that Lewis was on the right track: the Exclusion Principle can be thought of as an expression of the limits placed on the Hilbert space by Permutation Invariance. And when it comes to stellar collapse, the anti-symmetrization of the relevant wave function for the assembly constrains the distribution of electrons among the available states, such that they must occupy higher energy states, which leads to an apparent counter-gravitational 'pressure'. But it is this constraint that is doing the explanatory work.

The 'metaphysical dependence' to which Strevens alludes can now be understood as underpinning these constraints (French 2014), where it is the latter, and hence the associated symmetry, that 'makes things happen' (or not). If one is a realist about such matters, how one understands both the constraints and the dependence will depend on one's metaphysics: the structural realist, for example, will take the dependence to hold between the physical structure and the relevant entities, processes, and regularities. Of course, it isn't *necessary* to go down this route, as long as these constraints are understood as reflecting physical reality and an appropriate metaphysics is then given. The onus is then on the defender of the EIA to show that not only has the problem just been pushed back a step but that no such physical dependence holds and that these constraints must be understood as entirely mathematical. Given the availability of alternatives, the prospects for such a demonstration do not look good.

---

[23] This can of course be related to the previous answer (French 2014).
[24] Here the relevant counterfactuals help us track the dependencies referred to in Strevens' kairetic approach.

Our point, then, is, whether one chooses to go down the causal route and appeal to some conceptually accepted notion of 'degeneracy pressure' or take the relevant underlying symmetry as a feature of the structure of the world, or as a physical principle and accommodate it within either the D-N or Woodwardian frameworks, there is no need to open any door to mathematical structure in the role of *explanans*.

Nevertheless, someone keen to preserve some explanatory role for mathematics in this context might be tempted by the weaker claim that we mentioned at the beginning of this chapter and according to which the role of mathematics in science implies that certain physical objects and/or the associated quantities have some kind of 'hybrid' mathematical-and-physical nature. It is to that claim that we now turn.

This has received less attention than the Strong claim discussed above, but, again, relies on certain presuppositions about the interrelationships between science and mathematics. We shall consider the hybridity claim in the context of a case study involving spin and shall highlight its critical dependence on an argument to the effect that, in the absence of any apparent viable interpretation of that property, the only option is to adopt a form of realist stance towards the relevant mathematics. However, we shall suggest that there are other options 'on the table', as it were, and that the Weak claim can also be blocked in various ways. This will lead us into Chapter 9, where we shall examine a direct challenge to the framework we have presented here, in the form of so-called 'asymptotic explanations'. Here the concern is not with the existence of mathematical objects, but with the impact of certain limiting operations and associated mechanisms that have risen to prominence in the last fifty years or so. It has been argued that these cannot be accommodated within the framework presented here and that they force a re-evaluation of our view of explanation in general. We insist that, on the contrary, these mechanisms and the cases they feature in can be straightforwardly captured via the inferential framework plus partial structures and to this extent the latter demand for a re-evaluation of explanation is undermined.

8.4 The Weak Claim and the Hybridity of Spin

Early suggestions along these lines can be found in Heitler's reflections on the role of 'profound mathematical concepts', such as group theory, in quantum mechanics (as, again, detailed in previous chapters) which led him to conclude that the atom, for example, 'can hardly be thought of as something of a purely material nature' (Heitler 1963, p. 53).[25] The idea seems to be that it is not just that we have to accept wave–particle duality in one or other of the now very well-known ways, or give up some kind of substance-based ontology in the quantum context, as Eddington demanded,

---

[25] This is of course the same Heitler who worked with London and made the famous and provocative claim about eating chemistry with a spoon. He went on to do important work in quantum electrodynamics, among other things.

but that the role of the aforementioned concepts implies that the very nature of the atom is in some sense partly mathematical. Given that the kinds of symmetries touched on above are represented mathematically via group theory, one could invoke these examples as support for Heitler's view.

More recently, Morrison has argued that spin should be understood as 'hybrid' in that it possesses both mathematical and physical features and 'bridges' the mathematical and physical domains (Morrison 2007).[26] As she notes, spin effectively 'drops out' of the Dirac equation, in the sense that:

> the mathematical formalism of the Dirac equation and group theory require the existence of spin to guarantee conservation of angular momentum and to construct the generators of the rotation group.   (Morrison 2007, p. 546)

Although this provided a more complete theoretical treatment of spin than previous accounts, she regards it as 'essentially mathematical'. Such a view is reinforced by consideration of the group-theoretic analysis of spin that, as she recalls, helps to underpin our current understanding of this property (ibid., p. 552). Again, this is due to Wigner who extended the group-theoretic approach to elementary particles in his crucial and important work on the association of 'elementary physical systems' with representations of the Poincaré group (see Wigner 1935). This is the group of isometries of Minkowski space-time (the space-time of Special Relativity)[27] and there exists a 'unique correspondence' between these representations and the Lorentz invariant equations of quantum mechanics. Such irreducible representations, Wigner argued, in an echo of the sorts of considerations we touched on in Chapter 4, 'though not sufficient to replace the quantum mechanical equations entirely, can replace them to a large extent' (Wigner 1939, p. 151).[28] The labels of these representations can then be associated with the values of the properties of the relevant 'elementary systems', such as spin, charge, and mass. In other words, spin emerges from this analysis as just a group invariant characterizing the unitary representation of the Poincaré group associated with the wave equation.

Nevertheless, Morrison insists, it should not be regarded as wholly mathematical since, despite the problems with measuring single spins,[29] it can be regarded as a measurable feature of the world in some indirect sense (Morrison 2007). It is in this

---

[26] The two cases considered here—the Exclusion Principle and spin—can obviously be related via the spin-statistics theorem. Although there remains some doubt over what counts as an adequate proof of this theorem (see Sudarshan and Duck 2003), one could interpret it as grounding the relevant statistics in an understanding of spin, thus removing the need to appeal to symmetry as playing a fundamental explanatory role. Of course, this provides no succour at all to the defender of the EIA!

[27] Formally it is a 10-generator non-Abeliam Lie group that includes translations, rotations, and 'boosts', or transformations between reference frames moving with uniform velocity.

[28] It can give the change through time of a physical quantity corresponding to a particular operator, but not the relationships holding between operators at a given time.

[29] Problems that, as she says, enhanced its status as a truly non-classical feature of the world (Morrison 2007, p. 555).

sense, then, that spin can be said to 'bridge' the gap between the mathematical and the physical and 'emerges as a peculiar hybrid notion possessing both physical and mathematical features' (ibid., p. 547; also p. 552). And she takes this as symptomatic in that, 'the nature of both theory and experiment in contemporary physics has largely stripped us of the resources for a sharp division between the mathematical and the physical' (ibid., p. 555).

As a result, Morrison maintains, any attempt to give a realist interpretation of spin would be fundamentally misguided, including structural realist approaches (ibid., pp. 552–4; for a response see French 2014, pp. 109–11). Our concern here, however, is not with this latter issue but with that of the grounds for regarding properties such as spin as *hybrid* in the first place. It certainly cannot have anything to do with the 'Galilean' point about mathematics functioning as the language in which physics is written, since that concerns only the representational role touched on above and would reduce the claim of hybridity to triviality. It is in this Galilean sense that one might be inclined to agree, as noted previously in our discussion of Peressini's distinction between 'pure' mathematics and its formal analogue, that we do not have a mathematics-independent grasp of physics (or at least not the kind of physics we are concerned with here). But we take Morrison to be suggesting a 'deeper' sense of hybridity, as indicated by her use of the phrase '*essentially* mathematical' (italics added). One way of understanding this 'deeper' sense is in terms of a metaphysical claim: that certain properties, such as spin, are essentially both physical *and* mathematical, in some metaphysical sense (e.g. that the essence of spin involves some mathematical feature). We shall return to the issue of how we might make sense of this shortly (and as we shall see, we are doubtful that we can), but first let us pursue the possible grounds for suggesting such a position to begin with.

One such set of grounds concerns the problems—discussed extensively by Morrison (see also her 2015, pp. 234–8)—with measuring spin and, in particular, the spin of, for example, single electrons. As she notes, these problems could be a result of some physical restriction on the very possibility of such observations, akin to that which lies behind quark confinement for example, or it could simply be a matter of technological difficulties (ibid., p. 552). However, the lack of any currently acceptable, theoretical motivation for supposing the former, together with recent experimental developments, suggests the latter. Indeed, as she herself notes, the use of magnetic force resonance microscopy had got the relevant sensitivity down to around 100 electron spins at the time she considered this issue, but further work in this area has apparently demonstrated the ability to measure single spins (via the use of ultrasensitive cantilever-based force detection sensors to measure the force resulting from a precessing spin in a magnetic field, together with an increase in the relevant theoretical understanding of, for example, spin relaxation processes; see Rugar et al. 2004).

Alternatively, and perhaps more productively, we might consider the historical claim that Dirac's treatment of spin was 'essentially mathematical'. This kind of claim crops up at various places in the literature. Thus we recall that Steiner, again, described Dirac's discovery as 'magical' and as being made by operating only on the syntax of the relevant expressions (Steiner 1998, p. 163).[30] However, as we have emphasized, such claims, both in general and in this particular case, are based on a representation of the relevant history that downplays or leaves out altogether the motivations in the relevant developments in physics, together with the associated heuristic moves; once these are factored in, the discoveries start to look a lot less magical and mysterious (Kattau 2001). Returning to Dirac, we have presented his basically pragmatic attitude to the relevant mathematics in Chapter 7. There we were concerned with what we have called here the 'Strong' claim, and in particular with the dispensability of the delta function in the relevant physics but the arguments given there do not, of course, apply to supposedly hybrid properties, not least because they depend in part on the distinction between 'thick' and 'thin' epistemic access, where the criteria for the former, detailed previously, all seem to be satisfied by spin. Thus, spin obviously falls on the 'right' side of the quantifier/ontological commitment divide and our strategy of appealing to this distinction will not help us here. Nevertheless, our analysis further supports the above point that paying attention to the relevant heuristic moves helps dispel any magical mystery mongering and, as we shall see, there are ways of understanding spin without calling upon some notion of 'hybridity'.

One could also dismiss the above as merely historical and of little relevance to how we should understand spin in today's context. Here it is the group-theoretic framework that underpins that understanding, as Morrison notes, but how one gets from the deployment of that framework to the conclusion that spin is a hybrid property is not at all clear. Certainly there seems no obstacle to regarding that framework as representational or indexical in the manner suggested by Melia and Saatsi. Perhaps what we need is an appropriate understanding of properties in this context.

Morrison herself argues that the theoretical and experimental practices of physics have important implications for how we understand the notion of a physical property in general (2007, p. 548). This suggests that the argument is to be completed by a step involving a literal reading of these practices, to the effect that those aspects of the group-theoretical description that—based on a broadly classical understanding of properties[31]—one might have been prepared to regard as 'merely' mathematical, or physically surplus, in some sense, should be regarded as describing a property of the world, but one that is both physical and mathematical. This also suggests that there is a particular view of physical properties acting as a foil, which must be rejected in the face of this literal reading. Although this view is not made explicit, statements such as

---

[30] Morrison suggests that Steiner's account of the relationship between mathematics and physics (discussed in Chapter 1) might accommodate her conclusions; 2007, p. 548, note 29).

[31] An understanding that might lead one to erroneously regard spin in terms of a particle actually spinning like a top in space-time.

the following suggest that it is one on which properties like spin, (rest) mass, charge, etc. are regarded as monadic (and perhaps intrinsic):

we typically think of the electron as being identified by properties like charge, mass and spin with each of these having a value for the single entity in question.   (ibid., p. 548)[32]

The experimental problems with measuring spin together with its group-theoretical description are then taken to imply that this view must be abandoned.

However, even if this is correct,[33] it does not immediately yield the conclusion that the property must be conceived of as hybrid. First of all, the empiricist is going to remain untroubled by these sorts of considerations. For her, 'spin' will remain a theoretical term that may or may not refer to an actual unobservable property but the crucial issue is whether the theories that deploy are empirically adequate or not. As long as that feature is satisfied, it does not matter if we cannot give a coherent realist interpretation, although any such attempt will be welcomed because of its heuristic value.

Turning to the realist perspective, the situation is a little more complicated of course, but again, the move to hybridity can be resisted. Thus, alternative views of properties are available that are more accommodating than the above. In the case of spin in particular, Eddington, for example, urged that we take its group-theoretic description to be that of a 'pattern of interrelatedness of relations' (Eddington 1941, p. 278).[34] He saw his approach as 'rescuing' from that description what was essential for physical purposes, namely that the elements of the group are defined solely by their role in that group and he took this to require an expansion of our notion of (physical) property to embrace what he called 'patterns that weave themselves'. Those metaphysical positions that allow such an expansion are likely to be more accommodating to the group-theoretic description of spin without having to ascribe any hybridity to it. Again, current trends in philosophy have not served us so well in this respect: dispositionalist accounts of properties, in particular, appear to leave little room for the above kinds of considerations (see e.g. Bird 2007; see also French forthcoming). Nevertheless, alternatives are available in addition to structuralist approaches: according to Chakravartty's semi-realism, properties and relations are tied together in a holistic package that can also capture the above 'interrelatedness of relations' (see Chakravartty 2007).

That the realist needs an appropriate metaphysics of spin is further revealed by returning, in this context, to the indexing/representational approach. It might seem that, analogously to the use of physical rods of certain length to represent cicada life

---

[32] Earlier (ibid., p. 532) she refers to spin as an 'essential' property, although this is meant in an informal sense.

[33] We have already seen that the experimental issues are not decisive; although it could be argued that insofar as 'observations' of single spins involve a measurement context, they are not observations of spin *qua* quantum property.

[34] A view to which the structural realist would be sympathetic of course (see French 2014).

cycles, as suggested by Saatsi, one could argue that we can express rotations, say, via some kind of physical arrangement and in this case the relevant relation of combining elements might not even be that of actually performing successive rotations, obviating any need for a group-theoretic representation.[35] However, the response seems straightforward: when one considers the *representation* of the rotation group, the appropriate 'combining relation' is stated explicitly and if this relation were different, we would not be talking about the same group representation (see Eddington 1941, p. 270). The defender of hybridity might take this as again indicative of the role of mathematics in understanding the relevant property but it actually speaks to the significance of the relevant group transformations and the importance of constructing an appropriate metaphysical correlate.

Similarly, Melia's 'weaseling' move that we noted earlier—which suggests that the mathematics can be regarded as little more than heuristic scaffolding—needs to be treated with care in this context. For sure, certain elements might appear to have only a heuristic role in this case: we can introduce the components of spin, for example, by specifying them in a set of mutually orthogonal planes (corresponding to spin in the x-, y-, and z-directions) and after obtaining a group-multiplication table by taking the set of operations represented by rotations through $90°$ in each of the planes, we can effectively discard the planes and take this table (or the information it encodes) as representing the property (Eddington 1941, p. 279; see also French 2003). Here the planes perform only a heuristic role as scaffolding. However, in this case we are still left with the group table and so the defender of hybridity can argue that even with the scaffolding removed, the very fabric of the edifice so constructed is still interwoven with mathematics. Again, the response is to elaborate an appropriate metaphysics (such as that indicated by Eddington) that will then allow us to 'detach' this mathematics as representational and eliminate any need to appeal to a hybrid nature.

This suggests that in such cases, if we are to achieve the independent grasp on the mathematics that Peressini has suggested (recalling our discussion at the beginning of this chapter), we should either articulate an appropriate metaphysics of, for example, the relevant symmetry, if we are realists, or spell out how these features are merely tools, from the empiricist perspective. Unfortunately for the advocate of the EIA, there appear to be no obstacles to doing either.

Finally, let us return to the issue of the supposed hybridity itself. This, we suggest, is ontologically problematic.[36] First of all, it is not clear how the mathematics can be

---

[35] This is the suggestion made by Braithwaite in his critique of Eddington's philosophy of physics (French 2003).

[36] Note—just to be clear—that it is not simply a case of asserting that there are nominalistic facts about spin, say, but we can't say or think anything much about them without using mathematics, other than that they're not mathematical. That would be to maintain a form of the 'Galilean' understanding of hybridity touched on above and we are taking Morrison to be suggesting a more radical, but of course, more interesting, form. Furthermore, it is not at all clear that in cases such as spin, unlike that of the cicadas for example, we have grounds for asserting that there are such facts, unless it can be shown that the mathematics can be dispensed with.

essential to, or constitutive of, properties in this way, particularly since properties such as spin have causal impact (in the broad sense, setting aside issues to do with the nature of causality in quantum physics) and mathematics does not. Thus the idea would be that spin cannot be hybrid because it is causal and mathematical entities are not. Now, introducing causality here is famously problematic and those who agree with Heitler and Morrison might well point out that it is precisely in this context—namely that of quantum physics—that their view gains some traction! Insisting that spin is 'physical' simply begs the question but noting that it can be related to observable phenomena via precisely the sorts of moves that Morrison herself so nicely sets out (cf. French and Ladyman 2003) may not be sufficient either. It could be argued, for example, that it is not necessary for a hybrid property to effectively manifest all aspects of its hybrid nature on every occasion.

One could insist that it is the 'physical' aspect of spin that features in such moves and is manifested at the observable level or that it is this aspect that participates in causal relations, such as those involved in the above measurements but that one still requires the mathematical aspect to make conceptual sense of the property, for example. At this point one might turn the argument around and suggest that the onus is now on the defender of the hybrid view to provide a fuller account of what it is to be a hybrid in this case, if the two aspects can be separated so cleanly. We might try to draw on an analogy with cases of hybridity elsewhere. Thus, various conceptions of 'hybrid objects' crop up across a range of academic disciplines, from computer science to history of science. They all seem to feature the conjunction of disparate features, from the use of both continuous and discrete variables in the former case, to interpretations that combine elements of the animal world and the mineral or the physical and the chemical in the latter. But in all cases, both sets of features are taken to do useful work and it is this that grounds the 'hybridity'. In the case of spin, with the physical aspect cleaved off, as suggested above, only the latter aspect would do any empirical work. Of course, as just indicated, the defender of hybridity could maintain that the mathematical aspect does another kind of work, either conceptual or explanatory. However, if the latter, then we bring in the points raised earlier, namely that an appropriate account of explanation then needs to be spelled out. If the former, then an opponent of this view can appeal to the kinds of metaphysical accounts that also and perhaps better do the same conceptual work and the mathematics would be reduced to having only an indexing or representational role again.

A more apposite example, perhaps, than those above is that of 'impure' sets, which can be taken to exist where their members do. So, for example, one might say that where we have two cicadas, we also have the number two. As is well known, this may allow one to maintain that numbers, sets, etc. should not be taken as abstract in the sense of not being spatio-temporally located (see Rosen 2014 for some discussion and references). But, again, the point is that even if one were to grant this, it is the members of the set—the cicadas, for example—that wield any causal power and not the set—*qua* mathematical entity—itself.

A further concern with ascribing causal power to spin, conceived as hybrid, is that such ascriptions may be taken to underpin the identity of the properties concerned. But if spin is hybrid, its identity cannot be given by its causal powers alone, since its mathematical aspect cannot underpin or be associated with such powers. Again, of course, the issue of the role of causality in this context comes to the fore. A defender of the hybrid view might respond as follows:[37] insofar as the relevant causal powers are associated with or manifested via the kinds of moves Morrison spells out that relate the property to the empirical situation, they fall within the purview of, and underpin the identity of, the physical aspect of the property. But insofar as they are associated with or articulated in terms of the association of instances of the property with other instances, or with different properties entirely, in a sense that is obviously vastly more attenuated than how we understand causality outside of this context, they underpin the identity of the mathematical aspect. Thus we have a kind of 'double-identity' view that meshes with the hybrid nature of the property.

Again, to respond, the opponent of the hybrid view can appeal to one or other of the available metaphysical packages in order to underpin the identity of spin. The structuralist package, for example, does so by conceiving of the interrelationships both between instances of spin and between spin and other properties in terms of the relevant laws and symmetries, understood as features of the structure of the world (French 2014). The point is, some such account can be provided and that undercuts the motivation for ascribing hybridity in the first place.

A further option that might assuage this concern regarding identity conditions for properties would be to adapt Psillos's (2010) account of scientific models. Borrowing Dummett's phrase of 'physical abstract entities' to describe them, Psillos notes that models have physical properties ascribed to them; their identity (in part) is given by mathematical entities (such as phase spaces); and they are explanatorily relevant. The model of a simple harmonic oscillator, for example (ibid.), plays an important role both in explaining the behaviour of pendulums and in bringing under a single 'umbrella' a wide range of concrete entities; and it supports counterfactuals regarding, say, the results of changes to the length of a pendulum. Given these roles, and the commitment to such entities implied by reading theories literally, Psillos argues that causal inefficacy provides no grounds for denying reality to certain entities, and concludes: 'Causal inertia does not imply explanatory inertia' (ibid., p. 951).[38]

---

[37] Alternatively, one could retreat to a quidditistic view of properties, although that is generally regarded as deeply problematic and would represent a high cost.

[38] His aim in this work is not to defend the existence of mathematical objects but to argue that attempts to impose an 'austere' form of nominalist interpretation on scientific theories—in the sense that only concrete objects are admitted—would rule out many of those features of theories that give them their power.

Of course, one does not have to be a supporter of either the EIA or the hybridity claim to agree with that last point.[39] However, one can easily see how one might regard certain properties as 'physical abstract properties' in an analogous sense.[40] Thus, spin is obviously physical in at least certain respects, plays crucial explanatory and unificatory roles and, significantly, one can argue that its identity as a property is given in group-theoretical terms. However, it is then a further step to claim that this amounts to the identity conditions for the property being mathematically grounded.[41] One could, for example, insist again that group theory is acting in a representational capacity here and that it is the underlying structure—conceived of from the perspective of either structural realism or semi-realism, or indeed, some other ontological stance—that ultimately grounds the identity of this property. Of course, the availability of such alternatives does not provide a knock-down argument against mathematics performing such an identity-grounding role in this case and the supporter of the hybridity claim might wonder where the argumentative onus actually lies here. Nevertheless, one can see how the costs of the hybrid view (understood as involving more than the Galilean point) begin to mount up.

Thus consider, as a final comment, the point that, on such an account, properties could not be taken to be 'firmly rooted' in the spatio-temporal world as some views have it (Orilia and Swoyer 2016) but, depending on one's philosophy of mathematics perhaps, would have to be regarded as 'transcendental' and acausal or 'other-worldly' themselves. In this case, standard philosophical accounts of property instantiation will have to be revised to allow for the role of the mathematics.

Of course, such requirements are not impossible to satisfy;[42] our point here is just to indicate some of the costs associated with the hybridity claim. And given that there are alternative views of properties that may be more accommodating to the features associated with the group-theoretic description of spin, we suggest that as in the case

---

[39] Psillos himself notes that if 'non-mathematical abstract objects' such as models are explanatory, so are the mathematical entities from which they are, in part, constituted. Of course, here the issue arises again of what counts as explanatorily relevant: granted that models feature in scientific explanations, one could argue that they still only play a representational role and cannot function as the actual *explanandum*. Thus, adopting Strevens' account, one might insist that the models, *qua* models, cannot be difference-making.

[40] We would also reject the analogy, since it makes the mistake of reifying models: although we can represent various aspects of scientific practice via (set-theoretic) models, for example, this does not license us to claim that such models must exist, whether as 'physical abstract entities' or any other kind of entity.

[41] Note how this is different from the case of models, under Psillos' conception: as he notes, if these are considered to be abstract then there is little obstacle to their identity being partly constituted by other abstract objects. In the case of a (at least partly) physical property such as spin, one again bumps up against the issue of hybridity: How do the mathematically grounded identity conditions relate to the physically grounded ones in such a way as to fix the identity of spin as a property?

[42] One might adopt a partial instantiation account for example: when a property is instantiated only its physical aspect is instantiated such that it can be said to be 'rooted' in space-time. Since instantiation appears mysterious anyway—or, more politely perhaps, is claimed to be *sui generis*—it may well be flexible enough to accommodate all kinds of modifications (for critical discussion of standard accounts of instantiation, see Vallicella 2002).

of the Exclusion Principle, this move to an ontological construal of the role of mathematics is, at the very least, ontologically too costly.

8.5 Conclusion

We hope to have shown two things in this chapter. First, that when it comes to the EIA the possibility of mathematical entities acquiring a non-causal but explanatory role is not well motivated, even within the framework of an account of explanation that might be sympathetic to such a role. Second, that in the case of spin, the assertion of some kind of hybrid status also lacks strong motivation and comes with associated metaphysical costs.

The latter case can be taken to exemplify such assertions with regard to the kinds of properties one encounters in modern physics, given the role of group theory in affording us an appropriate conceptual grasp on the property and its widespread, if not ubiquitous, presence in the field. Spin offers perhaps the most striking case due to the well-known problems in conceiving of it in classical terms, as already noted, but given how charge is also presented in the quantum/relativistic context, one could also mount a hybridity claim in this case as well, which will be similarly undermined by considerations such as those presented above.[43]

In Chapter 9, we will pursue these themes further by examining the role of idealization in this context, and with specific regard to explanation, as well as the broad criteria of acceptability that, we would argue, any such explanatory account should meet. Our case study will be that of the phenomena of phase transitions in thermodynamics, and the role of the renormalization group, which has been held up as another example of mathematics playing a significant explanatory role. Again, we shall press our argument that once we have a clear framework for understanding representation, and an equally clear understanding of what is required of any explanation, such claims simply do not hold up.

---

[43] It may be that the claims of Franklin (1989) and Parsons (2008) with regard to 'quasi-concrete' objects might be similarly undermined.

9
Explaining with Mathematics?
Idealization, Universality, and the Criteria for Explanation

9.1 Introduction

It has been argued that the account we have been defending and developing here is in fact deficient in not appropriately accommodating certain forms of idealization that have recently come to the fore in the relevant philosophical literature and which are apparently crucial for explaining a range of significant physical behaviours (Batterman 2010). In particular, such idealizations depend on certain mathematical operations that cannot—or so it is claimed—be captured by the kind of set-theoretic approach that we favour. On the contrary, we insist, our account can in fact accommodate these idealizations in a perfectly straightforward manner.[1] Furthermore, we shall raise concerns as to the sense in which explanation is involved in the case considered. After revisiting the details of our account, we show that such claims either rely on a misconception of the partial structures approach, or can be straightforwardly accommodated by it.

9.2 Immersion, Inference, and Partial Structures

We recall that one of the core claims of our view is that it is by embedding certain features of the empirical world into a mathematical structure that it becomes possible to obtain inferences that would otherwise be extraordinarily hard (if not impossible) to obtain. Of course, as we have indicated, applied mathematics may have other roles (see also Bueno and Colyvan 2011), which range from unifying disparate scientific theories through helping to make novel predictions (from suitably interpreted mathematical structures) to providing explanations of empirical phenomena (again from certain interpretations of the mathematical formalism), an example of which we shall be exploring here (see again Pincock 2012).

---

[1] In previous work (see Batterman 2000b), Batterman has focused on the apparent gain in understanding that can be achieved by such idealizations. Although there has been some discussion of the relationship between explanation and understanding in the literature, in this chapter we shall primarily focus on the former, given the context in which Batterman raises his criticisms of the inferential conception of the application of mathematics.

However, all of these roles ultimately depend on the ability to establish *inferential relations* between empirical phenomena and mathematical structures, or among mathematical structures themselves. For example, when disparate scientific theories are unified, one establishes inferential relations between such theories, showing, for example, how one can derive the results of one of the theories from the other. Similarly, in the case of novel predictions, by invoking suitable empirical interpretations of mathematical theories, scientists can draw inferences about the empirical world that the original scientific theory wasn't constructed to make. Finally, in the case of mathematical explanations, inferences from (suitable interpretations of) the mathematical formalism to the empirical world are established, and in terms of these inferences, the explanations are formulated.

To accommodate this important inferential role, as we have emphasized, it is crucial to establish certain mappings in the form of partial isomorphisms (or partial homomorphisms) between the appropriate theoretical and mathematical structures, with further partial morphisms holding between the former and structures lower down in the hierarchy, all the way down to the empirical structures representing the appearances, at the bottom. Using these notions, as we have suggested throughout this book, we can provide a framework for accommodating the applicability of mathematics. The main idea, again, is that mathematics is applied by bringing structure from a mathematical domain (say, group theory) into a physical, but mathematized, domain (such as quantum mechanics). What we have, thus, is a structural perspective, which involves the establishment of relations between structures in different domains. Crucially, as we have emphasized, we typically have surplus structure at the mathematical level, so only *some* structure is brought from mathematics to physics: in particular, those relations which help us to find counterparts, at the empirical domain, of relations that hold at the mathematical domain. In this way, by "transferring structure" from a mathematical to a physical domain, empirical problems can be better represented and tackled.

It is straightforward to accommodate this situation using partial structures, as we have argued. The partial homomorphism represents the situation in which only some structure is brought from mathematics to physics (via the R_1- and R_2-components, which represent our current information about the relevant domain), although 'more structure' could be found at the mathematical domain (via the R_3-component, which is left open). Moreover, given the partiality of information, just *part* of the mathematical structures is preserved, namely that part about which we have enough information to match the empirical domain.

These formal details underpin our three-stage scheme, which we repeat here (see Chapter 2):

Step 1 (Immersion): A mapping is established from the physical situation to a convenient mathematical structure. The point is to relate the relevant aspects of the physical situation with the appropriate mathematical context. The former can be taken very broadly, and includes the whole spectrum of contexts to which mathematics is applied. Although the choice of mapping is a contextual matter, and largely dependent on the particular details of the application, typically such mappings will be partial, due to the presence, not least, of idealizations in the physical set-up. These can then be straightforwardly accommodated, and the partial mappings represented, via the framework of partial isomorphisms and partial homomorphisms.

Step 2 (Derivation): Consequences are drawn from the mathematical formalism, using the mathematical structure obtained in the immersion step. This is, of course, a key point of the application process, where consequences from the mathematical formalism are generated.

Step 3 (Interpretation): The mathematical consequences that were obtained in the derivation step are interpreted in terms of the initial physical situation. To establish an interpretation, a mapping from the mathematical structure to that initial physical set-up is needed. This need not be simply the inverse of the mapping used in the immersion step. In some contexts, we may have a different mapping from the one that was used in the latter. As long as the mappings in question are defined for suitable domains, no problems need emerge.

Thus, our account precisely emphasizes the point that "underlying both the purely representative aspects of (the mixed statements of) applied mathematics, and the explanatory aspects, is the idea that the proper understanding of applied mathematics involves some sort of mapping between mathematical structures and the physical situation under investigation" (Batterman 2010, p. 9). For us, the mapping is best characterized in terms of partial isomorphisms and homomorphisms holding between partial structures (although some other partial mappings can also be used).

We are now in a position to tackle recent claims that this kind of approach is deficient when it comes to capturing the kinds of mathematical devices that underpin the idealizations crucial to certain significant developments in physics.

9.3 Idealization and Surplus Structure

As we have emphasized, one of the virtues of the partial structures account is that it can easily accommodate idealizations in science (French and Ladyman 1998; da Costa and French 2003). Here two important questions can be asked in this regard: Can idealizations play an *explanatory* role in these cases? And if so, how? The standard answer to the first is they only *appear* to do so and hence part of the answer to the second has to do with how such idealizations can ultimately be dispensed with; that is, "we can tell a story about how they ultimately can be removed by paying more attention to details that are ignored or overlooked by more idealized models" (Batterman 2010, p. 15). However, it is claimed, in certain cases, no such story can be told, or so it would seem. Thus, in these cases, it is argued, the idealization plays an *ineliminable* explanatory role and furthermore, by virtue of the nature of the idealization, the mathematics that underpins it must play such a role as well.

Now, it is a good thing that it is no part of either the inferential account in particular or the partial structures approach in general that a less idealized model will necessarily be more explanatory than the given more idealized one![2] This can be seen from reflection on the general nature of the approach. If one were to insist that explanatory strength varies inversely with the degree of idealization (however strength and degree respectively are to be determined), one would have to supply some account of that relationship. However, such an account would have to draw on further resources that go beyond the purview of the inferential conception. In fact, the inferential conception is neutral on the particular relation between explanatory strength and idealization. In some cases, more idealized models are more explanatory; in other instances, the reverse is the case.

Having said that, we shall argue that there are difficulties in understanding the kinds of models that have been put forward in this context as explanatory *at all*. As will become clear, the central issue is not whether such models are *idealized*, but whether they are *explanatory*, given the role that mathematical devices play in such models.

Batterman maintains that the cases he is concerned with are at odds with structuralist mapping accounts such as ours because, first of all, "such idealisations trade on the fact that in many instances 'overly simple' model equations can better explain the most salient features of a phenomenon than can a more detailed less idealized model" (Batterman 2010, p. 17), and second, they involve limits that are *singular*, in the sense that the relevant object ceases to be either well defined or well behaved at a certain point (ibid.). A much-used example of this would be the function $f = 1/x$ which has a singularity on the real line at $x = 0$. Another (graphic) example, involving a limit, has been given by Berry (2002, p. 10): imagine you're eating an apple and you discover half a maggot; even more distressing would be to discover a third, or a quarter... and so on. Taking the limit, discovering no maggot at all should be the most distressing of apple-eating experiences! But of course, the reasoning fails because the limit is singular: a very small fraction of a maggot (where $f \ll 1$) is qualitatively very different from no maggot ($f = 0$). As Berry goes on to describe, such singular limits crop up again and again in modern physics. Here's another example that meshes with what we say below (ibid., p. 10): consider two spotlights, shining on the same area of a wall, say. In order to obtain the intensity of the combined beams, we need to take the amplitudes of the separate light waves, add them together, and then square the result to get the intensity. Taking account of the phases of the two waves, $\pm\phi$, say, the resultant intensity is given by $|e^{i\phi} + e^{-i\phi}|^2 = 2 + 2\cos 2\phi$, the values for which range between 0 and 4. This raises the question: Why then don't we see

---

[2] Mark Colyvan emphasized this point in a personal communication to Batterman (see Batterman 2010, p. 16).

interference fringes across that area of the wall? The explanation involves an appeal to a singular limit (ibid.): if the beams make an angle θ, then the interference fringe spacing will be $\lambda/2\theta$, where λ is the wavelength of the light. This vanishes in the geometrical limit of small λ. And as we approach that limit, $\cos 2\phi$ oscillates rapidly—indeed, infinitely fast as λ vanishes, corresponding to an 'essential' singularity in mathematical terms.[3] Given this rapid variation, $\cos 2\phi$ is replaced with its average value, namely 0, where this is justified by the finite resolution of the relevant detectors (our eyes, for example), the fact that the light is not monochromatic, and that there are phase variations between the two beams.[4] Hence we don't see the interference fringes and the visible resultant intensity is, in effect, the sum of the intensities of the two beams.

Now, as thus presented, the singularity appears to play a crucial role in the explanation of why we don't see the interference fringes. However, we would counsel caution! The mathematics is again only playing a representational role—one that is crucial of course for our understanding of interference phenomena in general but this particular device has to be abandoned in the limit and replaced. And this replacement is of course justified on pragmatic grounds, as just set out. We shall return to these points below.

Returning to Batterman's claims above, we maintain that the accommodation of singularities is in fact not a problem for our account, as we shall shortly see. As for the claim that models incorporating idealizations are often able to provide 'better' explanations than their de-idealized alternatives, we are happy to agree. The notion of a better explanation here presumably has to do with bringing out the relevant features of interest and depending on what those features are, adding further details in an effort to de-idealize the account may well obscure what is going on and lead to a less good explanation. There appears to be nothing here that would be at odds with the partial structures framework that we advocate.

More significantly, perhaps, there is the further claim that limiting operations of the kind involved in the explanations deployed in cases such as the above, "are simply not the sorts of gizmos which figure in a (partial) representation, the explication of which is the aim of the various mapping accounts" (Batterman 2010, p. 19). The criticism here appears to depend upon a limited understanding of the resources available to the partial structures approach. Certainly, the kinds of structures we deploy are set-theoretical, and insofar as a limiting mathematical operation is well defined, it can be characterized set-theoretically and hence represented within our framework.

Batterman goes on to insist that, "another slightly different way to see [his point] is by noting that there are no structures (properties of entities) that are involved in the limiting mathematical operations. [...] If the limits are not regular, then they yield various types of divergences and singularities for which there are *no* physical

---

[3] For a nice demonstration, see http://demonstrations.wolfram.com/AnEssentialSingularity/.

[4] As Berry remarks (ibid.), this is the classical equivalent of decoherence, which plays a central role in certain interpretations of quantum mechanics.

analogs" (2010, p. 19; italics in the original). This, in fact, raises a significantly different point, which challenges the interpretation step of the inferential conception. If there are no physical analogues corresponding to the divergences and singularities in the mathematical setting, the inferential conception cannot land back in the empirical set-up. Finally, Batterman notes that "one might [...] stretch terminology a bit and call the various divergences 'structures', but this won't help the mapping theorists as there are no possible physical structures analogous to such mathematical 'structures'" (2010, p. 19).

However, even if we grant that there are no possible physical structures corresponding to the relevant divergences, it is still possible to make room for the latter within our framework. One of the features of the account that we have repeatedly emphasized here is that it can accommodate the role of surplus mathematical structure, whereby a given physical structure can be related via partial homomorphisms (or some other partial morphism) to a suitable mathematical structure, which in turn is related to further mathematical structure, some of which can then in turn be interpreted physically (see Bueno 1997; Bueno, French, and Ladyman 2002; Bueno and Colyvan 2011).[5] We can represent such a surplus structure within the inferential conception by straightforward iteration: the initial mathematical model (Model 1) that is used to represent the original empirical set-up is itself immersed into another

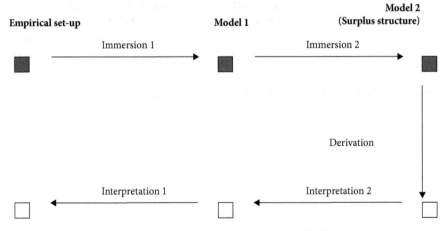

Figure 2 The Iterated Inferential Conception of Applied Mathematics

---

[5] In the example of the renormalization group to be discussed below, one takes the thermodynamic limit as the number of particles tends to infinity. In the purely mathematical sense, this is just a limit as n approaches infinity and can, of course, be represented set-theoretically. However, Batterman's further point is that the limit corresponds to an idealization that plays a significant explanatory role (2010, p. 7). Showing how one can represent such idealization in terms of surplus structure, without the mathematics alone playing an explanatory role, is one of the goals of this chapter.

model (Model 2), which gives us the surplus structure, and the results are then interpreted back into Model 1, which only then is interpreted into the physical set-up. Figure 2 illustrates the situation.

Even if there is no possible physical structure analogous to the surplus structure (where asymptotic reasoning takes place), it is perfectly possible for the intermediary structures—that is, Model 1 in Figure 2—to have a suitable physical interpretation. These intermediary structures ultimately link the surplus structure to the empirical set-up. In this way, as will become clear below, the formal framework we advance has suitable resources to accommodate the kinds of cases that Batterman has put forward.

In other words, we believe that the challenge can be met: the kinds of example presented can be accommodated as surplus structure, appropriately related to mathematical structures that are physically interpreted.[6]

Let us now look at some concrete cases, beginning with the now classic example of the rainbow as a useful illustration of what is involved.

9.4 The Rainbow

As is well known, Descartes proposed an early explanation of the formation of rainbows via the application of ray optics (essentially Snell's Laws). In particular, he showed that if rays from the same direction (that of the Sun) enter a raindrop, after reflection and refraction they tend to cluster together and exit in a particular direction, creating a significant concentration of light in that direction, known as the 'caustic' (http://www.ams.org/samplings/feature-column/fcarc-rainbows).[7] With the addition of the further understanding that light of different colours will refract in different directions when entering a given medium, generating different caustic rays, we have what might be called the standard explanation of rainbows. Note that this already involves both ray and wave optics as the latter is needed to explain refraction.

But this explanation cannot account for 'supernumerary' bows such as the faint violet and green bands sometimes observed on the inside of the main arc. Accommodating these involves, at base, an understanding of diffraction—and hence, again, wave optics—as the wavefront passing through the raindrop in effect doubles back on itself, producing the interference that results in the observed bands of colour (ibid.; see also Berry and Howls 1993). Such interference phenomena possess certain invariant features, having to do with the way the fringe spacings and intensities scale in the neighbourhood of the caustic. We shall return to this kind of invariance below as it underlies a significant feature that Batterman calls 'universality': namely

---

[6] We shall return to the issue of providing a physical interpretation below, when we consider Belot's (2005) contribution.

[7] It is called this because a similar phenomenon occurs when a magnifying glass focuses light on a piece of paper, with striking effects as every small child knows!

the way in which systems that are quite different at the micro-level exhibit similar macroscopic behaviour, under certain conditions (see Batterman 2002, especially section 6.3). In the case of the rainbow, these features can then be explained via catastrophe theory, which captures the structural stability involved (that leads to the formation of the bows even though the relevant wavefronts deviate from their ideal form), via the classification of those caustics whose topology survives perturbation. Without going into a lot of technical detail, the crucial relationship, from our point of view, is the isomorphism that holds between the theory of stable caustics and the relevant parts of catastrophe theory (Berry and Upstill 1980). It is this that underpins the embedding of the former into the latter in a way that fits nicely with the inferential approach we advocate. Indeed, the core point is that "the stable caustics have the same structure as the catastrophes" (ibid., p. 260).

Here the rays are represented by a multi-valued action function whose surfaces of constant action are the wavefronts of geometrical optics and whose branches meet on a caustic. This function, including its multi-valued nature, is represented in catastrophe theory by embedding it in a single-valued function with extra variables (Berry and Upstill 1980, p. 264). In the language of catastrophe theory, this latter function is a generating function in terms of which a "gradient map" can be defined from the spatial position of the point of observation to the relevant state variables. There is a nice analogy with the height of a landscape above a plane where the coordinates of the extrema (such as hilltops and saddles) correspond to the rays and the heights give the actions. As the observation point changes so do the extrema and when two or more extrema coalesce the gradient map is singular. Since rays coalesce on caustics, it can be concluded that caustics correspond to singularities of gradient maps (ibid., p. 267).

Thus, from this perspective, what we have is the empirical phenomenon, namely rainbows, first modelled in terms of ray optics. This corresponds to our immersion step 1. We then have an embedding within catastrophe theory, corresponding to immersion step 2, where the multi-valued function is embedded in a single-valued function with extra variables. In this context there is considerable surplus structure (e.g. there are numerous ways in which the branches of the multi-valued action function can be represented as the extreme values of the single-valued function). Interpretation 2 involves identifying the caustic as a *singularity* of the gradient map and interpretation 1 takes us back to geometrical optics. What is important is that the surplus structure of catastrophe theory can then be drawn upon to investigate the structural stability of these singularities (ibid., pp. 267–72). These investigations are then brought back over the structural relationship, as it were, to the model and its interpretation in terms of the phenomena concerned.

We can think of this situation as another example in which one mathematical structure (wave optics) is related to another (ray optics), via catastrophe theory, and understanding is achieved via this relationship, just as the theory of functions of a real variable is illuminated via its relation to the complex plane (Redhead 2004, p. 529). From this perspective, as we have indicated, catastrophe theory represents surplus structure, some of which comes to be physically interpreted so as to provide an account of the above universality. Now, Batterman's point is that this kind of structure, albeit surplus, nevertheless plays an explanatory role. However, as we shall argue in what follows, it remains unclear how that role can be spelled out *in his own terms*. We will canvass one possible way of doing this, and will argue that this can also be captured within our approach.

9.5 Accommodating Process and Limit Operations

Batterman's own account of the role of mathematics in physical theorizing hinges on the claim that "*to explain and understand the robustness of patterns and regularities, one sometimes needs to focus on places where those very regularities break down*" (2010, p. 20; italics in the original). As we'll see, it's the robustness of certain regularities to changes in the underlying physical material that functions as the *explanandum* in the cases that lie at the core of Batterman's account and it's the presence of certain singularities that is taken to be indicative of a breakdown in those regularities. Thus, consider another appearance of singularities in the mathematical modelling of certain situations, such as that of shock waves in a gas (see Wilson 2006; cited by Batterman 2010, p. 22). Shock waves form when a pressure front in the gas moves at supersonic speeds, so that, in effect, the gas in front of the wave can't react appropriately when the front arrives, leading to a dramatic increase in pressure over a very short distance—indeed, the thickness of such a front is of the order of the mean free path of the molecules of the gas (the average distance travelled between impacts). This justifies treating it as a plane in three dimensions, across which the properties of the gas are said to change 'almost instantaneously'. Modelling the gas as a fluid, we get a sharp discontinuity in the relevant properties, represented by a singularity in the mathematics, in the sense that at the specific place and time corresponding to the shock front, the relevant part of the gas will be attributed with two different velocities. (Indeed, this can be regarded as analogous to the kinds of phase transitions we shall discuss below.) We could, of course, abandon this model of the gas as a fluid and treat it as a collection of particles, with the faster molecules overtaking the slower ones. From this perspective the shock wave is treated as a boundary that determines the behaviour of the gas on either side. Thus, such singularities act as important signposts, in a sense, that help to highlight the interesting and potentially useful features of the situation under consideration.[8] As Wilson nicely puts it:

Insofar as the project of *achieving mathematical understanding* goes, singularities frequently prove our best friends, not our enemies.   (Wilson 2009, pp. 184–5; italics in the original)

---

[8] In general, of course, contradictions and inconsistencies in mathematics and science can serve as highly fruitful heuristic 'signposts'; see Post (1971) and da Costa and French (2003, chapter 5).

Indeed, as Batterman emphasizes, it is through the investigation of singularities in mathematical limiting operations that we begin to understand the *effectiveness* of mathematics in these situations. However, he maintains—and this is a crucial point—"it is an approach that is completely orthogonal to structuralist/mapping accounts that take explanations necessarily to involve static representational maps" (2010, p. 21). Here he draws again on Wilson who notes that the above friendly aspect of singularities only applies to mathematical understanding and that when it comes to modelling the physical situation, they tend to be eschewed.[9] Such modelling-based or, broadly, representational accounts are thus deficient in that they miss "in many cases, what is explanatorily relevant about idealisations; namely, that they often involve *processes* or limiting *operations*" (2010, p. 10; italics in the original). On the contrary, however, we believe that the many useful and illuminating examples given in this debate do not in fact clash with our approach (as already indicated), not least because insofar as his putative explanations involve non-static processes and limiting operations, these can both be accommodated in our framework.

Of core significance to this whole discussion is Batterman's point that many examples regarding the role of mathematics in science share a common feature in that they invoke a mathematical *entity* that supposedly plays the explanatory role. Here we might recall the examples given in Chapter 8, where the entities concerned are numbers (and we raised the concern as to whether any purported support for their existence could 'spread' to other such entities). As we saw there, given that these entities are typically regarded as abstract, defenders of the claim that mathematics does play such a role have typically turned to non-causal accounts of explanation in science in terms of which they articulate their claim. As Batterman says, and here we think he's probably right, this feature arises out of the central concern with the indispensability argument, enhanced or otherwise, and grounds for positing the existence of such entities in general. As he makes clear, however, this is not his concern and, moreover, his focus is on those 'examples of mathematical explanations of physical phenomenon [*sic*] that do *not* require that one associate a mathematical entity or its properties with some physical structure had by the system of interest' (2010, p. 4; italics in the original). In these cases, what are involved are not mathematical entities but mathematical *operations*, such as are involved in taking a certain limit.

Now, these operations and mathematical "processes" in general are simply func-tions, transformations, mappings, and so forth, and these can be, and indeed explicitly have been, incorporated into the partial structures framework. In the context of the latter, the distinction between objects and operations is a bit of a red herring. Given the set-theoretic context of our account, functions, transformations, and mappings are

---

[9] Again, we would emphasize that even in such contexts, inconsistencies are not always simply or straightforwardly dismissed.

just particular kinds of relations, and thus can be immediately represented.[10] From that perspective they are mathematical objects too, just like numbers. But the fact that certain contributors to the debate over the indispensability argument have concentrated their attention on rather simplistic examples involving numbers and properties, such as primeness, should not be taken as indicative either of those who have looked at more complex cases, such as those presented here, or of any supposed deficiency in the partial structures account.

Hence it is not the case that mapping-based accounts have focused solely on 'static' relationships, insofar as this can be clearly understood. Of course, there is a trivial sense in which a partial homomorphism between a mathematical structure and the relevant physics is 'static' in that that particular partial homomorphism does not change with time (how could it?). And of course we can represent changes in the relationships between given structures as we proceed through the stages of theory development by deploying different partial homomorphisms at each stage. Indeed, the notion of partial structure was developed precisely to accommodate the open-endedness of theory change, as we have said, and the way in which new information often requires adjustments in the relevant structures. It seems to us that there is no other sense of 'dynamical' relationship that would be appropriate here.

With this point out of the way, we can now examine the more significant claim that in some important examples of explanations in physics, certain mathematical operations feature *ineliminably* in the *explanans*, requiring us to grant mathematics the kind of explanatory role that Batterman suggests.

9.6 Renormalization and the Stability of Mathematical Representations

Let us consider another very familiar example: the transition of water from a solid phase, to liquid, to vapour. Each phase corresponds to a different state of aggregation with very different properties. And, of course, in this case there is a striking change in the long-range orientational order that characterizes ice as a crystalline solid (and here we might recall our brief discussion of what led Wigner to group theory, in Chapter 4). Following Landau, such a change can be understood as a form of symmetry breaking, characterized by a change in a certain 'order parameter'. In the case of water, as it undergoes the transition from liquid to steam, say, this is given by the mass density. Such parameters can then be used to describe the nature and extent of the symmetry breaking (see Kadanoff 2009 for a useful introduction).

---

[10] Our framework inherits all the advantages of the set-theoretic underpinnings of mathematics. In particular, with regard to the ontology of mathematics, only sets need to be assumed. Other kinds of mathematical objects, such as functions, relations, and numbers, can all be represented in terms of sets. And for those who are inclined towards a nominalist understanding of mathematics, it is still possible to provide a fictionalist reading of the framework as well (see Bueno 2009, 2013a).

Examples such as the above are classified as 'first-order' phase transitions and involve the absorption or loss of a fixed amount of energy per volume—as heat is added, the temperature of the system remains constant until all parts of the system have completed the transition. During that transition, however, certain thermodynamical properties of the system—the density, say—change abruptly (analogously to the case of the shock wave touched on above). With so-called 'second order', or 'continuous', phase transitions, on the other hand, the change in certain physical quantities is, as the name suggests, continuous but the first derivative with respect to the relevant order parameter will be discontinuous. Consider the case of ferromagnetism, which is explained in terms of two of the features of quantum mechanics that we have already considered, namely spin and the Pauli Exclusion Principle: a ferromagnetic substance (such as iron) is one that can be magnetized by an external magnetic field and can retain its magnetism. It is the electron's spin, together with its orbital angular momentum, that gives rise to its magnetic dipole moment that, like the spin, can be oriented 'up' or 'down'. In a filled 'shell' or orbital in an atom, the Paul Exclusion Principle will ensure that each 'up' spin is paired with a 'down' spin, so the overall magnetic dipole of the atom will be zero. However, if the outer 'shell' is only partially filled, then the atom will have an overall non-zero magnetic dipole moment. If one of this atom's neighbours also has a partially filled outer 'shell' then a lower energy state can be achieved if the spins of those outer-shell electrons line up in parallel; again this is due to the Exclusion Principle: if the spins are parallel, the spatial distribution of the electrons' charge is greater (thinking semi-classically, one might visualize the electrons as being further apart) and the electrostatic contribution to the overall energy is reduced (the difference being referred to as the 'exchange energy'). This energy difference is much greater than that associated with the interaction between the dipoles, which would tend to keep them anti-parallel and so the dipoles of such magnetic materials tend to be aligned. Of course, a lump of iron does not display such spontaneous magnetization as the effect of the Exclusion Principle is short-range and so the material divides into magnetic domains, within each of which the dipoles line up together. When a strong enough external field is applied the walls between these domains effectively break down and the dipoles line up throughout the material, creating a permanent magnet.

Now, if the lump of iron is then heated up, the thermal motion of the atoms will interfere with the alignment of the dipoles. At a certain temperature, known as the Curie temperature, this alignment is destroyed and the permanent magnetization disappears (although the material can still become magnetic in an external field). At this point there is a second-order or continuous phase transition as the properties of the material change dramatically and it loses its spontaneous magnetization.[11] In this case the energy changes continuously but the first derivative of the energy with

---

[11] Below this point there is spontaneous symmetry breaking as the dipole moments become aligned.

respect to the temperature (the relevant order parameter) is discontinuous—and mathematically we have another singularity.[12] That's all very interesting in and of itself, of course, but what is even more interesting is that systems that are quite different in nature exhibit very similar behaviour near such singularities. Indeed, the details of the systems can then be effectively ignored and the similarities can be captured by partitioning the systems into 'universality classes'. Furthermore, the relevant features of this universal behaviour can then be described via a mathematical device known as the 'renormalization group' and it is this, it is claimed, that plays an ineliminable explanatory role in such cases.

It is important to be clear on what the *explanandum* is in such cases: it is not the phenomena associated with the transitions themselves; rather it is the remarkable *similarity* in the behaviour of these very different systems—that is, systems with different molecular constitutions—near this kind of phase transition, as captured by the 'universality' of the critical phenomena (that is, in the region of the transition concerned).[13] This behaviour of the systems is described in terms of certain order parameters that scale in a certain way, namely as a specific power law (Batterman 2010, pp. 5–7).

The relevant 'why question' then, is why do these parameters for very different systems all scale in this way? Thus, the *explanandum* is not a simple regularity per se but rather has to do with the *robustness* of that regularity. The framework of the explanation is then as follows (for a handy summary, see Reutlinger 2017): we begin with the Hamiltonian of the system, capturing the interaction energy between its components. As the system undergoes a phase transition, it becomes impossible to keep track of the relevant interactions and so a transformation is applied that re-describes the characteristic length of the interactions. This amounts to a kind of coarse-graining and it is such transformations that form the renormalization group. This application yields a new Hamiltonian which, together with the original, can be represented in an abstract mathematical space representing all the possible Hamiltonians. Repeated applications of the renormalization group generate further such Hamiltonians which 'flow' towards certain fixed points in this Hamiltonian space. These fixed points describe possible system behaviour[14] and different systems whose Hamiltonians 'flow' to the same fixed point will then exhibit the same behaviour and can be allocated to the same universality class.

Thus, the *explanans* invokes this notion of a 'renormalization group', which mathematically captures the 'scale invariance' of certain systems, in terms of which they appear the same, or exhibit 'self-similarity', at different distance scales (a useful

---

[12] We recall that the superfluid transition in liquid helium is another phase transition, as is Bose–Einstein condensation in general, as discussed in Chapter 5.

[13] This can be demonstrated experimentally in terms of their possessing the same characteristic number, or 'critical exponent', related to the relevant order parameter.

[14] Which can be characterized by the above 'critical exponent'.

overview is given by Wilson 2008).[15] By means of this device, then, one can demonstrate that different systems can be described by the same fixed-point inter-action, where the fixed points emerge as invariants of the renormalization group (and here we might recall our discussion from Chapter 4). Broad classes of physical Hamiltonians, corresponding to these different systems, can then be attributed to the same universality class and hence the above 'universality' is thereby explained.[16]

So, the central idea is to demonstrate, and thus explain, the equivalence of the behaviour of different systems by representing those systems in an abstract Hamil-tonian space in which there exist points that are fixed, and hence invariant, under an imposed renormalization group transformation, and towards which the universality class of the given systems will 'flow', in a sense.[17] Again, certain singularities and divergences prove to be significant, in particular the divergence of the relevant correlation length (which offers a measure of how ordered a system is) in the thermodynamic limit (i.e. as the number of particles in the systems tends to infinity, with the particle density held fixed). This divergence turns out to play a crucial role in this demonstration as it underpins the loss of distinguishability of the systems.[18]

Here is a more concrete way of seeing what is going on, using Kadanoff's spin block approach (see Kadanoff 2009). Recall our example of some ferromagnetic material and imagine that it is modelled via a lattice of spins (Batterman 2011).[19] Let's proceed in reverse, as it were: at a suitable high temperature (above the critical temperature) the spins are randomized, and thus the correlation length (which gives a measure of the order of the system) is small (thus the system is disordered). As the temperature decreases, the spins and hence the magnetic dipoles begin to align and the spatial extent of a 'block' of such correlated spins increases, with the size of the block then offering a measure of the correlation length. At the critical temperature, the correlation length diverges, a phase transition takes place, and the material becomes ferromagnetic. Near the critical temperature, the correlation length becomes very large, a singularity emerges in the mathematics, and the relevant governing equations cannot be solved.

---

[15] As Kenneth Wilson notes:

> The renormalization group is a nonlinear transformation group of the kind that occurs in classical mechanics. The equations of motion of a classical system with time-independent potentials define transformations on phase space which form a group. The finite trans-formations of the group are the transformations induced by a finite translation in time; the infinitesimal transformation is defined by the equations of motion themselves.
>
> (Wilson 1971, p. 3175)

[16] For further details, see Batterman (2011).

[17] But only in a sense—what we have here are shifts in certain parameters in an abstract, mathematical space.

[18] And here we might identify a further theme, running through from Chapter 4, namely the role of this crucial idealizing move of rendering systems indistinguishable.

[19] Such a model might be an Ising model for example (see http://scienceworld.wolfram.com/physics/IsingModel.html).

The renormalization group is then applied in order to transform an intractable problem involving such large correlation lengths into a more tractable problem involving reduced lengths, thereby reducing the number of coupled degrees of freedom (again, see Batterman 2011, p. 1042). Thus, in the above example, we group the spin into 'blocks', and introduce new 'block variables' that describe the average behaviour of the block; that is, we replace the grouped spin with a so-called 'block spin'. The relevant lengths are then transformed so that these new spins sit on the same lattice sites as the old, and then the new spin variables are likewise transformed so that the new system is as much like the old as possible. In effect this "blocking" allows us to define the relevant features of the theory at large distances in terms of aggregates of features at shorter distances. This sequence can be represented in terms of a series of transformations between the relevant Hamiltonians, where these represent the systems and characterize the kinds of interactions between the degrees of freedom (e.g. between the spins) as well as any effects of external fields. Performing the renormalization transformation yields a sequence of Hamiltonians describing systems with the same lattice spacing, but for which the correlation lengths get smaller and smaller (Batterman 2011, p. 1044).

If we then consider the abstract space whose coordinates are the parameters appearing in the various Hamiltonians of the systems, each point in such a space will correspond to a possible such Hamiltonian. In the case of the lattice system, with all parameters except the temperature fixed, as the temperature approaches the critical point, the point representing the system moves about in the space of Hamiltonians, where the path that point makes is called 'the physical line'. The space can then be divided up into Hamiltonians of constant correlation length, where a 'critical surface' can be defined that corresponds to infinite correlation length.

Under the renormalization group transformations, every point in the space gets mapped to another, yielding a trajectory issuing from that point. A trajectory generated from a transformation on a point on the physical line on the critical surface remains on that surface, whereas points off the physical line yield trajectories that diverge from the critical surface, intersecting surfaces that correspond to successively lower correlation lengths (Batterman 2011, pp. 1044–5). The fixed points are those points in this space that are their own trajectories: i.e. that represent a state of the system that is invariant under the renormalization transformation τ. Finding these fixed points means solving the fixed-point equation: $\tau(H^*) = H^*$.

In other words, one must determine the fixed-point Hamiltonian that is independent of any choice of initial Hamiltonian. It is by reference to the properties of such fixed points that universal behaviour is explained: "More precisely, [this behaviour] is related to the stability of the fixed points and to how the renormalization group transformation τ maps points in the neighbourhood of the fixed points" (Batterman 2011, p. 1045). Crucially, that the renormalization group is involved in this manner is taken to be indicative of the explanatory role of mathematics, since, it is claimed, there is *no physical correlate* to the transformations described, unlike, say, the cases of permutation symmetry and atomic or nuclear structure, as discussed previously, or that of the Poincaré group and Minkowski space-time.

Now, this is all remarkable, Nobel prize-winning work. But consider: first of all, the renormalization group is primarily invoked because of an issue of *tractability*.[20] It allows us to recast a difficult problem into terms that we can get to grips with (and again, we are reminded of the Wigner programme in quantum mechanics, as described in Chapter 4). In this sense, *qua* mathematical structure, it functions as a device, but the central question is: In what sense do such devices contribute to explanations of physical phenomena?[21] Consider the case of the delta function, for example, as discussed in Chapter 7. As we saw there, this was explicitly dispensable and treated only as a mathematical device with no suggestion that it played a genuine role in any relevant explanation (not least because it was inconsistent!). This is perhaps a more extreme case, but the point remains: certainly, such devices allow for certain simplifications, and they may highlight or make room for the exemplification of certain features of interest. However, although they may play an important role in allowing us to perform the relevant derivations and make the relevant inferences, it should not be immediately assumed that they have an *explanatory* role, not least because the delta function example indicates how such an assumption might lead us astray. The spin block technique is only one of many such devices deployed in this and related contexts of course; others introduce explicitly fictional spaces of fractional dimensions, or again, explicitly fictional particles with very large masses, or Wilson's 'lattice regularization' (see Wilson 1974), where a space-time constructed out of hyper-cubical lattices is constructed. In all such cases, the question remains: What is doing the explanatory work?

Of course, neither the inferential conception on its own, nor, more generally, the partial structures account, settle that issue. It all depends on how the relevant mathematics that is used to represent the empirical set-up is *interpreted*. That mathematics can be interpreted in realist terms or not, and it can also be interpreted as playing an explanatory role or not. Realists about mathematics will take the derivation and the interpretation steps of the inferential conception as involving reference to independently existing mathematical entities, as we discussed in Chapter 8 (see Colyvan 2001), whereas nominalists about mathematics will resist such a move. On their view, mathematical objects are either not referred to at all, since these objects do not exist (see Field 1989), or are referred to, but the notion of reference does not require the existence of the corresponding entities (see Azzouni 2004).

---

[20] This is not to deny that, *as appropriately interpreted*, its success might support the realist inference that it represents something fundamental about the systems to which it is applied. This is precisely what we wish to emphasize. Of course, anti-realists will resist the realist implications of the move, while acknowledging its heuristic significance. (Once again, we are grateful to one of the referees for urging us to make this explicit.)

[21] Again, the issue is whether and in what sense the renormalization group as a mathematical structure can feature as the *explanans*, not whether that which the realist takes it to represent can serve in that role.

Furthermore, those who hold that mathematics *does* play such an explanatory role owe us an account of the nature of explanation involved in the relevant examples from scientific practice. Expressing it as neutrally as possible, any such account must be able to tell us how the mathematics does the relevant explanatory *work*. The most obvious way of getting a handle on that would be to show how the mathematics and the relevant physical phenomena are related in a manner that goes beyond the representation of this relation via deduction or other formal devices. One might, for example, suggest that the phenomena in question are *grounded* in the mathematics, in some sense. That, of course, immediately invites the request to cash out that sense and articulate what it is that this 'grounding' consists in. One way of doing that—perhaps the most obvious way—would be to say how it is that the relevant physical phenomenon is 'brought about' by the mathematics. Furthermore, this 'bringing about' needn't be explicated via appeal to causal factors, which of course would raise obvious concerns about compatibility with mathematics as the *explanans*—one might invoke certain structural commonalities, for example, and argue that the relevant structural features of the phenomena are grounded in the corresponding features of the mathematical structure putatively serving as the *explanans*.

9.7 Structural Explanations

We have already met an example of a broadly 'structural' explanation that might seem, at first sight, to be amenable to Batterman's approach, but which, in fact, precisely reveals the concern we have, namely the explanation of the halting of the collapse of a white dwarf star, discussed in Chapter 8. We recall that here the *explanans* is Pauli's Exclusion Principle that determines how many electrons can occupy the relevant energy state. The core of the explanation consists in the claim that the gravitational attraction on the mass element is balanced by the difference in what is sometimes called the 'Pauli pressure', or 'degeneracy pressure' across the mass shell created by the occupancy of the energy states. But, of course, Pauli's Principle itself is not 'merely' mathematical—it is a statement or description of a physical feature of the world. As we have emphasized in previous chapters, it can be understood as a consequence of the requirement that the relevant wave function for an assembly of fermions be asymmetric, which in turn can be interpreted in terms of the action of permutation symmetry, understood as capturing either a physical constraint or 'meta-law' or a fundamental feature of the physical structure of the world (see French 2014; French and Saatsi 2018).

Here is another example that might seem even more conducive to Batterman's analysis, insofar as it involves an *explanans* that is understood to be *fictional*, in a sense.[22] This concerns the semi-classical phenomenon of 'wave-function scarring'

---

[22] Bokulich (2008a) explicitly refers to Batterman's work in this context. A useful comparison might also be made with Lange (2013). For additional development of Lange's approach, see Lange (2017).

and the central idea is to invoke certain classical structures in order to explain quantum phenomena. The *explanandum* here involves what is sometimes known as 'quantum billiards': consider a particle bouncing chaotically around a 'stadium'-shaped enclosure. It transpires that certain, rare, trajectories repeat themselves and when these systems are treated from the perspective of quantum mechanics, the probability density of the relevant wave functions is strongly localized around these rare periodic orbits. The explanation of this phenomenon then appeals to classical structures that are known not to exist. In essence, it hinges on the fact that the classical evolution of Gaussian wavepackets[23] can arbitrarily approximate solutions of the time-dependent Schrödinger equation in the limit as $h/2\pi \rightarrow 0$ (see Bokulich 2008a, pp. 222–3).

According to Bokulich (2008a, 2008b), these strictly fictional structures can be regarded as genuinely explanatory and she gives a model-based account of their explanatory power. The core idea is taken from Morrison (1999), who articulated the explanatory power of models in terms of their exhibition of certain kinds of structural dependencies. As Bokulich notes, this is too general as it stands, and fails to distinguish genuinely *explanatory* models from those that merely save the phenomena. So, she enhances Morrison's approach by appealing to Woodward's now classic rendering of the dependence of the *explanandum* on the *explanans* in counterfactual terms: we consider the difference it would make to the *explanandum* if certain (relevant) factors in the *explanans* had been different in various possible ways. Thus on this view, explanatory power is cashed out in terms of supplying an answer to '*what-if-things-had-been-different*' questions (see e.g. Pexton and Saatsi 2012). However, she leaves aside Woodward's (2003) interventionism, according to which something causes, and therefore explains the occurrence of, a particular phenomenon if by intervening on that something, some (relevant) feature of that phenomenon would change. That would obviously reduce Bokulich's account to a causal framework, which would be inappropriate in this situation (Bokulich 2008a, p. 226).[24] As she notes, the exhibition of the structural elements means that there is a sense in which the elements of the model can be said to 'reproduce' the relevant features of the *explanandum*, and satisfaction of Woodward's counterfactual condition means the model should also be able to give information about how the target system *would* behave, if the structures represented in the model were changed in various ways. In addition, there is a justificatory step in which the domain of applicability of the model is specified and assurance is given that the *explanandum* falls within that domain. It is via such a justification that genuine explanations are distinguished from those that merely save the phenomena.

---

[23] The graph of which is a bell-shaped curve.

[24] This extension of the Woodwardian account is also followed by Saatsi in his account of 'geometric' explanations, as noted in Chapter 8 (see Saatsi forthcoming).

Bokulich then goes on to argue that the relevant periodic classical orbits genuinely explain the phenomenon of wave-function scarring both because the relevant pattern of counterfactual dependence is exhibited and because top-down justification is supplied by a theoretical device (Gutzwiller's periodic orbit theory) that specifies how the classical trajectories can be deployed to model the relevant features of the quantum dynamics. Furthermore, it is not the case that the classical trajectories *cause* the scarring; rather, as she emphasizes, what we have here is a case of 'structural explanation', in which:

> the *explanandum* is explained by showing how the (typically mathematical) structure of the theory itself limits what sorts of objects, properties, states, or behaviours are admissible within the framework of that theory, and then showing that the *explanandum* is in fact a consequence of that structure.   (Bokulich 2008a, p. 229)

Now note, first of all, that there seems to be nothing here that conflicts with either the partial structures approach in general or the inferential conception in particular. Certainly, one could appeal to the former to characterize the relevant models and in particular to capture the way in which such semi-classical models capture only certain features of what is ultimately a quantum mechanical phenomenon. However, Bokulich allows mathematical structures to limit the appropriate behaviour of the system in question, yielding the relevant phenomenon as a consequence. This could be taken to exemplify the sense in which that phenomenon might be understood as *grounded* in the mathematics, as suggested above and, as we indicated, it is in this respect that Bokulich's analysis appears to be conducive to Batterman's account. Here, we suggest, one must step carefully. If it were to be asserted that a mathematical structure, *qua* a piece of mathematics, sets such limits, our concern would arise again (we recall again the discussion of the role of the Pauli Exclusion Principle in Chapter 8): In what sense can mathematics limit the behaviour of physical systems (beyond the obvious point that such systems cannot behave in logically impossible ways)? And likewise, we would want to know in what way a physical phenomenon could arise as a *consequence* of a piece of mathematics. But we do not think that is what Bokulich intends. As an example of structural explanation, she gives Hughes' explanation of the invariance of the speed of light as a consequence of the structure of Special Relativity and the underlying space-time (2008a, p. 229).[25] Thus it is not purely mathematical structure that serves as *explanans*, but rather physical structure that, of course, is characterized or described mathematically.[26]

---

[25] Hughes (1989, p. 199) argued that one can explain the invariance of the speed of light by appealing to the structure of Minkowski space-time, which imposes certain constraints on the admissible coordinate systems and transformations of the theory. Obviously, insofar as the kinds of symmetries characterized group-theoretically likewise impose certain kinds of constraints, one might view explanations that invoke such symmetries as structural in this sense.

[26] Consider, also, Bokulich's discussion of what she calls 'classical structures' (2008a, p. 233), which suggests that she takes these structures to be physically interpreted in terms of classical mechanics rather than as being purely mathematical. Recall our point made previously, in Chapter 1, note 32.

And this is certainly the case when it comes to the explanation of wave-function scarring, with the caveat that the physical structure in this case is strictly fictional, since the classical orbits do not actually exist (but would be physical had they existed). Now, of course, the fictional nature of the model needs to be explored further, since one might have the concern that if it were entirely fictional in all respects, it could not perform any explanatory work. As Bokulich states, the relevant quantum wave functions exhibit a 'pattern of dependence' on the associated classical trajectories (2008a, p. 230) and it is the latter that are obviously fictional, since the particles are not, of course, classical.[27] Thus, in the case of the classical models that Bokulich considers, it is the *relationship* between the classical elements that represents the relevant features of the quantum phenomena, and it is this relationship that is captured by Morrison's structural dependencies. Of course, insofar as any idealized model involves strictly false elements, as far as the partial structures account is concerned, these will be placed in the R_2, but the model cannot consist entirely of those, otherwise it would be pragmatically useless. It might be thought that fictional models, of the sort discussed by Bokulich here, raise problems for this tripartite classification. If such models are thought of as consisting entirely of false elements, then it becomes unclear how, on our account, they could be useful in performing an explanatory role. However, if we accept the above point that the quantum wave functions exhibit a pattern of dependence on the classical trajectories and that although the elements representing the classical trajectories are strictly false, the structural relationships between them faithfully represent the quantum behaviour of the system, we can claim to have higher-order structural dependencies that would be placed in the R_1 within our framework, which is where the 'insights' afforded by the model are represented. Furthermore, the literature on wave-function scarring suggests that we could also identify both first- and higher-order relationships that should fall under the R_3 of our scheme. Of course, there is more to be said on this issue, but we hope we have indicated how even in cases of fictional models, the partial structures approach can capture what is going on.

Equally obviously, more could also be said about the explanatory role of fictional models in general (see e.g. the essays collected in Suárez 2009) but our point is twofold: first, Bokulich gives a clear explanatory schema in terms of Morrison's and Woodward's accounts, and second, we do not have in this case an example of the mathematics itself performing an explanatory role—rather it is certain physical relationships that do all the work in the *explanans*.

Certainly, then, structural explanations such as those exemplified by the cases above do not straightforwardly support Batterman's argument. In both cases—the

---

[27] Thus, in our terms, the physical structure that constrains the system is ultimately quantum in nature. That one can partially represent this classically, in the sense that Bokulich indicates, is of course a surprising result and it is this representation that, she argues, requires the introduction of fictional models which, as we indicate here, can be accommodated within our framework. Again, we are grateful to Manuel Barrantes for encouraging us to be clearer on this.

explanation of the halting of white dwarf collapse and that of wave-function scarring—something over and above merely citing the relevant mathematics is required.[28] And we must confess, we are hard put to conceive of other ways of explicating the way in which physical phenomena, or features of such phenomena, might be grounded in mathematics—certainly the onus is on the advocate of such a view to come up with further examples.

Furthermore, even if the relevant mathematical feature that is introduced as part of the putative *explanans* is ineliminable (as in the case of the renormalization group but unlike that of the delta function), it still may be non-explanatory. So, for example, the axiom of choice—at least in one of its formulations—is ineliminable from the proof that every set is well-ordered, but such a proof does not constitute an explanation, given that we cannot exhibit the well-ordering in question. Here we are just dealing with mathematical proofs and explanations, of course. In the cases Batterman presents, we have the additional concern as to how it could be that an ineliminable piece of mathematics can thereby account for how some particular physical phenomenon comes about. Again, one might appeal to the corresponding causal factors or structural features. Neither of these is what Batterman has in mind, but then it remains unclear in what sense the mathematics is explaining anything.

9.8 Explanation and Eliminability

If derivation is taken to yield explanation, then the derivation of the fixed points in explanations that deploy the renormalization group, as outlined above, counts as an explanation. But it is clearly implausible to take mere derivation as sufficient for explanation, as the example just mentioned of the axiom of choice makes clear. In his earlier work, Batterman himself appears to hold that derivation from *ineliminable* mathematics is sufficient, and presents the case studies in a naturalistic manner as evidence that scientists themselves are deploying a form of explanation not covered by standard philosophical accounts.[29] And he has argued that case studies such as

---

[28] As we saw in Chapter 8, one of the central issues in recent discussions concerns the claim that despite the acausal nature of mathematics, it may still be explanatorily indispensable. As should be clear, we are not here assuming that explanations need to be causal. Nevertheless, we maintain that some physical interpretation must be given to the mathematics; otherwise, it is simply radically indeterminate what the mathematics states about the physical world (with the possible exception of cardinality considerations).

[29] Again, it may be that Batterman and we have different senses of explanation in mind here (Bokulich too emphasizes how we must take the explanatory practices of scientists seriously when constructing philosophical accounts of explanation; 2008a, pp. 223–4). However, no matter the strength of one's naturalistic inclinations, some caution must be exercised in taking scientists' own reflections on their practice at face value. Of course, some of these reflections may be more philosophically informed than others. It may be that they attach the word 'explanation' to the kinds of moves at issue here (as indeed they do), but this may be no more than a convenient label signifying the kind of deductive relationship we have concerns about. Indeed, they may switch almost in the same breath from describing such moves as 'techniques' or even 'tricks' and then refer to them as explanatory. Hence, the importance of being clear about the sense of explanation that is in question. (We are grateful to Juha Saatsi for pressing us on this issue.) However, we are not suggesting that one should adopt a blanket scepticism with regard to such claims, nor are we suggesting that the naturalistic project of attempting to construct a theory of explanation

those considered here reveal fundamental theories in science to be explanatorily inadequate (2002; Batterman and Rice 2014), since in order to understand the phenomena involved, concepts must be imported from less fundamental theories (as in Bokulich's case of wave-function scarring described above). Thus, it is claimed, the standard deductive-nomological (D-N) and causal-mechanical accounts of explanation cannot accommodate the role of asymptotic reasoning in the explanation of universality, since the description of the relevant behaviour is not to be obtained on the basis of 'from-first-principle' solutions to the relevant equations, but is 'deeply encoded' in them and revealed only via asymptotic analysis.

The suggestion that derivation from *ineliminable* mathematics is sufficient for explanation can be challenged in two ways. First, we can reject the claim that the mathematics Batterman discusses is indeed ineliminable. Second, even if we grant that the relevant mathematics has that character, we can still contest that derivation from such mathematics is *sufficient* for explanation. We consider each of these responses in turn.

With regard to the first response, care must be taken not to simply follow the physicists in taking certain mathematical structures as making an ineliminable contribution to our understanding of the relevant phenomena, when in fact consideration of the mathematics may show that these structures in fact make only a heuristic and thus entirely eliminable contribution. Indeed, one could follow Belot (2005) in imagining a 'great intuitive analyst' who is asked to construct the asymptotic approximate solutions of a given partial linear differential equation, and who in effect 'decodes' these solutions. Various results could be obtained but, Belot maintains, at this point the analyst has only a *mathematical* understanding of the problem and in order to transform this into an explanation with physical content, a physical interpretation needs to be given which, crucially, will be in terms of the more fundamental theory. At best the less fundamental structures act as what Belot calls "mathematical crutches" that enable us to make the relevant inferences that *can* be accommodated within a form of the D-N account. By taking their explanatory role seriously, it is alleged, Batterman is effectively guilty of reifying the mathematics.

This is a charge also made by Redhead (2004) who notes, first of all, that with regard to the above explanation of universal behaviour, the purported explanation offered by the renormalization group does not seem to be accommodated by causal accounts, since such behaviour is precisely characterized in terms of an insensitivity to detailed causal mechanisms. Furthermore, the putative explanation is not accommodated by the account of explanation via unification of Friedman and others either (see Friedman 1974), since universality has to do with the similar behaviour exhibited by systems with very different causal mechanisms, rather than bringing together

that tracks scientists' own understanding is ill founded. Essentially all we are saying here is that until and unless it is made clear how (uninterpreted) mathematical structures can interact with or be appropriately related to physical systems, one should exercise caution about such claims.

apparently diverse behaviour. Second, however, and crucially for what we argue here, Redhead suggests that the asymptotic analysis of universality should be left as a purely mathematical exercise, along the lines of using complex analysis to illuminate the theory of functions of a real variable. This exercise would then be conceived of as taking place within 'surplus mathematical structure', and understood in that fashion the moves Batterman points to can be seen as related to similar moves involving surplus structure throughout the history of modern physics that we have highlighted throughout this book.[30]

Batterman has responded by repeating that the D-N account, for example, is actually incapable of accommodating his examples because (as we recall) these involve as *explanandum* not a particular instance of a pattern or regularity but the existence of the pattern or regularity itself (Batterman 2005). And he insists, again, that his form of 'asymptotic explanation' is very different from all standard accounts of explanation. Furthermore, he maintains that, contrary to Belot's assertion, in the example considered with the great intuitive analyst, concepts from the less fundamental theory must be appealed to in order to provide at least part of the physical interpretation. The issue at stake here, however, is not about fundamentality, but about explanation. In what sense, exactly, is the invocation of ineliminable mathematics explanatory?

But even if we grant that the relevant mathematics is ineliminable, as we noted, this does not guarantee that it is thereby *explanatory*. As we have repeatedly emphasized, the basic requirement here is that we understand how the phenomena in question are grounded, in some sense, in the *explanans* or how the latter leads, in some sense, to those phenomena obtaining, not simply that they do in fact obtain.[31]

This is the central issue. It is not enough, as we have said, to appeal to straightforward deduction of the relevant results in explicating this grounding or 'leading to' or bringing about. Alternatively, such understanding may be provided through the identification of suitable physical interpretations of the relevant mathematical results, as we noted in the cases of white dwarf collapse in Chapter 8 and wave-function scarring above.[32] Identifying such suitable physical interpretations is clearly a major component of the inferential conception.

Of course, this is precisely what Batterman rejects, and it might be thought that we are in danger of begging the question here, given that he is arguing for a new form of

---

[30] In this context, it is worth mentioning again Fraser's argument that the need to account for certain phenomena (specifically that associated with spontaneous symmetry breaking) does not provide unequivocal support for the explanatory indispensability of the thermodynamic limit, which is central to Batterman's discussion, since such phenomena may be explained via the properties of large but finite systems (Fraser 2016).

[31] This requirement may be satisfied by a causal account, but such an account is not necessary. As we shall indicate below, symmetry constraints may also satisfy the requirement. In this sense, at least, we are not begging the question against Batterman.

[32] Furthermore, not all deductions in the 'derivation step' will be permissible and their exclusion will also effectively contribute to our understanding. We shall return to this point below.

explanation (which we will come to shortly). Certainly, however, it is not enough to simply reject extant accounts of explanation and yet insist that the relevant mathematical features are still explanatory, for they may well be merely an instance of useful surplus structure. Something further is required to render that surplus structure explanatory. And although in the case of the spin blocking technique touched on above, some of the physicists involved in developing and applying this technique do state that it helps provide 'understanding' of the phenomenon of universality, the sense in which the relevant reasoning can be understood as explanatorily forceful remains unclear. Of course, one can add explanatory force to the surplus structure by interpreting it in realist terms. But given the highly idealized nature of the device, realist interpretations of the latter will not be straightforward.[33] This highly idealized nature is, of course, recognized in the literature, and in this context, the method is simply described as a 'technique'. This suggests to us that, as in the example of wave-function scarring, it is appropriately located within the R_3-components of the partial structures approach.

9.9 Requirements for Explanation

The core of our disagreement with Batterman now becomes clear. For us, the relevant techniques sketched above involve significant surplus structure. For Batterman, the claim that they provide understanding and explanation means that the structure *cannot* be surplus. This further exemplifies the need for a clear explication of the sense of explanation that is being appealed to. Without such an explication, it is not clear how one can articulate the difference between a mathematical *description* and a mathematical *explanation*.

Consider the example of a stone thrown into the air. At one point in time, the velocity that features in the mathematical equation describing the stone's movement has value zero. Does the fact that the variable in the relevant equation has such a value provide an explanation of why the stone is at rest, or does it simply offer a mathematical description of the phenomenon in question? Presumably, no one would consider the fact that a variable has value zero to be by itself an explanation of a physical phenomenon. A suitable physical interpretation, which identifies the relevant physical processes responsible for the production of the phenomena in question, is needed in order to yield a satisfactory explanation. Unless Batterman can give a relevant account of explanation, he is unable to answer these questions appropriately.

---

[33] Due to the role of the renormalization group in this technique, it might be possible to understand this as representing a structural feature of the world, as in the case of the Exclusion Principle mentioned above. However, such a possibility requires considerable further elaboration, and, of course, in that case it would not be the mathematics that explains the phenomenon, but this structural feature.

Of course, it might be insisted that in the case of the explanation of universality for example, something different is going on, and we do have mathematical explanations without any corresponding physical interpretation. However, in precisely these cases, it is not clear whether all that we have are very elaborate mathematical structures that only describe the relevant phenomena. It is suggested that fundamentally different behaviour in the mathematical limit can be explanatory of physical regularities where that limit is not reached. Without an explication of what counts as explanatory here, it is not clear whether the behaviour being referred to is the result of nothing more than giving too much epistemic weight to these mathematical structures.

It would seem that the only grounds for even considering them to be explanatory in the first place are the statements of the physicists who may be investing these structures with greater epistemic significance than is warranted. As part of the practice of physics, nothing is really changed by referring to some structures as being explanatory or being just descriptive. However, a philosophical account of the practice has to be more careful, of course. The distinction between explanation and description is significant, and it cannot be blurred just because physicists play fast and loose with the terminology.

In particular, it needs to be shown that these sorts of examples meet some basic requirements that all explanations satisfy before it becomes clear that we are dealing with a truly new sort of explanation. Here we shall set out a set of conditions that are in fact met by extant accounts of explanation. We suggest that if Batterman's asymptotic reasoning proposal does not meet them, it is unclear in what sense it qualifies as a form of explanation at all:

(a) *Explanations are typically tied to understanding.* Exactly what kind of understanding is involved in the production of the phenomena Batterman considers? How is that understanding different from a mere description of the phenomena in question?

In the case of the renormalization group, Batterman insists that:

the explanation for the universality of critical phenomena requires singularities; in particular, the divergence of the correlation length. Without this, we have no understanding of how physically diverse systems can realize the same behavior at their respective critical points.

(2010, p. 18)

Thus, the mathematics is regarded as essential for our understanding of the phenomenon described by the *explanandum*. However, it remains unclear what this understanding consists in or how it differs from mere mathematical description.[34]

---

[34] There has been, of course, some discussion of the role of understanding in explanation. But the details are not relevant for our argument here (see De Regt et al. 2009).

(b)   *Explanations have a certain structure*, which varies according to different views, ranging from arguments to answers to why questions. Exactly what is the structure of these explanations that, according to Batterman, indispensably involve these mathematical devices? Presumably, it is such that the explanations are different from the mathematical descriptions of the phenomena that physicists articulate; otherwise, once again the distinction between explanation and description is lost.

An obvious option at this point would be to say that explanations involving asymptotic reasoning are ultimately answers to a why question, such as: Why do physically diverse systems realize the same behaviour at their respective critical points? However, if one takes up this option then what we end up with is not quite the radically new, unaccounted-for type of explanation it is advertised as being. Perhaps a better option would be to say that the structure of explanations based on asymptotic reasoning is this: the issue to be explained is the stability of the diverse physical systems rather than a particular empirical regularity. And in order to explain that stability we need to invoke the relevant singularities, which play a deductive role in accounting for stability. However, if the singularities in question explain the stability in virtue of deductive relations that can be established between them, then, once again, it is hard to see in what sense the proposed type of explanation is new—it is just a form of deductive explanation from some sort of general principles (which need not be physical laws).

(c)   *Explanations indicate the epistemic significance of the items that are invoked in explanatory contexts*. Why should we give epistemic significance to these particular features, such as the fixed points in the parameter space or the thermodynamic limit in general? In virtue of what exactly should these items receive that sort of epistemic warrant?

It might be replied that we should grant such significance to the divergence of the relevant correlation length, for example, precisely because it provides a unificatory account of "how physically diverse systems can realize the same behavior at their respective critical points" (2010, p. 18). However, contrast this with the significance given to the height of the flagpole in the explanation of the length of its shadow: advocates of causal views of explanation argue that it is precisely the causal role that allows us to grant this significance and without it we lose the crucial asymmetry. We are not, of course, saying that such a view must be adopted, but at least it should be indicated how such significance can be conferred upon the divergent correlation length if the appropriate explanatory asymmetry is to be maintained.

(d)   *Explanations typically involve the distinction between* explanandum *and* explanans. However, in the cases considered here, it is unclear how to draw that line. Of course, the asymptotic limit, the fixed points, etc., and the description of the stability and universality to be explained can be distinguished *mathematically*.

But then so can all kinds of pairs of mathematical elements to which we would not ascribe the terms '*explanans*' and '*explanandum*'. We are owed an account of the distinction in this particular case.

Thus, what we are offering here is a challenge to the account of explanation via asymptotic reasoning. If this account is truly explanatory, it needs to be shown how it meets at least these four criteria. We believe that appropriate explanations of the halting of white dwarf collapse and wave-function scarring can certainly satisfy these criteria.[35] But what about the explanation of universality?

We recall Batterman's insistence that most accounts of explanation in science (even causal, non-covering accounts) assume that explanation involves *subsumption* of the *explanandum* under some regularity and hence cannot handle those cases where, by virtue of singularities such as indicated here, there are no such regularities and no laws governing the world. The claim is that it is by examining such cases that we come to explain the very regularities that hold elsewhere, and this explanation 'will involve a demonstration of the stability of the phenomenon or pattern under changes in various details' (2010, p. 21).

However, this is not clear. One can, after all, explain low-level empirical regularities and not just their instances from higher-level laws. Furthermore, there is nothing in *our* framework that states that the relevant relations must be law-like (although they may be, of course). So as it stands, there is nothing in Batterman's insistence above that precludes an account based on the mappings we have in mind from accommodating his examples.

9.10 Attempts at Explanation

More recently Batterman and Rice have articulated a 'minimal models' account according to which renormalization group explanations are explanatory by virtue of showing that the *explanandum* is completely independent of all the relevant micro-details (Batterman and Rice 2014). The core idea is that one constructs an explanation by demonstrating that certain factors are irrelevant with regard to accounting for the *explanandum* and those features that remain will be the relevant ones, to be included in the *explanans* (ibid., p. 363). Thus, what is explanatory about the models constructed by physicists to account for the kind of behaviour considered above is that they provide a 'back story' about why various details that distinguish the different systems from one another are essentially irrelevant. This both delimits the relevant universality class and underpins the stability of the behaviour of the systems under significant changes in the microphysical details:

The renormalization group strategy, in delimiting the universality class, provides the relevant modal structure that makes the model explanatory: we can employ complete caricatures—minimal

---

[35] With regard to the former, again see French and Saatsi (forthcoming).

models that look nothing like the actual systems—in explanatory contexts because we have been able to demonstrate that these caricatures are in the relevant universality class. As such, one might as well use those minimal models for computational ease. The backstory guarantees that, with respect to continuum scale behaviors, it will reproduce the behaviors of real systems. (ibid., p. 364)[36]

In particular, '... what accounts for the explanatory power of this model is not that it correctly mirrors, maps onto, or otherwise accurately represents the real systems of interest' (ibid.). Hence they reject any such account that meshes with the kind of structural framework that we are advocating here.

However, Lange has argued that Batterman and Rice do not, in fact, succeed in establishing this latter claim (Lange 2015). In particular, he suggests that on their own account, the system's macroscopic behaviour is explained by its possessing the property of being such that it is attracted to a certain fixed point in the abstract phase space when it repeatedly undergoes a renormalization group transformation (ibid., pp. 299–300). But since all systems in that universality class possess this property, this appears to be precisely the sort of 'common feature' that Batterman and Rice decry. So, what is going on here?

The answer is that Batterman and Rice have already made a prior decision as to what features can serve in the relevant *explanans*, whereby the aforementioned property, and those to do with conservation and symmetry in general, are ruled out: 'We think it stretches the imagination to think of locality, conservation, and symmetry as causal factors that make a difference to the occurrence of certain patterns of fluid flow' (Batterman and Rice 2014, p. 360). Note the grounds for this decision: even if one were to agree that it might 'stretch the imagination' to understand symmetry, for example, as a *causal* factor that brings about the macroscopic behaviour being explained, there are other ways of conceiving of the 'difference making' capabilities of symmetry, as we have indicated here and in Chapter 8—one may take the relevant principles to be Wignerian meta-laws and adopt an extended form of D-N account, or one may reify them as part of the structure of the world, or, taking a less overtly realist tack, one could situate them in a broadly pragmatic approach to explanation and so on.

Lange himself insists that '[p]lenty of causal explanations appeal precisely to such properties to explain macrobehaviors' (2015, p. 300), citing Herschel's explanation of certain macroscopic features of diffusion phenomena[37] via an appeal to the symmetries of the microphysical (and causal) laws.[38] As Lange says, there seems to be no

---

[36] And they also apply this account to the examples of fluid mechanics and Fisher's sex ratio model in biology.

[37] Namely the fact that the concentration of the material after being released is distributed along a Gaussian curve. As Lange notes, Maxwell subsequently used the same argument to show why the velocities of gas molecules likewise form a Gaussian distribution.

[38] Specifically, the rotational symmetry, reflecting the fact that the laws privilege no particular direction in space.

reason to deny such symmetries the status of 'common features' that minimal models might appeal to in their explanatory role. Of course, Batterman and Rice might insist that the property we are talking about when the renormalization group transformations are applied—namely the property of being such that the system is attracted to a certain point in phase space under repeated application of these transformations— is different in some way from these other kinds of symmetries, perhaps because it is not represented 'in' the model itself but only manifests when the transformations are applied. However, as Lange suggests (2015, p. 302), this is to confuse metaphysics with epistemology: even before the transformation is applied, the system can be said to possess the property of being such and such, although of course, in order to *know* whether the system has that property, the transformation must be applied. And of course, what we actually apply the transformation to is our model of the system, so in the relevant sense, that model can be said to represent that particular property (in our terms, the property might initially be included under the R_3, to be moved to the R_1 as we discover that the system does in fact possess it). But again, as Lange notes, this is exactly the same for symmetries, which manifest themselves as invariance under the given transformations.[39]

The core point is that it is by virtue of sharing such 'common features' with the target systems that the relevant models can be said to explain, and in this sense such models can be said to 'mirror' their target systems, contrary to Batterman and Rice's claim (Batterman and Rice 2014, p. 364; Lange 2015, p. 300). And of course, we would insist that such 'mirroring' or representation in general is always only partial, to be 'meta-represented', as it were, by philosophers of science in terms of the partial structures framework.[40]

Furthermore, at the core of Batterman and Rice's 'minimal models' account is the argument that all systems in a given universality class are alike in their macroscopic behaviour, *where this includes both the target system and the minimal model.* But then, as Lange points out, this means that the target system and the minimal model are treated alike, whereas in the relevant explanation they should be treated differently (Lange 2015). He asks: 'How does the minimal model's macrobehavior acquire explanatory priority over the target system's macrobehavior?' (ibid., p. 295) and

---

[39] This forms the basis of a rejoinder to those colleagues and students who have insisted that this application of the renormalization group cannot be embraced by structural realism! (Having said that, such a suggestion requires further elaboration, as we noted in note 33.)

[40] Batterman and Rice go on to argue that citing these 'common features' is just to push the question back one step, since one must now explain why all these different systems possess such features (Batterman and Rice 2014, p. 374). Lange maintains—and we agree—that these are different questions (2015, p. 304). In particular, one might, if one were in realist mode, explain why all these systems have the same features by appealing to 'the structure of the world' (cf. the question why no system can travel faster than the speed of light). Of course, this is not to deny that at some point the explanatory buck may just have to stop: Why do we only ever seem to observe the symmetric and anti-symmetric representations of the permutation group, corresponding to bosons and fermions, respectively? Perhaps because that is just the way this world is (see French and Rickles 2003)!

notes that Batterman and Rice have no answer. In other words, their account fails to meet our criterion (d) above: it does not offer any grounds for giving the *explanans* (the minimal model) explanatory priority over the *explanandum* (the target system). As Lange says, reproducing the behaviour of real systems is not the same as *explaining* that behaviour.

As an alternative, Reutlinger (2017) has suggested that the role of the renormalization group with regard to explaining the universality of critical phenomena can be accommodated within the kind of counterfactual account of explanation that was touched on previously.[41] We recall that the core element of such accounts is that explanations are conceived of as answering 'what if things had been different?' questions. The relationship between *explanans* and *explanandum* is then considered to be explanatory if the following conditions are met (ibid.):

Veridicality: the *explanans*, auxiliary conditions, and *explanandum* are all taken to be either approximately true or at least well confirmed.

Implication: the *explanans*, together with auxiliary conditions, logically imply either the *explanandum* (or strictly, the relevant statements are so related) or a statement of the relevant conditional probability.

Dependency: the *explanans* supports at least one counterfactual of the form: had the auxiliary conditions been different, the *explanandum* (or the relevant conditional probability) would have been different.

Reutlinger then shows that renormalization group explanations of universality meet these conditions. Identifying the key explanatory elements as the renormalization group transformations, the relevant Hamiltonians, and the 'flow' of those Hamiltonians, and noting the crucial claim that systems belong to the same universality class, if reiterating renormalization group transformations demonstrates that such systems 'flow' to the same fixed point in the abstract space, he argues that renormalization group explanations clearly exhibit the requisite structure insofar as the *explanandum* is the occurrence of universality, the *explanans* consists of the relevant Hamiltonians (with the laws of statistical mechanics lurking in the background), and the auxiliary conditions are the renormalization group transformations and the Hamiltonian 'flow'.

Furthermore, and more importantly, veridicality is satisfied, or so Reutlinger claims, because the *explanandum* and *explanans* are clearly both true, or approximately so, or,

---

[41] Insofar as this would fall under what Batterman and Rice call a 'common features account' they would dismiss it on the grounds that presenting the relevant common features is 'clearly insufficient to explain the behaviors' (Batterman and Rice 2014, p. 361) of the target systems. However, as Lange notes, such an account need not and typically would not, of course, say that simply giving these common features would be sufficient: 'it can require that an explanation show how the common features result in the macrobehavior. It might, for instance, require an explanation to trace the way in which causes or laws possessing the common features give rise to the macrobehavior—thereby using the minimal model to describe processes occurring in the target system' (Lange 2015, p. 299). Or it might, for example, track the dependence of the macrobehaviour on the relevant microphysical details via the kind of counterfactual analysis sketched here.

presumably if one is inclined towards anti-realism, well confirmed. There is an obvious issue as to whether the auxiliary conditions can similarly be regarded as approximately true/well confirmed, given that they involve the kinds of idealizations that are precisely at issue in our challenge to Batterman above. We'll come back to this shortly. Implication also holds because the *explanans* plus auxiliary conditions imply that systems with different Hamiltonians will display the same macroscopic behaviour. And, finally, dependency is satisfied because the *explanans* supports counterfactuals of the following form (ibid.):

There is a physically possible Hamiltonian H^* such that: if (1) a physical system had the original Hamiltonian H^* (instead of its actual original Hamiltonian H), if (2) H^* were subject to repeated [renormalization group] transformations, and if (3) we determined the resulting flow of the Hamiltonians to a fixed point, then a system with original H^* would be in a different universality class than a system with original Hamiltonian H.

Crucially, as Reutlinger notes, the *explanandum* counterfactually depends on only some changes in the *explanans*, which is all that the condition requires. He also points out that Batterman himself had previously noted that such explanations show how membership of the universality class is dependent on such features as the symmetry properties of the order parameter (a significant observation as far as we are concerned of course!) and the spatial dimensions of the system (Batterman 2000a, p. 127).

Now the point of our going through the above is not so much to advocate Reutlinger's account per se as to indicate that *here* we do have a definite and clearly articulated explanatory account that offers some hope of meeting our criteria above. But it might be objected that this hope is dashed precisely because of the role of the idealizations that some (presumably Batterman) would insist undermine the veridicality condition of the above account.[42] We recall that for Batterman, the renormalization group explains universality by virtue of exploiting the divergence in correlation length that results from taking the thermodynamic limit. That limit, and the resulting singularities, are hence, in his eyes, explanatorily ineliminable. Thus, as Saatsi and Reutlinger (forthcoming) emphasize, the renormalization group explanation is 'all about' the fixed points of the relevant transformations and, from our perspective, if

---

[42] Reutlinger doesn't think so, citing Norton (2012), who maintains that the idealization of going to the thermodynamic limit is in fact dispensable. Although we are sympathetic to this line, we agree with Saatsi when he points out that Norton's focus is on the explanation of the occurrence of phase transitions, rather than on the explanation of the universality of macroscopic behaviour of systems undergoing those phase transitions (forthcoming, pp. 2–3). Similarly, Shech (2013) argues that the apparent paradox here—that finite systems undergo phase transitions, while on the basis of the explanation, only infinite systems can undergo phase transitions—can be dissolved by distinguishing between representation as denotation and faithful representation. Thus, if we retreat from the view that such representations should be completely faithful, we don't need to be ontologically committed to such features as the thermodynamic limit. Now of course, we can easily accommodate 'partial faithfulness' of representations within our account, but, again, that does not really tackle the issue of the explanatory indispensability of this limit.

these are effectively left as mathematical devices playing an indispensable explanatory role, then Batterman's position appears to remain intact.

This issue of the eliminability of these devices has been discussed in the literature, of course (see Menon and Callender 2013) but not, according to Saatsi and Reutlinger, in the context of a clear account of the nature of the renormalization group explanation. This is what they seek to provide and the central feature of this explanation that they highlight is the network of dependencies between the critical exponents and the spatial dimensionality of the system, certain qualitative features of the relevant Hamiltonians and, in addition, the dimensionality of the spin parameter (Saatsi and Reutlinger forthcoming, pp. 8–10). The explanation hinges on the fact that this network follows from the relevant laws of statistical mechanics and it is these dependencies, of course, that the counterfactual account of explanation tracks (for details, see ibid.). Crucially, from our point of view, this dependence is 'directed', in the sense that the critical exponents depend on the microphysical details of the system as captured by the relevant Hamiltonian, and on the spatial dimensions and the dimensionality of the spin parameter, and not the other way around (ibid., p. 23). But what is significant for our discussion is the claim that within this analysis, the fixed points play an indispensable *but instrumental* role. What this means can be spelt out as follows.

As Saatsi and Reutlinger emphasize, the renormalization group analysis reveals the 'broad structural features' displayed by the 'flows' in the abstract space of parameters. These flows are attracted to the fixed points that underpin the (topological) structure in the way that the different Hamiltonians 'coarse grain' to yield similar macroscopic behaviour (ibid., p. 14). In this context, the fixed points are indispensable in the following sense:

> The theoretical resources involving the fixed points and their basins of attraction are indispensable (in the current state of physics at least) for grasping this topological structure exhibited by the space of parameters, and for studying its repercussions on those models that . . . can be taken to faithfully represent a system of interest approaching a critical point.   (ibid.)[43]

Nevertheless, the aim of this analysis is to give us information about the explanatory network of dependencies noted above, which have to do with how these universality classes depend on actual features of the systems that do not concern the fixed points themselves. In this sense, the fixed points—which are not 'physical' in that they do not pertain to any finite system—are 'instrumentally indispensable' for capturing the features via which these dependencies can be exhibited (ibid., p. 22).

As Saatsi and Reutlinger go on to articulate in detail, we need to refer to the fixed points in order to talk about the 'flow' in the neighbourhood of such a point, where

---

[43] But, of course, what is observed in the lab is the behaviour of finite systems, to which not all points in this space of parameters correspond. Thus, 'these points are best construed as mathematical approximations of properties of sequences of corresponding finite models, having these points as limits' (ibid., p. 16).

the correlation length is large enough for the renormalization group analysis to be valid, and thereby perform the kinds of calculations that demonstrate the above dependencies. However, such reference is 'ontologically innocuous'[44] since, in effect, the point is just there to help pin down the neighbourhood, as it were (ibid., p. 22). Second, we need to refer to these points in order to express the explanatorily relevant feature shared by all the Hamiltonians in a given universality class; i.e. that when renormalized, they all end up in the vicinity of the same fixed point (ibid.). Thus, although the fixed points are indispensable in these senses, they are not ontologically committing. Furthermore, there is nothing in the values of the critical exponents that depends on them—the fixed points are not variables that we can associate with different states of the *explanandum* (ibid., p. 23)—and so from our perspective, they do not play an explanatory role (within the context of the counterfactual account).

With all of this under our belts, let us return to our four criteria. First of all, the requirement that the renormalization group explanation provide understanding appears to be met, not least because of the counterfactual dependence between the critical exponents in terms of which the universality can be expressed and the various physical features of the systems as listed above. We can also concede that reference to the fixed points, and the associated thermodynamic limit, contributes to that under-standing, while acknowledging that we do not need to be ontologically committed to such points, while Reutlinger's veridicality criterion can also be met.

Second, with these developments and extensions of the counterfactual account, the relevant explanatory structure is made clear and, in particular, one can see how this is different from a mere mathematical description of the relevant phenomenon. Again the role of the dependencies is crucial (as of course it is in the case of the collapsing white dwarf case) and that of the fixed points is indeed no more than that of enabling us, via the mathematics, to get a fix on the universality phenomenon being accounted for. Third, and more importantly perhaps, we can ascribe epistemic significance to the relevant features of the *explanans* which, to repeat, are the microphysical details of the system as represented via the relevant Hamiltonian, its spatial dimensions, and the dimensionality of the spin parameter. These, of course, are as explanatorily 'kosher' as the flagpole mentioned above, whereas the signifi-cance of the fixed points, say, is, as Saatsi and Reutlinger point out, instrumental only. And finally, of course, the distinction between *explanans* and *explanandum* is now clear and given, not just in mathematical terms, but by referring to the relevant physical features that actually do the explanatory work.

Thus, with this nuanced account of the role of such devices as the fixed points, together with the application of the counterfactual account of explanation, we have

---

[44] This notion of being 'instrumentally indispensable' might usefully be compared with Melia's 'weasel-ing' manoeuvre discussed in Chapter 8 (see Melia 2000). Just as in that case, we introduce the mathematics for certain purposes but even though it is indispensable in order to achieve those aims, we are not required to commit ourselves to it, ontologically speaking.

what we need: a framework for explaining universality that does not attribute crucial explanatory roles to mathematical devices such as fixed points and the thermodynamic limit and thereby meets our four demands.

Indeed, we would argue that adherence to our four criteria rules out the possibility of mathematical explanations of physical phenomena in general. Thus, consider the classic example of the bridges of Königsberg: there are seven bridges connecting two islands and the mainland. The question then is whether it is possible to cross all the bridges exactly once, given that each island must be accessed via a bridge and each bridge, once accessed, must be crossed to its end (see e.g. Pincock 2012, pp. 51–3).[45] The answer of course is that it is not, a result that Euler proved by framing the problem in graph-theoretic terms: if parts of Königsberg are taken to correspond to nodes in a graph and the bridges are taken to correspond to the edges, then Euler showed that there is a path through such a graph if and only if either all the nodes in the graph are connected to an even number of edges, or exactly two nodes in the graph (one of which we take as our starting point) are connected to an odd number of edges (see Pincock 2012, p. 51). Since the actual arrangement of bridges in Königsberg cannot be represented by such a graph, there is no way of crossing all the bridges exactly once, given the stated conditions.

Now, we will concede that this account meets our criteria (a) and (b). There is a sense in which a form of understanding is provided, and one can charitably argue that the above satisfies some kind of 'why' question. However, when it comes to the issue of epistemic significance, the problem arises as to how a pure mathematical structure can even in principle account for physical relations. There is a fundamental difference, of course, between an abstract structure and a physical configuration.[46] And given that, as is acknowledged by both sides of the debate, the onus is on the advocates of the role of mathematics to articulate the sense in which entities from such different ontological categories can be appropriately related so as to support the claimed explanatory structure.

So, consider: the crucial mathematical features in the case of the bridges are that either all parts of Königsberg are connected by an even number of bridges or exactly two parts of Königsberg are connected by an odd number of bridges. But neither of these conditions is satisfied in the case of Königsberg. Once the physical situation is fully understood, one can see that none of the relevant four parts of Königsberg (namely, two islands and the two parts of the mainland) are connected by an even number of bridges (it is not the case that to every bridge in there is exactly one bridge

---

[45] We acknowledge that one can give more general versions of this problem and that these can be addressed as problems in topology, for example. However, these do not amount to the problem we are considering here which is about the bridges of Königsberg!

[46] This is just to repeat the point, made in note 32 of Chapter 1, that physical structure is concrete in a way that mathematical structure is not. Again, we shall not spell this out metaphysically except to note that the former, but not the latter, can be related to experimental outcomes.

out), nor is it the case that exactly two parts of Königsberg are connected by an odd number of bridges (there are four relevant parts and not two). In this context, we understand the evenness and oddness of the number of bridges in concrete terms. The number of bridges is even as long as to every bridge in there is a bridge out, and that number is odd just in case for every bridge in we have a bridge out plus one. This use of arithmetic can be dispensed with, in familiar ways, in terms of logic alone. In the end, the issue rests on the particular bridges and their configurations.

Note the general nature of the move we are suggesting, which is exemplified in our earlier cases of the white dwarf and universality: in each case, there is a physical alternative to the proposed mathematical *explanans* and given the aforementioned ontological difference, the onus is on the advocate of the explanatory role of mathematics to articulate the sense in which elements of one category can play an explanatory role with regard to those of another.

It should now be clear that criterion (d) is also violated in this case, and indeed all such cases in general. If one were to maintain that a distinction can be made between the *explanans* and *explanandum* by virtue of the fact that the former is mathematical and the latter physical, one would run up against the issue just outlined. Alternatively, suppose that the account insists that there is a dependence relation that holds between the *explanans* and the *explanandum* (such as is maintained in counterfactual accounts of explanation as we have sketched above, following Reutlinger). Obviously in this case, the onus is on the advocate of mathematical explanation to articulate the nature of that dependence, even if this is construed non-causally so as to embrace very general forms of metaphysical dependence. And it would be curious, to say the least, if such a person were to maintain that the physical world was in some sense metaphysically dependent upon mathematical structures.[47]

Of course, one might be tempted to eschew dependence and these counterfactual accounts entirely, and adopt some form of D-N view, perhaps. In that case, clearly the distinction between *explanans* and *explanandum* is indeed drawn—so, for example, the mathematical element might feature as certain mathematical laws in the premises—but at the cost of allowing for the physical to 'explain' the mathematical, since the relevant asymmetry has been lost on this view, as the previously mentioned example of the flagpole demonstrates. In this case, a distinction can be drawn—and of course, D-N accounts in general meet our criteria—but yields an account that falls prey to an obvious objection.

Finally, if the distinction between *explanans* and *explanandum* is not drawn at all then (d) cannot be met.

---

[47] One might point to Pythagoreanism in this context, but note that such a view does not so much insist on some form of dependence between the mathematical and the physical as it *reduces* the latter to the former. As a result, the very distinction between the physical and the mathematical, and hence between the *explanandum* and *explanans* in the case of mathematical explanations of physical phenomena, is lost.

9.11 Interpretation and Idealization

Returning, then, to our framework, Batterman's rejection of the role of mappings in explanation and of the inferential conception, in particular, is ungrounded, and his examples can be accommodated by this account and the partial structures approach in general, as we have tried to indicate here.

In particular, it is with regard to the role of *interpretation* that the inferential conception goes beyond a mere mapping account. In focusing only on the mappings, Batterman is correct in noting that:

one might ask why simply having a partial mapping between some aspects of the physical situation and an appropriate mathematical structure accounts for the *explanatory* role that idealisations can play in applied contexts. So far what we have is a framework in which we can get some kind of partial representation of the full actual structure. [...] *Prima facie*, it seems we have no reason to believe that simply having an appropriate (partial) mapping is explanatory. Indeed, what is the argument that such a partial representation itself plays an explanatory role?   (2010, p. 14; italics in the original)

He goes on to glean an answer from Bueno and Colyvan's (2011) discussion of economic theory that emphasizes the ranking of idealizations. Thus, he suggests, the less idealized the account, the more explanatory it is. However, as we've already said, the demand for less idealized accounts forms no part of our framework, nor does any imposed ranking, and furthermore this does not strike us as a reasonable account of either idealization or explanation. As we noted above, what we require is an account of why the results in question obtain, and this in turn will be provided by the relevant suitable interpretation, where that suitability is contextually dependent. Of course, different theories of explanation will capture the relevant context dependence in different ways, and our account does not preclude the adoption of any of these theories.[48]

We also note that in whatever way the physical interpretation is understood, Belot's account, mentioned previously, fits nicely within our framework (Belot 2005). Thus we can think of his analyst as embodying the immersion into the mathematical structure that he or she is presented with, which includes the relevant partial linear differential equation. Various inferences are then drawn regarding the asymptotic behaviour, but then crucially a physical interpretation needs to be provided so these results can acquire physical import, along the lines indicated by Saatsi and Reutlinger above.

What is more interesting, for us at least, than the details of the physical interpretation, is Batterman's insistence that he is not reifying the relevant mathematics, but that in the cases of interest this mathematics, and in particular infinite idealizations,

---

[48] Thus, again, one could give an account of why a particular phenomenon occurs in causal-mechanical terms, although this is not necessarily an account that we would adopt.

must be appealed to in our explanatory practices. Again, he is not concerned with indispensability-type arguments but with the putative role of mathematical operations in explanation. Here, and later in his response, he seems to agree with Redhead's point about the surplus mathematics, but insists that the role of this surplus is explanatorily ineliminable. However, as we have already noted, ineliminability and explanatory capacity are very different things.

9.12 Conclusion: Explanation and the Inferential Conception

Let us be clear: the inferential conception does not offer an account of explanation per se, but rather it provides a framework in terms of which certain kinds of explanations can be articulated (as we have tried to indicate above). Nevertheless, and given our criteria (a)–(d) above, certain kinds of permissible moves on the mathematical side are going to be ruled out when it comes to the explanation of physical phenomena. Thus, referring back to the diagram of our iterated inferential conception above, when it comes to the 'derivation' step, certain kinds of moves here should be excluded. So, for example, in addition to those moves that do not track the relevant explanatory asymmetries, derivations from inconsistent premises which would be trivial within classical logic would also be ruled out. Likewise, 'brute computation strategies' might also be deemed to be unacceptable. Consider the case of the Königsberg bridges again: we could, for example, simple try out all possible routes over the bridges and thereby conclude that none of them are Eulerian. But that would only show *that* a Eulerian path is impossible in this case, and not *why* it is impossible. For that, as we noted above, we need to understand how the *explanans* leads, in some sense, to the relevant results obtaining. That understanding, in turn, may be provided through the identification of suitable physical interpretations of the relevant mathematical results, as we also said, but a contribution to such understanding may also be obtained by insisting that certain moves in the derivation step, as just indicated, should be excluded on the grounds that they do not contribute to the explanation.[49]

Thus, we certainly agree that the phenomena that Batterman has emphasized stand in need of explanation. Where we disagree is with the claims that (i) the inferential conception cannot accommodate such phenomena, and (ii) the relevant mathematics itself plays an explanatory role.

With regard to (ii), there is something curious about taking surplus mathematical structure to be ineliminable in scientific explanations. Consider an alternative example: that of the application of group theory to quantum mechanics, as discussed

---

[49] We are, once again, grateful to Manuel Barrantes for pressing us on this point and giving us the example of the brute computational strategy in the Königsberg bridges example.

in Chapter 4. There the application crucially involved surplus structure and in particular what Weyl called the 'bridges' between different parts of the mathematics—between the representations of the symmetry and unitary groups, for example. But when it comes to the use of group theory as a framework for quantum statistics, and the explanation of, say, the behaviour of liquid helium involving Bose–Einstein statistics (Bueno, French, and Ladyman 2002), or the role of the Exclusion Principle in accounting for the halting of the collapse of white dwarf stars, as discussed in Chapter 8, we don't take the symmetrization or anti-symmetrization of the relevant wave functions as merely pure mathematics playing an ineliminable explanatory role. Rather, we take permutation symmetry to represent a fundamental feature of physical reality, if we are realists, or as supporting a possible physical interpretation if we are not. Here it is the relevant physical symmetry, however construed, that is doing the explaining, not the mathematics by which it is represented. Likewise, even if the move to the Hamiltonian space and the invocation of the renormalization group in the above sketch is not eliminable in the way that Batterman (2002) indicates for the case of wave and geometrical optics, one can argue that the actual explanatory work will be done by the physical interpretation, as we emphasized above. Here the interpretation will be of the renormalization group and the way the latter represents the scale invariance associated with certain systems. In this case, again, a form of structural explanation can accommodate this example in which this scale invariance is understood in terms of certain features of the physical structure of the world.

Furthermore, and now considering (i), the above sketch of spin blocking can be straightforwardly accommodated by the inferential account underpinned by partial structures: the phenomena to be represented concern the stability of certain properties of diverse systems. This is characterized mathematically, and abstract systems are employed to focus on and reveal the relevant property. This, in turn, is represented via appropriate Hamiltonians in Hamiltonian space. The relevant relations or mappings between the physical system and the mathematics, and between the mathematics, conceived of as surplus structure, can then be represented in terms of partial homomorphisms. With the application of the renormalization transformations one can make certain derivations—a crucial feature of our framework, of course—and one completes the process by reinterpreting the results obtained (concerning the fixed points) in terms of the relevant physical properties.[50]

Similarly, the case of asymptotic reasoning in optics that Batterman (2002) has identified can also be accommodated via the iterated inferential conception plus partial structures. The phenomena in question are modelled via the introduction of an immersion step into a new model in which the mathematical asymptotic phenomena can be exhibited and derived. In turn these results are interpreted back into the original model and they in turn are interpreted in a physically relevant manner.

---

[50] While acknowledging, of course, Saatsi and Reutlinger's point regarding the instrumental indispensability of such fixed points.

Let us also again recall the rainbow example (ibid.). Here, neither wave nor ray theories of optics are capable of providing an appropriate explanation on their own. Instead, Batterman argues, features of both must be appealed to in order to construct an asymptotic 'borderland', which is effectively a model incorporating such features and through which an explanation can be provided via an embedding into the structure of catastrophe theory, for example, as we indicated above. Partial structures are precisely capable of capturing this kind of piecemeal construction whereby certain elements of the wave and ray theories are partially immersed, via partial morphisms, into an asymptotic model from which the derivation of the relevant results is obtained, and which in turn are interpreted in terms of the physical set-up. Asymptotic reasoning and partial structures need not be in conflict after all.

It might be objected (Batterman 2010, p. 14), that the mere existence of a partial mapping between a mathematical structure and an empirical set-up is not enough to guarantee an explanatory role for the mathematics. On the inferential conception, the objection goes, these mappings are seen to be explanatory because they suggest that the mathematical theory approximates a more complex structure that would be an exact analogue to the physical structure. However, if there can be partial mappings to mathematical structures that cannot be physically interpreted, then the ability of the inferential account to come up with such mappings would not account for the explanatory role of the mathematics used.

On the contrary, what we have shown above is that we don't need to come up with a *direct* physical interpretation of the relevant mathematical structures to accommodate Batterman's examples, since the relevant structures act as surplus structure related to those structures that are physically interpreted. As noted above, we don't think that mathematical structures *on their own* have such an explanatory role. Nevertheless, their role in the kinds of piecemeal constructions indicated above can be captured within our account using partial structures (and surplus structures when needed).

Thus, we agree with Batterman that his examples shed new light on the practice of science and are significant for our understanding of crucial aspects of scientific reasoning. By bringing together the inferential conception and the partial structures framework, we are in a position to account for the nature and significance of the phenomena involved, as well as to offer an understanding of them within a unitary account of scientific practice.

10

Conclusion
Between Lazy Optimism and Unbridled Opportunism

10.1 Introduction

As we indicated at the end of Chapter 1, our overall aim in this work has been to steer a middle way between opportunism and 'lazy' optimism with regard to the application of mathematics. On the one hand, we have tried to argue that applicability is not simply a case of finding a mathematical structure to fit the phenomena under consideration, since these phenomena themselves must be represented in such a way that will involve significant idealizations. But, on the other, we have insisted that it is not just a matter of wrestling uncooperative mathematics into place, since with these idealizations in place, the 'special circumstances' via which mathematics is able to say something useful about physical behaviour are effectively constructed. The case studies we have presented here can then be understood as setting out a series of such special circumstances by which mathematics can be applied to physical systems and our hope is that by presenting them in such detail we can blow away the air of mystery that is often taken to surround this application process.

The question then arises: Are there any cases in which such circumstances do not arise? If the answer were 'yes', then our approach would founder. Thus Pincock, in his detailed critical analysis of Steiner's account, records our earlier discussion of the role of group theory in the articulation of isospin (French 2000), noting the way in which the application of group theory was grounded in physical reasoning, but then suggests:

That said, there remain many cases where a discovery was the result of deploying a mathematical analogy based on no independent scientific motivation. (Pincock 2012, p. 187)

But, of course, as they say, the devil is in the details (appropriating Batterman 2011) and we would need to see such details before we could comment. Again we ask, are there any such cases? Well, here's a possible candidate: the discovery of the Ω^- particle (here we draw extensively on Bangu 2008, 2012).

10.2 The Ω^- Case

In short, the history is as follows: with the claim that certain elementary particles possessed the property of 'strangeness', which is conserved under both electromagnetic and 'strong' nuclear interactions, the relevant group was expanded from SU(2), which, we recall, captures isospin, to SU(3), as we discussed in Chapter 4.[1] Again, this yields certain irreducible representations, the dimensionalities of which correspond to the cardinality of associated hadron multiplets. Briefly, considering the decuplet of spin 3/2 baryons, it was immediately apparent that although nine of the mathematical nodes in the decuplet corresponded to known particles, the tenth at that time did not. Gell-Mann and Ne'eman then reasoned that if nine of the ten nodes were effectively 'occupied', the tenth had to be too, and they were able to specify the properties of this new proposed particle and thus to suggest to the experimentalists where to look for it. The rest, as they say, is history—with again Nobel prizes all round!

Now Bangu (2008, 2012) takes the reasoning here to be essentially the same as Dirac deployed in the case of the positron but, crucially, different from that used in other predictions of new entities, such as Leverrier and Adam's prediction of the existence of Neptune or Pauli's prediction of the existence of the neutrino. In both these latter cases there was an experimentally observed anomaly—a perturbation in the orbit of Uranus and an apparent failure of energy conservation, respectively—but there was nothing along these lines that motivated the positing of either the positron or the Ω^-. Instead, Bangu argues, the reasoning in these cases involved, at least implicitly, the following principle:

If Γ and Γ' are elements of the mathematical formalism describing a physical context, and Γ' is formally similar to Γ then, if Γ has a physical referent, Γ' has a physical referent as well. (In Bangu 2008 this is termed the 'Reification Principle', but in his 2012 it is referred to as the 'Identification Principle'.)

By implicitly deploying this principle, Gell-Mann and Ne'eman were able to specify the physical characteristics of the new particle and enable the experimentalists to search for it. But what is the status of this principle? As stated, it seems to represent a new and non-naturalistic methodological component in modern science, and as Bangu notes, Steiner certainly seems to take it that way, lumping it under 'Pythagorean expectations' (Steiner 1998, p. 162). On Steiner's view, this case would seem to precisely fit Pincock's speculation of possible counterexamples to our approach.

However, let us recall our discussion of the positron example. In that case, Colyvan appears to be advocating something similar to the above principle, and using its role

---

[1] With the reduction of protons and neutrons to assemblies of quarks, and the associated property of 'colour' charge, the strong force can be viewed as a manifestation of the fundamental colour force and SU(3) can then be taken to be the associated local gauge group (for a useful introduction to colour, see Greenberg 2009).

in the relevant prediction to leverage an indispensability claim. But, as we saw, it was not the case that Dirac simply relied on a formal similarity in the mathematics. In fact, he grounded his prediction in relevant physical considerations, albeit, of course, in the context of the existence of the relevant surplus mathematical structure.

Now Bangu disagrees: in considering the passage from Dirac that we cited earlier, where Dirac insists that one can arbitrarily exclude negative energy solutions in classical physics but not in the quantum case, since perturbations will generate transitions into the negative energy states, Bangu argues that all that this supports is the claim that quantum physics can accommodate such negative energy states, not that there is a positive reason to believe there exist such states (2012, p. 91).[2] And he asks 'what prompted him [Dirac] to even think about such entities? Where did he get the idea from? The answer is, I believe, clear: the mathematical formalism' (ibid., p. 192). As with Dirac and the positron, so with Gell-Mann, Ne'eman, and the Ω^-.

We believe we have answered these questions in Chapter 7, and the answer does not lie with the mathematical formalism, at least not wholly. Of course, the surplus mathematical structure needs to be present in the first place, not least to offer the scope for the physical reasons to come into play! But to insist that Dirac 'got the idea' from this surplus structure seems to us to be a case of misplaced origin. Indeed, we think that such questions are not helpful; at least, we would argue that, typically, the location of where scientists 'get their ideas' cannot be so easily pinned down. A variety of factors enter into the mix and this is certainly true in Dirac's case. As we saw, it was only after he had a physical interpretation of his equation which allowed him to identify the transition process that he took the negative energy solutions seriously and what guided him was not the mathematics on its own, but in fact valence theory. Similarly, when it comes to giving a positive reason to believe in such states, the mathematics alone clearly does not give that—after all, it just supplies the structure that, in the classical context, is dismissed as 'surplus'. It is only when situated in the context of quantum physics and with the physical reason sketched in Chapter 7 that we have the positive reasons to posit such states.

As we have repeatedly suggested, what we are attempting to do here is to map out the complex set of heuristic factors involved in the application of mathematics to modern physics. Bangu dismisses such attempts on the grounds that they require a sharp distinction between the contexts of discovery (where heuristic factors come into their own) and justification, cashed out in the following terms:

First, the physicists, studying the world, notice an intriguing physical phenomenon or entity. Second, they seek the correct theoretical treatment of that phenomenon or entity (i.e., they seek the equation, or the theory describing it). In an attempt to find the theory, physicists are free to use any hint they think might help them discover the equation. Once pure inspiration (or a

---

[2] Bangu also acknowledges the heuristic aspect of Dirac's prediction, but insists that Colyvan was right to emphasize the significant role of the mathematical formalism (2012, p. 91, note 197).

good analogy with a previous situation, or any other trick) has delivered a candidate equation, the process of discovery comes to an end and the physicist enters the context of justification, in which experiments are run to test it, etc.   (ibid., p. 252)

However, as he says, the Ω^- case does not fit this scheme, first because the physicists involved did not begin by noticing some 'intriguing' phenomenon or entity and second, because the order seems to have been reversed, in that they began with the relevant formalism and only then looked for something that it described. In particular, he argues, heuristics typically gives us the equation for some entity, x, not x itself, whose positing requires some further premise as represented by the Reification/Identification Principle above. Indeed, insofar as this principle was crucial to the prediction and the observation of the Ω^- particle and was taken to confirm the SU(3) symmetry, the principle falls into the domain of justification (see also Bangu 2012, pp. 88–91).

We agree that the Ω^- case does not fit the heuristic scheme as characterized above. However, we would reject this characterization as overly restrictive. In particular, it is simply not the case that all scientific discoveries begin with some intriguing phenomenon or entity. As has been well documented (see e.g. Post 1971), the situations in which scientific discoveries are made are rich and complex, involving a variety of heuristic factors, from theoretical flaws or even outright inconsistencies, to thought experiments and, yes, even odd phenomena. Nor is it the case that the heuristic moves and factors deployed in such situations only yield the relevant equations and not entities. Indeed, the whole point of Redhead's highlighting of mathematical surplus structure was to illustrate its heuristic role where that role covered such cases as the kinetic theory of matter and the existence of atoms, for example (Redhead 1975, p. 88). We also do not take there to be a sharp distinction between the contexts of discovery and justification (see da Costa and French 2003, chapter 4); indeed, we accept that the two may be intertwined in such complicated ways as to render any such sharp distinction impossible. Thus, we have a broader conception of heuristics than is captured in the above scheme, one that allows a role for mathematical structure but only in the context of the relevant physical reasoning.[3]

Of course, returning to the Ω^- case, the tenth node in the baryon decuplet—the apex, in fact—can likewise be regarded as surplus mathematical structure which, under the action of certain physical reasoning, can be interpreted so as to predict the existence of a new particle. The core of this reasoning is nicely set out in a passage from Ne'eman (given in Bangu 2008, p. 243) who begins by noting that the

---

[3] Bangu also rejects the suggestion that what is going on in such cases is interpretation of the mathematical formalism, or at least not as understood in the sense that one might interpret the 'F' in $F = ma$ as 'force' (2008, p. 255, 2012, p. 92). The difference, he maintains, is that in the positron and Ω^- cases we have *reification* of a formalism. However, it is not clear to us where the difference lies. After all, those who believe in the existence of forces will argue that what we have in the case of interpreting $F = ma$ is likewise a reification of what the term 'F' stands for.

overarching framework provided by the so-called 'Eight-Fold Way'—so named because it organized baryons and mesons into octets[4]—required that the spin 3/2 particles fit into either a decuplet or a family of 27.[5] Certain observations ruled out the latter (basically it should have allowed for certain particles which were not detected), leaving the decuplet as the only contender. As Ne'eman put it, referring to 'the creators of the eightfold way' (such as Gell-Mann):

> They saw the pyramid being completed before their very eyes. [...] Only the apex was missing, and with the aid of the model they had conceived, it was possible to describe exactly what the properties of the missing particle should be!   (cited in Bangu 2008, p. 243)

Thus, as Bangu notes, part of the physical reasoning was the observation that the nine known spin 3/2 baryons 'fitted' into the ten-dimensional representation of the SU(3) group (ibid.; also 2012, p. 81). Given that the tenth node, at the apex, *qua* surplus structure, was of course formally similar to the other nine, there was good reason to expect that it too could be interpreted as representing a yet to be discovered spin 3/2 baryon.

Now Bangu does not deny that physical reasons played some role in the prediction, but he insists that 'the reasoning leading to the prediction of the characteristics of the omega minus particle proceeded by an analysis of the (*mathematical*) *description* of a physical system, as opposed to an analysis of the *physical interactions* within the system (like in the neutrino and Neptune cases)' (2008, p. 254; italics in the original). Again, we agree but would insist that such an analysis can be and indeed has been a crucial part of scientific discovery in the modern era, with the relevant mathematical description corresponding to what Redhead has identified as surplus structure, which would be dismissed as such (as in the case of negative energy solutions and classical physics) unless situated in an appropriate interpretational context, in which relevant physical reasons can be given for the corresponding interpretation (as in the case of quantum theory and the negative energy solutions).[6] When it comes to the Ω^- example, the immediate context was the Eight-Fold Way and the interpretation of the relevant nodes in the various octet representations of SU (3) as baryons and mesons, but the broader context, of course, was the resurgence of interest in group theory, following the rediscovery of Wigner's work and its subsequent application to elementary particle physics in general. Thus, it was not simply a case of observing that the nine known spin 3/2 baryons fit into the scheme and hence there should be a tenth, but of drawing on indirect evidence provided by the success of a major programme in physics. Given that, and given the ruling out of the only

---

[4] That is, eight-dimensional representations of SU(3).

[5] SU(3) allows multiplets of 1, 8, 10, and 27.

[6] Thus, we disagree with Nambu's overenthusiastic remark that Gell-Mann's and Ne'eman's work was based only on the mathematical symmetry (cited in Bangu 2008, p. 254). As we noted in Chapter 9, we should be careful not to let our naturalistic inclinations lead us to take everything physicists say at face value!

alternative, as described by Ne'eman, the physical interpretation of the surplus structure represented by the apex of the decuplet and the consequent postulation of a new particle was well grounded—in the relevant physics!

Thus, the Ω^- case, we think, can be accommodated within our approach. And this, together with what we have covered in this work, gives us good reason to doubt Pincock's claim above. Of course, the role of the mathematics *qua* surplus structure is crucial, but we do not think there are any examples of scientific discoveries based only on mathematical analogies with 'no independent scientific motivation'.

10.3 Partial Structures as a Meta-Level Device

The other aspect of our account that we have tried to make clear is the following: we agree that the way that mathematics and physics are brought together does not support the meta-level optimistic view that the relationship can be directly captured via the framework of isomorphisms, within the semantic approach. Nevertheless, there are other formal devices that can be pressed into service in order to illuminate that relationship for us—philosophers of science. It is important to emphasize again that last point since there has been some confusion in the literature about what the role of such devices should be. The core issue, as we see it, is whether they should be taken as merely *representational*, in the sense that we philosophers of science, and others, may use them to represent salient features of scientific practice, or whether they should be understood as actually *constitutive* of the theories being considered. Let us sketch the contours of this issue.

As we hinted at in Chapter 2, a debate has recently broken out with regard to the semantic approach in general which has focused on its presumptive role in underpinning the *identity* of theories. Halvorson, for example, insists that, 'according to the semantic view, a theory *is* [italics added] a class of models' (2012, p. 190)[7] and the aim of his critique is to show that 'it is impossible to formulate good identity criteria for theories when they are considered as classes of models' (ibid., p. 190; see also p. 201).[8] However, framing the debate over the viability of the semantic approach in this way leads to the possibility of question begging over what counts as 'the same' theory to begin with. To demonstrate that the semantic approach identifies theories that should be regarded as distinct, Halvorson's strategy is to syntactically formulate two theories, show that they are inequivalent by the standard criterion of definitional equivalence, and then point out that the relevant sets of models are isomorphic and hence the theories must be counted as the same according to the semantic approach, but contrary to how they 'should' be understood.

---

[7] Indeed, the whole thrust of his paper is encapsulated in section 4, which is entitled 'Identity Crisis for Theories'.

[8] Basically, by demonstrating how certain proposals for defining an isomorphism fail.

However, the question of what 'is' the theory is precisely what is in dispute in this debate. To maintain that a theory 'is' its syntactic formulation in terms of which it can be shown to be inequivalent to another, which the semantic approach renders as equivalent, is precisely to beg the question (Glymour 2013, p. 287). Furthermore, the examples of 'theories' presented in this debate are either 'toy' logic cases or taken from mathematics where, in both cases, clearly articulated formulations can be given in terms of which the equivalence, or not, of the theories can be explicitly demonstrated via some standard technical device. This can then be contrasted with the relevant relationship obtained via the appropriate device at the level of classes of relational structures. In such cases the identity criteria of the theories concerned *can* be made clear, at one level or another. But this is typically not the case when it comes to examples of *scientific* theories. Should Newton's theory of mechanics or Maxwell's theory of electrodynamics or Einstein's General Relativity be identified with certain syntactic formulations? To do so would clearly beg the question against the semantic approach, and in these cases, we don't have the clearly articulated formulations presented in the debate. Instead, what we have in practice (at least when it comes to physics) are some equations, interpreted of course, written down in various texts, in various languages, mathematical and 'everyday', sometimes 'expressed' or presented in quite different ways, some models perhaps, taken as exemplars or as 'filling out' the abstract principles, and so on. We could, of course, attempt to construct a syntactic formulation of any or all of these various features, along the lines of the Received View, but to insist that *that* formulation *is* the theory we are concerned with and that in such terms the semantic approach misidentifies it, is, of course, to beg the question. And equally, the proponent of the Received View can say the same if we were to articulate the criteria of theory identification in model-theoretic or 'semantic' terms!

Furthermore, although scientists themselves make reference to specific theories and models, we should be careful not to take such talk as indicative of something whose identity we can definitively pin down. Consider classical mechanics, for example: Do we take the theory to be delineated by Newton's equations, or the Hamiltonian or Lagrangian formulations? And before the reader cries that they are all equivalent, there is an ongoing debate regarding that very point (North 2009; Curiel 2014). Indeed, the question 'what is classical physics?' requires a nuanced discussion in which the distinction becomes subject to both temporal and geographical variation (Gooday and Mitchell 2013). And, of course, classical continuum mechanics is famously inhomogeneous, presenting a complex patchwork of models, principles, approximations, and so forth (Wilson 2014). Perhaps one might insist that at least some cases are clear: we might recall that famous line that Maxwell's theory is just Maxwell's equations! However, leaving aside that, what we all learned in our physics textbooks are actually Heaviside's equations, not Maxwell's (which are better suited for the quantum context), and it has been argued that the debate over whether Maxwell's theory is formally inconsistent or not turns precisely on the issue of how the theory is delineated (Vickers 2013).

The situation gets even worse when we move to the quantum context. Indeed, this very book is precisely about, in large part, the role of mathematics in shaping the newly developing quantum theory. Thus, taking an overview of the material covered in Chapters 4–7, during the latter part of the 1920s physicists faced a situation in which there were various apparent alternatives in play, including not only Schrödinger's wave mechanics and Heisenberg's matrix mechanics, of course, but also Dirac's 'general science of non-commuting quantities' and Weyl's group-theoretic approach. Now one can show that certain equivalences hold: Dirac's 'transformational' approach is mathematically the same as Weyl's, and as we have discussed, von Neumann demonstrated that Schrödinger's mechanics and Heisenberg's are equivalent, in the sense that they may be considered to be alternative representations on an underlying separable Hilbert space. Following that, for many commentators, including physicists as well as philosophers of physics, it is the latter, with its representation of states as vectors (or more generally, rays) in that Hilbert space that provides *the* formulation 'of' quantum mechanics. However, many, especially physicists themselves, would also agree that Dirac's approach, with its 'bra' and 'ket' formalism, offers certain pragmatic advantages.

As we have also noted, von Neumann himself became dissatisfied with the Hilbert space approach, and attempted to delineate an entirely new framework capable of accommodating probability attributions for quantum systems with an infinite number of degrees of freedom. As for Dirac, the delta function became mathematically 'kosher' through the work of Schwartz, as we have discussed, and the former's approach in general was put on a sound mathematical footing via what is known as 'rigged' Hilbert space (see Kronz and Lupher 2012). Interestingly, it is claimed that this 'formulation' ('of quantum mechanics') can handle a broader range of phenomena than that of separable Hilbert space (ibid.; see also Antoine et al. 2009). And just as von Neumann criticized Dirac for his lack of rigour in using the delta function, so Weyl admonished Jordan's presentation of canonical variables in the new matrix mechanics as 'mathematically unsatisfactory and physically unfeasible' (Scholz 2007, p. 255), offering group theory as a way of yielding 'deeper insight into the true state of affairs' (ibid.).[9]

All of these quantum revolutionaries were obviously seeking to disseminate what each thought were the basic principles of quantum mechanics, drawing on pragmatic considerations as well as concerns about mathematical rigour and outright consistency. In effect, each was attempting to mould or shape quantum physics in a certain way.[10] But then, given this situation of competing mathematical and physical principles and a

[9] In particular, Weyl was able to obtain Schrödinger's formulation from a group-theoretic basis: 'In the end, the Schrödinger characterization of a free particle turned out to be nothing but a well-chosen basis description of the irreducible ray representation of the non-relativistic kinematical group R^{2n}' (Scholz 2007, p. 257).

[10] As Kragh (2013) puts it, with regard to Dirac's *Principles*: 'He wanted to *shape* a theory which had not yet found its final shape' (italics in the original).

still emerging consensus as to which were foundational, to insist that there *is* a clearly delineated theory whose representations in terms of the syntactic or semantic approach can then be compared, would seem to be hopeless, or even foolish.[11] Whatever one might think of the above claims, regarding the delineation of classical mechanics or the inconsistency of Maxwellian electrodynamics, here we are precisely concerned to illustrate the role played by mathematics in the emergence of this fundamentally important new physics and the point is that we don't have the nice clear and clean examples that the above debate has centred on. What we have is something a lot more complex and a lot messier, in the context of which the articulation of criteria of theory identification is, at the very least, a much less straightforward and much more contentious business. As a result, we, as philosophers of science, then have to decide on what basis we are going to select those features of this complex practice, with its interwoven mathematical and physical principles that we then focus on. One answer—drawn from the recent developments of the semantic approach—is to centre any such focus on the representational role of the relevant scientific models.

Thus, van Fraassen, in responding to this debate, writes that when a scientist presents a theory 'she provides a class of models for the representation of those phenomena' (2014, p. 277). The characteristics of representation in general by which we may pick out scientific representations from the melee that is scientific practice in any field have been discussed in Chapter 3. As we saw there, it is standard to draw appropriate comparisons with representation in art: 'we properly speak of a model of combustion or of the San Francisco Bay in the way we speak of a painting of fire or of the Giaconda' (ibid.).

Given, then, that scientific models are primarily representations, in what sense may they also be mathematical structures in the way that the semantic approach proposes? The answer is straightforward: 'A model is a mathematical structure in the same sense that the *Mona Lisa* is a painted piece of wood' (ibid.). In other words, both the representational content of the painting and the actual painted piece of wood are what make the *Mona Lisa* the artefact that it is, and similarly, there is more to a model, as a scientific artefact, than the relational structure in terms of which we can define embeddability, isomorphism, and so on. In particular, if we restrict our considerations to the former, and take a model to be a structure plus an interpretation which maps expressions in some language to elements of that structure, so that sentences may come out true under such an interpretation, we stand to overlook the representational aspect that is so crucial in the scientific context.

---

[11] One can further pursue this point: which version of *quantum field theory* should we take as canonical, the rigorously axiomatized one that doesn't apply to any actual system, or the one that does but is likewise a hodgepodge of models and techniques at best, inconsistent at worst (see the debate between Fraser 2011 and Wallace 2011)? There is more to say about this, of course, but a strong claim can be made that any insistence that science itself presents clearly identifiable and delineable theories breaks down once close attention is paid to the history and the practice.

Thus, we could, rather perversely perhaps, adopt a kind of Received View stance towards the philosophy of art and rationally reconstruct the *Mona Lisa* in terms of a mapping from certain natural language expressions to features of the painted piece of wood, such that certain statements made in art books, say, come out true under that interpretation, but this is just as un-illuminating when it comes to artistic practice as adopting the above stance towards scientific representations (ibid., p. 278). In both cases, *when it comes to this representational role*, it is more natural to point to the painting or some particular feature of scientific practice and say 'that *is* the *Mona Lisa*/Newtonian mechanics (respectively)'.

The upshot then is that given that a scientific model is a representation, 'it does not follow that the identity of a theory can be defined in terms of the corresponding set of mathematical structures without reference to their representational function' (ibid., p. 278). And if we focus only on such structures while ignoring the representational function then of course we will identify putative theories that are distinct—but we always knew that, as the well-known examples of the equations describing gas diffusion and temperature distributions over time demonstrate (ibid., p. 279). It is only by appreciating their distinct representational functions that we can see that they are not the same, even if the relevant mathematical structures are.

A further issue in this debate, then, is how we should understand the semantic approach. Here we will simply note the point made some years ago but apparently forgotten in this context: 'It seems to be a popular misconception of the semantic view that it says *nothing* but the following about theories: theories *are* (with 'is' of identity) just structures (models)' (French and Saatsi 2006, p. 552; italics in the original; see in contrast Halvorson 2012, p. 204). Dropping that misconception opens up the possibility of developing a more nuanced approach to how we, philosophers of science, should represent, for our own purposes, the elements of practice that we are concerned with.

Returning from these general issues to the specific ones we are concerned with here, when it comes to the application of mathematics to physics and given the highly fluid and contested nature of the situation in which that application takes place, we believe that the framework of partial structures offers perhaps a particularly useful and appropriate device for such purposes, not least because, as we have said, it captures the open-ended nature of these developments. However, this claim has also been criticized.

In particular, Suárez and Cartwright (2008) raise what they see as two significant problems with the partial structures account:

Problem 1:   The representations involved capture almost none of the interesting features of how the kinds of theory change we see in science actually occur, including what kinds of motivations the scientists had, what techniques they were familiar with, what information they had to hand, etc. Suárez and Cartwright "see no way to express these common assumptions as relations between set-theoretical structures" (2008, p. 74).

But, of course, to try to represent set-theoretically the relevant scientists' *motivations* would be an entirely misconceived endeavour, akin to the attempt by those set-theoretic structuralists mentioned in Chapter 2 to accommodate sociological factors in theory change by stipulating let S be a set of scientists! Here we bump into one of the major fault lines in science studies, and it can be posed in the form of a question: To what extent should the scientists' intentions, motivations, and psychological states feature in one's account of their practices? Cartwright and Suárez appear to regard it as a flaw in our framework that it does not feature such aspects directly. But, of course, that they are not represented set-theoretically does not mean that they do not feature in one's philosophy of science at all. Indeed, they can be accommodated indirectly, through the appropriate meta-level (that is, at the level of the philosophy of science) representation of the relations that hold between theories, between theories and data, etc., together—and this is crucial, as we have repeatedly emphasized—with the heuristic moves that establish such relations. The factors that Cartwright and Suárez mention certainly do feature in our account, but indirectly, via the heuristic factors that they manifest in practice, for example.[12] Furthermore, we would maintain that we can directly capture the *information* that scientists have available to them, in the form of these relations with other theories and models and background information in general.

Problem 2:   The application of partial isomorphisms is inappropriate because the theories themselves do not contain any partial relations.

This simply falls under the wheels of the point made above: we are not suggesting that scientific theories themselves *in the sense of those features identified as such in scientific practice* (rather than the reconstructions of philosophers) 'contain' (in whatever sense) partial relations, because we do not regard theories *as* partial structures. It seems obvious that scientists themselves, in presenting, discussing, disputing, etc. their theories, do not write them down in the terms set out in Chapters 2 and 3. They use natural language plus, of course, mathematics, as well as graphs, diagrams, and other representational devices, *but*, and this is crucial, they also (typically) acknowledge that their theories are open to further development, via being extended within a domain, or even across domains. This open-ended character of theories, or of scientific practice more generally, obviously cannot be captured directly by or in terms of the mathematics used. It could be stated using natural language—by, for example, simply asserting 'This theory is open to further development'—but such a statement would not be a part of the theory, but *about* the theory. It is we, philosophers of science, who need a way of capturing this

---

[12] Not for nothing did da Costa and French allocate an entire chapter to the discussion of heuristic factors when they set out the partial structures approach in book form (da Costa and French 2003). In his otherwise critical analysis of this approach, Pincock explicitly notes this point, for example (Pincock 2005, p. 1253).

open-ended character at the level of our representations of scientific practice and, furthermore, for our own ends and purposes. Partial structures allow us—that is, philosophers of science—to do that.

Thus, we take such devices to be deployed at the meta-level and that deployment to be justified on basically pragmatic grounds at that level: that is, on grounds to do with whether they help us, philosophers of science, achieve our aims, whatever they may be! In particular, we do not take such devices to be *constitutive* of the theories, models, or more generally features of practice that we happen to be analysing.

And as we also emphasized in Chapter 1, as far as our case studies here are concerned, any such formal device should be able to accommodate and illuminate three core features of the application of mathematics to physics:

  (i) the openness to further developments of both mathematical and scientific theories;
 (ii) the nature and role of surplus structure on both the mathematical and physics sides; and
(iii) the kinds of idealizing moves that are made and intervening representations that are introduced in order to bring the mathematics and physics together.

We think the partial structures approach does the job in this regard. Others may disagree, as indeed they have. But any proposed alternative needs to get the balance right: too little formal framework and the account degenerates into a mere recitation of one case study after another; too much and we lose sight of the practices we are trying to explicate. The line we have tried to push throughout this book is that the framework of partial structures offers a kind of 'Goldilocks' account in this respect.

Furthermore, in terms of this framework, we have identified and discussed three roles associated with the application of mathematics: representational, unificatory, and explanatory. With regard to the first, we noted how the partial structures approach offers a rigorous 'backbone' for an account of scientific representation in this context that, in particular, accommodates the role of mathematics. Specifically, we considered four different features of this representational role:

(a.1) *Applying genuinely new mathematics*: Here we considered the application of new mathematics in the sense of being new to the particular context of application, and the kinds of heuristic moves, idealizations, and analogies that are required to bring the relevant mathematical and physical structures together, using, as our case study, the introduction of group theory into quantum mechanics (see Chapter 4).

(a.2) *Appropriately representing physical phenomena*: The introduction of new mathematics from the 'top down' as it were, may not be sufficient, as the relevant physical phenomena need to be represented in a certain 'bottom-up' way in order for the mathematics to be appropriately applied, as in the case of the application of Bose–Einstein statistics to the behaviour of liquid helium (Chapter 5).

(a.3) *Applying problematic mathematics*: Sometimes, the application of new mathematics leads to concerns over consistency and suchlike, as in Dirac's introduction of the delta function. In such cases, certain dispensability strategies are typically deployed as we mapped out in Chapter 7.

(a.4) *Interpreting mathematics*: Finally, the introduction of certain mathematics on its own is typically not enough to account for physical phenomena, and an appropriate interpretation is required in order for the phenomena to be accommodated, as we noted also in Chapter 7 in the context of Dirac's equation and in Chapter 9 with regard to the phenomenon of universality in thermodynamics.

When it comes to the unifying use of mathematics, we considered how it can be used to bring different domains together as in the case of von Neumann's search for a unified mathematical formulation of quantum mechanics, in which the representation of probability, quantum states, and their dynamics along with an appropriate underlying logic could all hang together (as discussed in Chapter 6). The purported explanatory role of mathematics is more problematic of course. Here it is claimed that certain mathematical structures are explanatory of physical phenomena, with examples ranging from the life cycles of cicadas, to the bridges of Königsberg, to certain universality phenomena in statistical physics. However, we have argued that mathematics plays no such explanatory role (Chapters 8 and 9).

In that respect, the conclusion of this book may seem rather modest. And this modesty is replicated along other lines, as we have tried to steer alternative paths between optimism and opportunism with regard to the application of mathematics in scientific practice and between overly formal and informal meta-level representations of that practice. Nevertheless, we hope that we have also illuminated the different 'dimensions' of that application and the richness of the relevant history, as well as further illustrating the power and usefulness of partial structures.

References

Achinstein, P. (1968): *Concepts of Science: A Philosophical Analysis*. Baltimore: Johns Hopkins University Press.

Anderson, C.D. (1932a): "The Apparent Existence of Easily Deflectable Positives", *Science 76*: 238–9.

Anderson, C.D. (1932b): "Energies of Cosmic-Ray Particles", *Physical Review 41*: 405–12.

Anderson, C.D. (1966): "Interview with C. Weiner, June 30, 1966", transcript in Niels Bohr Library, American Institute of Physics, New York.

Anderson, C.D., and Anderson, H.L. (1983): "Unraveling the Particle Content of Cosmic Rays", in Brown, L.M., and Hoddeson, L. (eds.), *The Birth of Particle Physics*. Cambridge: Cambridge University Press, pp. 131–54.

Antoine, J.-P., Bishop, R.C., Bohm, A., and Wickramasekara, S. (2009): "Rigged Hilbert Spaces in Quantum Physics", in Greenberger, D., Hentschel, K., and Weinert, F. (eds.), *Compendium of Quantum Physics*. Dordrecht: Springer, pp. 640–50.

Azzouni, J. (1997): "Thick Epistemic Access: Distinguishing the Mathematical from the Empirical", *Journal of Philosophy 94*: 472–84.

Azzouni, J. (1998): "On 'On What There Is'", *Pacific Philosophical Quarterly 79*: 1–18.

Azzouni, J. (2000): "Applying Mathematics: An Attempt to Design a Philosophical Problem", *The Monist 83*: 209–27.

Azzouni, J. (2004): *Deflating Existential Consequence: A Case for Nominalism*. New York: Oxford University Press.

Baker, A. (2003): "Does the Existence of Mathematical Objects Make a Difference?", *Australasian Journal of Philosophy 81*: 246–64.

Baker, A. (2005): "Are There Genuine Mathematical Explanations of Physical Facts?", *Mind 114*: 223–38.

Baker, A. (2009): "Mathematical Explanation in Science", *British Journal for the Philosophy of Science 60*: 611–33.

Baker, A. (2012): "Science-Driven Mathematical Explanation", *Mind 121*: 243–67.

Balzer, W., Moulines, C.U., and Sneed, J. (1987): *An Architectonic for Science: The Structuralist Program*. Dordrecht: Reidel.

Bangu, S. (2006): "Steiner on the Applicability of Mathematics and Naturalism", *Philosophia Mathematica* (III) *14*: 26–43.

Bangu, S. (2008): "Reifying Mathematics? Prediction and Symmetry Classification", *Studies in History and Philosophy of Modern Physics 39*: 239–58.

Bangu, S. (2012): *The Applicability of Mathematics in Science: Indispensability and Ontology*. Houndmills: Palgrave Macmillan.

Batterman, R.W. (2000a): "Multiple Realizability and Universality", *British Journal for the Philosophy of Science 51*: 115–45.

Batterman, R.W. (2000b): "A 'Modern' (= Victorian?) Attitude towards Scientific Understanding", *The Monist 83*: 228–57.

Batterman, R.W. (2002): *The Devil in the Details: Asymptotic Reasoning in Explanation, Reduction, and Emergence.* New York: Oxford University Press.

Batterman, R.W. (2005): "Response to Belot's 'Whose Devil? Which Details?'", *Philosophy of Science 72*: 154–63.

Batterman, R.W. (2010): "On the Explanatory Role of Mathematics in Empirical Science", *British Journal for the Philosophy of Science 61*: 1–25.

Batterman, R.W. (2011): "Emergence, Singularities, and Symmetry Breaking", *Foundations of Physics 41*: 1031–50.

Batterman, R.W., and Rice, C. (2014): "Minimal Model Explanations", *Philosophy of Science 81*: 349–76.

Belot, G. (2005): "Whose Devil? Which Details?", *Philosophy of Science 72*: 128–53.

Belousek, D. (2000): "Statistics, Symmetry and the Conventionality of Indistinguishability in Quantum Mechanics", *Foundations of Physics 30*: 1–34.

Bergia, S. (1987): "Who Discovered the Bose-Einstein Statistics", in Doncel, M.G., Hermann, A., Michel, L., and Pais, A. (eds.), *Symmetries in Physics (1600–1980): Proceedings of the 1st International Meeting on the History of Scientific Ideas.* Barcelona: Bellaterra, pp. 221–48.

Berry, M.V. (2002): "Singular Limits", *Physics Today 55*: 10–11.

Berry, M.V., and Howls, C. (1993): "Infinity Interpreted", *Physics World 6* (June): 35–9.

Berry, M.V., and Upstill, C. (1980): "Catastrophe Optics: Morphology of Caustics and their Diffraction Patterns", in Wolf, E. (ed.), *Progress in Optics XVIII.* Amsterdam: North-Holland, pp. 257–346.

Beth, E. (1949): "Towards an Up-to-Date Philosophy of the Natural Sciences", *Methodos 1*: 178–85.

Bird, A. (2007): *Nature's Metaphysics: Laws and Properties.* Oxford: Oxford University Press.

Birkhoff, G., and von Neumann, J. (1936): "The Logic of Quantum Mechanics", *Annals of Mathematics 37*: 823–43. (Reprinted in von Neumann, J. (1962): *Collected Works, vol. IV. Continuous Geometry and Other Topics* (edited by A.H. Taub). Oxford: Pergamon Press, pp. 105–25.)

Blackett, P.M.S. (1933): "The Positive Electron", *Nature 133*: 917–18.

Blackett, P.M.S., and Occhialini, G.P.S. (1933): "Some Photographs of the Tracks of Penetrating Radiation", *Proceedings of the Royal Society of London A139*: 699–720.

Bokulich, A. (2008a): "Can Classical Structures Explain Quantum Phenomena?", *British Journal for the Philosophy of Science 59*: 217–35.

Bokulich, A. (2008b): *Reexamining the Quantum-Classical Relation: Beyond Reductionism and Pluralism.* Cambridge: Cambridge University Press.

Bonolis, L. (2004): "From the Rise of the Group Concept to the Stormy Onset of Group Theory in the New Quantum Mechanics: A Saga of the Invariant Characterization of Physical Objects, Events and Theories", *Rivista Del Nuovo Cimento 27*: 4–5.

Boolos, G. (1985): "Nominalist Platonism", *Philosophical Review 94*: 327–44.

Borel, A. (2001): *Essays in the History of Lie Groups and Algebraic Groups.* Providence, RI and London: American Mathematical Society and London Mathematical Society.

Born, M. (1937): "The Statistical Mechanics of Condensing Systems", *Physica 4*: 1034–44.

Braithwaite, R. (1962): "Models in the Empirical Sciences", in Nagel, E., Suppes, P., and Tarski, A. (eds.), *Logic, Methodology and the Philosophy of Science: Proceedings of the 1960 International Congress.* Stanford: Stanford University Press, pp. 224–31.

Brezis, H., and Browder, F. (1998): "Partial Differential Equations in the 20th Century", *Advances in Mathematics 135*: 76–144.

Brush, S.G. (1983): *Statistical Physics and the Atomic Theory of Matter*. Princeton: Princeton University Press.

Bub, J. (1981a): "What Does Quantum Logic Explain?", in Beltrametti, E., and van Fraassen, B.C. (eds.), *Current Issues in Quantum Logic*. New York: Plenum Press, pp. 89–100.

Bub, J. (1981b): "Hidden Variables and Quantum Logic: A Sceptical Review", *Erkenntnis 16*: 275–93.

Bueno, O. (1997): "Empirical Adequacy: A Partial Structures Approach", *Studies in History and Philosophy of Science 28*: 585–610.

Bueno, O. (1999): "Empiricism, Conservativeness and Quasi-Truth", *Philosophy of Science 66* (Proceedings): S474–85.

Bueno, O. (2000): "Empiricism, Mathematical Change and Scientific Change", *Studies in History and Philosophy of Science 31*: 269–96.

Bueno, O. (2003): "Is It Possible to Nominalize Quantum Mechanics?", *Philosophy of Science 70*: 1424–36.

Bueno, O. (2005): "Dirac and the Dispensability of Mathematics", *Studies in History and Philosophy of Modern Physics 36*: 465–90.

Bueno, O. (2006): "Representation at the Nanoscale", *Philosophy of Science 73*: 617–28.

Bueno, O. (2009): "Mathematical Fictionalism", in Bueno, O., and Linnebo, Ø. (eds.), *New Waves in Philosophy of Mathematics*. Houndmills: Palgrave Macmillan, pp. 59–79.

Bueno, O. (2013a): "Nominalism in the Philosophy of Mathematics", in Zalta, E.N. (ed.), *Stanford Encyclopedia of Philosophy* (Fall 2013 edition). Stanford: The Metaphysics Research Lab, Center for the Study of Language and Information (CSLI), Stanford University. URL: http://plato.stanford.edu/archives/fall2013/entries/nominalism-mathematics/.

Bueno, O. (2013b): "Putnam and the Indispensability of Mathematics", *Principia 17*: 217–34.

Bueno, O. (2014): "Computer Simulations: An Inferential Conception", *The Monist 97*: 378–98.

Bueno, O. (2016): "Belief Systems and Partial Spaces", *Foundations of Science 21*: 225–36.

Bueno, O., and Colyvan, M. (2011): "An Inferential Conception of the Application of Mathematics", *Noûs 45*: 345–74.

Bueno, O., and French, S. (1999): "Infestation or Pest Control: The Introduction of Group Theory into Quantum Mechanics", *Manuscrito 22*: 37–86.

Bueno, O., and French, S. (2011): "How Theories Represent", *The British Journal for the Philosophy of Science 62*: 857–94.

Bueno, O., French, S., and Ladyman, J. (2002): "On Representing the Relationship between the Mathematical and the Empirical", *Philosophy of Science 69*: 452–73.

Bueno, O., French, S., and Ladyman, J. (2012a): "Empirical Factors and Structural Transference: Returning to the London Account", *Studies in History and Philosophy of Modern Physics 43*: 95–103.

Bueno, O., French, S., and Ladyman, J. (2012b): "Models and Structures: Phenomenological and Partial", *Studies in History and Philosophy of Modern Physics 43*: 43–6.

Bunge, M. (1978): "Review of W. Stegmüller, *The Structure and Dynamic of Theories*", *Mathematical Reviews 55*: 333.

Callender, C., and Cohen, J. (2006): "There is No Special Problem about Scientific Representation", *Theoria 55*: 7–25.

Cao, T. (2003): "Structural Realism and the Interpretation of Quantum Field Theory", *Synthese* *136*: 3–24.

Carnap, R. (1939): *Foundations of Logic and Mathematics*. Chicago: University of Chicago Press.

Carson, C. (1996): "The Peculiar Notion of Exchange Forces I: Origins in Quantum Mechanics, 1926–1928", *Studies in History and Philosophy of Modern Physics 27*: 23–45.

Cartwright, N., Shomar, T., and Suárez, M. (1995): "The Tool Box of Science: Tools for Building of Models with a Superconductivity Example", in Herfel, W., Krajewski, W., Niiniluoto, I., and Wójcicki, R. (eds.), *Theories and Models in Scientific Processes*. Amsterdam: Rodopi, pp. 137–49.

Chakravartty, A. (2007): *A Metaphysics for Scientific Realism*. Cambridge: Cambridge University Press.

Chakravartty, A. (2009): "Informational versus Functional Theories of Scientific Representation", *Synthese 72*: 197–213.

Chayut, M. (2001): "From the Periphery: The Genesis of Eugene P. Wigner's Application of Group Theory to Quantum Mechanics", *Foundations of Chemistry 3*: 55–78.

Coleman, A.J. (1997): "Groups and Physics: Dogmatic Opinions of a Senior Citizen", *Notices of the American Mathematical Society 44*: 8–17.

Collier, J.D. (1992): "Critical Notice: Paul Thompson, *The Structure of Biological Theories* (1989)", *Canadian Journal of Philosophy 22*: 287–98.

Colyvan, M. (1999): "Causal Explanation and Ontological Commitment", in Meixner, U. and Simons, P. (eds.), *Metaphysics in the Post-Metaphysical Age: Papers of the 22nd International Wittgenstein Symposium*, vol. 1. Kirchberg: Austrian Ludwig Wittgenstein Society, pp. 141–6.

Colyvan, M. (2001): *The Indispensability of Mathematics*. Oxford: Oxford University Press.

Colyvan, M. (2010): "There is No Easy Road to Nominalism", *Mind 119*: 285–306.

Colyvan, M. (2015): "Indispensability Arguments in the Philosophy of Mathematics", in Zalta, E.N. (ed.), *The Stanford Encyclopedia of Philosophy* (Spring 2015 edition). Stanford: The Metaphysics Research Lab, Center for the Study of Language and Information (CSLI), Stanford University. URL: http://plato.stanford.edu/archives/spr2015/entries/mathphil-indis/.

Contessa, G. (2007): *Representing Reality: The Ontology of Scientific Models and Their Representational Function*. PhD dissertation. London School of Economics and Political Science.

Cox, R.T., and Carlton, C.E. (2003): "A Comment on Gene Introgression versus *en masse* Cycle Switching in the Evolution of 13-Year and 17-Year Life Cycles in Periodical Cicadas", *Evolution 57*: 428–32.

Cummins, R. (1991): *Meaning and Mental Representation*. Cambridge, Mass.: MIT Press.

Curd, M. (1980): "The Logic of Discovery: An Analysis of Three Approaches", in Nickles, T. (ed.), *Scientific Discovery, Logic, and Rationality*. Dordrecht: Reidel, pp. 201–19. (Reprinted in Brody, B., and Grandy, R. (eds.) (1989): *Readings in the Philosophy of Science* (2nd edition). Englewood Cliffs, N.J.: Prentice-Hall, pp. 417–30.)

Curiel, E. (2014): "Classical Mechanics Is Lagrangian; It Is Not Hamiltonian", *British Journal for the Philosophy of Science 65*: 269–321.

da Costa, N.C.A., Bueno, O., and French, S. (1998): "The Logic of Pragmatic Truth", *Journal of Philosophical Logic 27*: 603–20.

da Costa, N.C.A., and Chuaqui, R. (1988): "On Suppes' Set-Theoretical Predicates", *Erkenntnis 29*: 95–112.

da Costa, N.C.A., and Doria, F.A. (1991): "Undecidability and Incompleteness in Classical Mechanics", *International Journal of Theoretical Physics 30*: 1041–73.

da Costa, N.C.A., and Doria, F.A. (1995): "Undecidability, Incompleteness and Arnold's Problems", *Studia Logica 55*: 23–32.

da Costa, N.C.A., and French, S. (1989): "Pragmatic Truth and the Logic of Induction", *The British Journal for the Philosophy of Science 40*: 333–56.

da Costa, N.C.A., and French, S. (1990): "The Model-Theoretic Approach in the Philosophy of Science", *Philosophy of Science 57*: 248–65.

da Costa, N.C.A., and French, S. (2000): "Models, Theories and Structures: Thirty Years On", *Philosophy of Science 67* (Proceedings): S116–27.

da Costa, N.C.A., and French, S. (2003): *Science and Partial Truth*. New York: Oxford University Press.

Daly, C., and Langford, S. (2009): "Mathematical Explanation and Indispensability Arguments", *The Philosophical Quarterly 59*: 641–58.

Davis, P. (1987): "When Mathematics Says No", in Davis, P., and Park, D., *No Way: The Nature of the Impossible*. New York: Freeman, pp. 161–177.

Dawid, R. (2013): *String Theory and the Scientific Method*. Cambridge: Cambridge University Press.

Demopoulos, W., and Friedman, M. (1985): "Critical Notice: Bertrand Russell's *The Analysis of Matter*: Its Historical Context and Contemporary Interest", *Philosophy of Science 52*: 621–39.

De Regt, H., Leonelli, S., and Eigner, K. (eds.) (2009): *Scientific Understanding: Philosophical Perspectives*. Pittsburgh: University of Pittsburgh Press.

Diederich, W. (1996a): "Structuralism as Developed within the Model-Theoretic Approach in the Philosophy of Science", in Balzer, W., and Moulines, C.U. (eds.), *Structuralist Theory of Science: Focal Issues, New Results*. Berlin: De Gruyter, pp. 15–21.

Diederich, W. (1996b): "Pragmatic and Diachronic Aspects of Structuralism", in Balzer, W., and Moulines, C.U. (eds.), *Structuralist Theory of Science: Focal Issues, New Results*. Berlin: De Gruyter, pp. 75–82.

Dirac, P.A.M. (1926): "On the Theory of Quantum Mechanics", *Proceedings of the Royal Society A* 112: 661–77.

Dirac, P. (1928a): "The Quantum Theory of the Electron", *Proceedings of the Royal Society of London A117*: 610–24.

Dirac, P. (1928b): "The Quantum Theory of the Electron. Part II", *Proceedings of the Royal Society of London A118*: 351–61.

Dirac, P. (1930): "A Theory of Electrons and Photons", *Proceedings of the Royal Society of London A126*: 360–5.

Dirac, P. (1931): "Quantized Singularities in the Electromagnetic Field", *Proceedings of the Royal Society of London A133*: 60–72.

Dirac, P. (1958): *The Principles of Quantum Mechanics* (4th edition; 1st edition published in 1930). Oxford: Clarendon Press.

Dirac, P. (1978): *Directions in Physics*. New York: Wiley.

Doncel, M.A., Hermann, A., Michel, L., and Pais, A. (eds.) (1987): *Symmetries in Physics (1600–1980): Proceedings of the 1st International Meeting on the History of Scientific Ideas*. Barcelona: Bellaterra.

Donini, E. (1987): "Einstein and a Realistic Conception of Light-Matter Symmetry, 1905–1925", in Doncel, M.G., Hermann, A., Michel, L., and Pais, A. (eds.), *Symmetries in Physics (1600–1980): Proceedings of the 1st International Meeting on the History of Scientific Ideas.* Barcelona: Bellaterra, pp. 105–12.

Douven, I. (2011): "Abduction", in Zalta, E.N. (ed.), *The Stanford Encyclopedia of Philosophy* (Spring 2011 edition). Stanford: The Metaphysics Research Lab, Center for the Study of Language and Information (CSLI), Stanford University. URL: http://plato.stanford.edu/archives/spr2011/entries/abduction/

Downes, S.M. (1992): "The Importance of Models in Theorizing (a Deflationary Semantic View", *PSA 1992*, vol. 1: 142–53.

Duck, I., and Sudarshan, E.C.G. (1997): *Pauli and the Spin-Statistics Theorem.* Singapore: World Scientific.

Eco, U. (1998): *Faith in Fakes: Travels in Hyperreality.* London: Vintage Books.

Eddington, A. (1941): "Discussion: Group Structure in Physical Science", *Mind 50*: 268–79.

Einstein, A. (1924): "Quantentheorie des Einatomigen Idealen Gases", *Sitzungsberichte der Preussischen Akademie der Wissenschaften 22*: 261–7.

Einstein, A. (1925): "Quantentheorie des Einatomigen Idealen Gases", *Sitzungsberichte der Preussischen Akademie der Wissenschaften 23*: 3–16.

Falconer, I. (1987): "Corpuscles, Electrons and Cathode Rays: J.J. Thomson and the 'Discovery of the Electron'", *British Journal for the History of Science 20*: 241–76.

Feyerabend, P. (1975): *Against Method.* London: Verso.

Feynman, R.P. (1953): "The λ-Transition in Liquid Helium", *Physical Review 90*: 1116–17.

Field, H. (1980): *Science without Numbers: A Defense of Nominalism.* Princeton: Princeton University Press.

Field, H. (1989): *Realism, Mathematics and Modality.* Oxford: Basil Blackwell.

Franklin, J. (1989): "Mathematical Necessity and Reality", *Australasian Journal of Philosophy 67*: 286–94.

Fraser, D. (2011): "How to Take Particle Physics Seriously: A Further Defence of Axiomatic Quantum Field Theory", *Studies in History and Philosophy of Modern Physics 42*: 126–35.

Fraser, J. (2016): "Spontaneous Symmetry Breaking in Finite Systems", *Philosophy of Science 83*: 585–605.

French, S. (1989): "Identity and Individuality in Classical and Quantum Physics", *Australasian Journal of Philosophy 67*: 432–46.

French, S. (1997): "Partiality, Pursuit and Practice", in Dalla Chiara, M.L., Doets, K., Mundici, D., and van Bentham, J. (eds.), *Structures and Norms in Science.* Dordrecht: Kluwer Academic, pp. 35–52.

French, S. (1999): "Models and Mathematics in Physics: The Role of Group Theory", in Butterfield, J., and Pagonis, C. (eds.), *From Physics to Philosophy.* Cambridge: Cambridge University Press, pp. 187–207.

French, S. (2000): "The Reasonable Effectiveness of Mathematics: Partial Structures and the Application of Group Theory to Physics", *Synthese 125*: 103–20.

French, S. (2003): "Scribbling on the Blank Sheet: Eddington's Structuralist Conception of Objects", *Studies in History and Philosophy of Modern Physics 34*: 227–59.

French, S. (2003a): "A Model-Theoretic Account of Representation (Or, I Don't Know Much about Art...But I Know It Involves Isomorphism)", *Philosophy of Science 70*: 1472–83.

French, S. (2014): *The Structure of the World: Metaphysics and Representation.* Oxford: Oxford University Press.

French, S. (forthcoming): "Doing Away with Dispositions", forthcoming in Meincke, A. (ed.), *Dispositionalism. Perspectives from Metaphysics and the Philosophy of Science.* Springer Synthese Library. Dordrecht: Springer.

French, S., and Krause, D. (2006): *Identity in Physics.* Oxford: Oxford University Press.

French, S., and Ladyman, J. (1997): "Superconductivity and Structures: Revisiting the London Account", *Studies in History and Philosophy of Modern Physics 28*: 363–93.

French, S., and Ladyman, J. (1998): "A Semantic Perspective on Idealisation in Quantum Mechanics", in Shanks, N. (ed.), *Idealization IX: Idealization in Contemporary Physics.* Amsterdam: Rodopi, pp. 51–73.

French, S., and Ladyman, J. (1999): "Reinflating the Semantic Approach", *International Studies in the Philosophy of Science 13*: 103–21.

French, S., and Ladyman, J. (2003): "Remodelling Structural Realism: Quantum Physics and the Metaphysics of Structure: A Reply to Cao", *Synthese 136*: 31–56.

French, S., and Rickles, D. (2003): "Understanding Permutation Symmetry", in Brading, K. and Castellani, E. (eds.), *Symmetries in Physics: Philosophical Reflections.* Cambridge: Cambridge University Press, pp. 212–38.

French, S., and Saatsi, J. (2006): "Realism about Structure: The Semantic View and Non-linguistic Representations", *Philosophy of Science (Proceedings) 78*: 548–59.

French, S., and Saatsi, J. (2018): "Symmetries and Explanatory Dependencies", in J. Saatsi and R. Reutlinger (eds.), *Explanation Beyond Causation.* Oxford: Oxford University Press, pp. 185–205.

Friederich, S. (2015): "Symmetry, Empirical Equivalence, and Identity", *British Journal for the Philosophy of Science 66*: 537–59.

Friedman, M. (1974): "Explanation and Scientific Understanding", *Journal of Philosophy 71*: 5–19.

Frigg, R. (2006): "Scientific Representation and the Semantic View of Theories", *Theoria 21*: 49–65.

Frigg, R., and Hartmann, S. (2012): "Models in Science", in Zalta, E.N. (ed.), *The Stanford Encyclopedia of Philosophy* (Fall 2012 edition). Stanford: The Metaphysics Research Lab, Center for the Study of Language and Information (CSLI), Stanford University. URL: http://plato.stanford.edu/archives/fall2012/entries/models-science/.

Gavroglu, K. (1995): *Fritz London: A Scientific Biography.* Cambridge: Cambridge University Press.

Gell-Mann, M. (1987): "Particle Theory from S-Matrix to Quarks", in Doncel, M.G., Hermann, A., Michel, L., and Pais, A. (eds.), *Symmetries in Physics (1600–1980): Proceedings of the 1st International Meeting on the History of Scientific Ideas.* Barcelona: Bellaterra, pp. 473–97.

Giere, R. (1979): *Understanding Scientific Reasoning.* New York: Holt, Rinehart & Winston.

Giere, R. (1988): *Explaining Science: A Cognitive Approach.* Chicago: University of Chicago Press.

Giere, R. (1994): "The Cognitive Structure of Scientific Theories", *Philosophy of Science 61*: 276–96.

Glymour, C. (1980): *Theory and Evidence.* Princeton: Princeton University Press.

Glymour, C. (2013): "Theoretical Equivalence and the Semantic View of Theories", *Philosophy of Science 80*: 286–97.

Goles, E., Schulz, O., and Markus, M. (2001): "Prime Number Selection of Cycles in a Predator-Prey Model", *Complexity 6*: 33-8.

Gooday, G., and Mitchell, D. (2013): "Rethinking 'Classical Physics'", in Buchwald, J., and Fox, R. (eds.), *The Oxford Handbook of the History of Physics*. Oxford: Oxford University Press, pp. 721-64.

Goodman, N. (1976): *Languages of Art*. Indianapolis: Hackett.

Grattan-Guinness, I. (1992): "Structure-Similarity as a Cornerstone of the Philosophy of Mathematics", in Echeverria, J., Ibarra, A., and Mormann, T. (eds.), *The Space of Mathematics*. Berlin: de Gruyter, pp. 91-111.

Greaves, H., and Wallace, D. (2014): "Empirical Consequences of Symmetries", *British Journal for the Philosophy of Science 65*: 59-89.

Greenberg, O.W. (2009): "Color Charge", *Scholarpedia 4*(11): 6933.

Hacking, I. (1989): "Extragalactic Reality: The Case of Gravitational Lensing", *Philosophy of Science 56*: 555-81.

Hales, T.C. (2001): "The Honeycomb Conjecture", *Discrete Computational Geometry 25*: 1-22.

Halperin, I., and Schwartz, L. (1952): *Introduction to the Theory of Distributions*. Toronto: Toronto University Press.

Halvorson, H. (2012): "What Scientific Theories Could Not Be", *Philosophy of Science 79*: 183-206.

Halvorson, H. (2013): "The Semantic View, If Plausible, Is Syntactic", *Philosophy of Science 80*: 475-8.

Hartmann, S. (1995): "Models as a Tool for Theory Construction: Some Strategies of Preliminary Physics", in Herfel, W., Krajewski, W., Niiniluoto, I., and Wójcicki, R. (eds.), *Theories and Models in Scientific Processes*. Amsterdam: Rodopi, pp. 49-67.

Heisenberg, W. (1926a): "Mehrkörperproblem und Resonanz in der Quantenmechanik", *Zeitschrift für Physik 38*: 411-26.

Heisenberg, W. (1926b): "Über die Spektra von Atomsystemen mit zwei Elektronen", *Zeitschrift für Physik 39*: 499-518.

Heisenberg, W. (1932): "Über den Bau der Atomkerne, I", *Zeitschrift für Physik 77*: 1-11.

Heitler, W. (1963): *Man and Science*. Edinburgh: Oliver and Boyd.

Hellman, G. (1989): *Mathematics without Numbers: Towards a Modal-Structural Interpretation*. Oxford: Clarendon Press.

Hendry, R., and Psillos, S. (2007): "How to Do Things with Theories: An Interactive View of Language and Models in Science", in Brzeziński, J., Klawiter, A., Kuipers, T.A.F., Łastowski, K., Paprzycka, K., and Przybysz, P. (eds.), *The Courage of Doing Philosophy*. Amsterdam: Rodopi, pp. 59-115.

Herfel, W., Krajewski, W., Niiniluoto, I., and Wójcicki, R. (eds.) (1995): *Theories and Models in Scientific Processes*. Amsterdam: Rodopi.

Hesse, M. (1963): *Models and Analogies in Science*. Oxford: Oxford University Press.

Hilbert, D., Nordheim, L., and von Neumann, J. (1927): "Über die Grundlagen der Quantenmechanik", *Mathematische Annalen 98*: 1-30. (Reprinted in von Neumann, J. (1961): *Collected Works, vol. I. Logic, Theory of Sets and Quantum Mechanics* (edited by A.H. Taub). Oxford: Pergamon Press.)

Huggett, N. (1995): "What are Quanta, and Why Does it Matter?", *PSA: Proceedings of the Biennial Meeting of the Philosophy of Science Association 1994*, vol. 2: 69-76.

Hughes, R.I.G. (1989): "Bell's Theorem, Ideology, and Structural Explanation", in Cushing, J.T., and McMullin, E. (eds.), *Philosophical Consequences of Quantum Theory: Reflections on Bell's Theorem*. Notre Dame, IN: University of Notre Dame Press, pp. 195–207.

Hughes, R.I.G. (1997): "Models and Representation", *Philosophy of Science 64*: S325–36.

Hughes, R.I.G. (2010): *The Theoretical Practices of Physics*. Oxford: Clarendon Press.

Husserl, E. (1954): *The Crisis of European Sciences and Transcendental Phenomenology* (translated by D. Carr). Evanston: Northwestern University Press.

Jacob, M. (1998): "Antimatter", in Goddard, P. (ed.), *Paul Dirac: The Man and His Work*. Cambridge: Cambridge University Press, pp. 46–87.

Judd, B.R. (1993): "Applied Group Theory 1926-1935", in Wightman, A.S. (ed.), *The Collected Works of Eugene Paul Wigner. The Scientific Papers*, vol. 1. Berlin: Springer-Verlag, pp. 17–33.

Kadanoff, L. (2009): "Phases of Matter and Phase Transitions: From Mean Field Theory to Critical Phenomena", unpublished manuscript. URL: http://jfi.uchicago.edu/~leop/RejectedPapers/ExtraV1.2.pdf.

Kahn, B., and Uhlenbeck, G.E. (1938a): "On the Theory of Condensation", *Physica 4*: 1155–6.

Kahn, B., and Uhlenbeck, G.E. (1938b): "On the Theory of Condensation", *Physica 5*: 399–414.

Kattau, S. (2001): "Kabbalistic Philosophy of Science" (Review of Mark Steiner's *The Applicability of Mathematics as a Philosophical Problem*), *Metascience 10*: 22–31.

Kragh, H. (1990): *Dirac: A Scientific Biography*. Cambridge: Cambridge University Press.

Kragh, H. (2013): "Paul Dirac and *The Principles of Quantum Mechanics*", in Baudino, M., and Navarro, J. (eds.), *Research and Pedagogy: A History of Quantum Mechanics through Its Textbooks* (edition Open Access). URL: http://edition-open-access.de/studies/2/11/.

Kronz, F., and Lupher, T. (2012): "Quantum Theory: von Neumann vs. Dirac", in Zalta, E.N. (ed.), *The Stanford Encyclopedia of Philosophy* (Summer 2012 edition). Stanford: The Metaphysics Research Lab, Center for the Study of Language and Information (CSLI), Stanford University. URL: http://plato.stanford.edu/archives/sum2012/entries/qt-nvd/.

Kuhn, T.S. (1978): *Black-Body Theory and the Quantum Discontinuity, 1894–1912*. Oxford: Oxford University Press.

Landau, L.D., and Lifshitz, L.M. (1958): *Quantum Mechanics* (revised edition: 1977; reprinted in 2003). Oxford: Pergamon Press.

Lange, M. (2013): "What Makes a Scientific Explanation Distinctively Mathematical?", *British Journal for the Philosophy of Science 64*: 485–511.

Lange, M. (2015): "On 'Minimal Model Explanations': A Reply to Batterman and Rice", *Philosophy of Science 82*: 292–305.

Lange, M. (2017): *Because without Cause: Non-Causal Explanations in Science and Mathematics*. New York: Oxford University Press.

Lewis, D. (1986): "Causal Explanation", in Lewis, D., *Philosophical Papers*, vol. 2. Oxford: Oxford University Press, pp. 214–40.

Liggins, D. (2012): "Weaseling and the Content of Science", *Mind 121*: 997–1005.

London, F. (1935): "Macroscopical Interpretation of Supraconductivity", *Proceedings of the Royal Society (London) A152*: 24–34.

London, F. (1937): "A New Conception of Supraconductivity", *Nature 140*: 793–6; 834–6.

London, F. (1938a): "The λ-Phenomenon of Liquid Helium and the Bose-Einstein Degeneracy", *Nature 141*: 643–4.

London, F. (1938b): "On the Bose-Einstein Condensation", *Physical Review 54*: 947–54.

London, F. (1948): "On the Problem of the Molecular Theory of Superconductivity", *Physical Review 74*: 562–74.

London, F. (1950): *Superfluids: Macroscopic Theory of Superconductivity*, vol. I. New York: Wiley.

London, F. (1954): *Superfluids*, vol. 2. New York: John Wiley and Sons.

London, F., and London, H. (1935): "The Electromagnetic Equations of the Supraconductor", *Proceedings of the Royal Society (London) A149*: 71–88.

Mackey, G.W. (1978): *Unitary Group Representations in Physics, Probability, and Number Theory*. Reading, Mass. The Benjamin/Cummings Pub. Co.

Mackey, G.W. (1993): "The Mathematical Papers", in Wightman, A.S. (ed.), *The Collected Works of Eugene Paul Wigner. The Scientific Papers*, vol. 1. Berlin: Springer-Verlag, pp. 241–90.

McMullin, E. (1985): "Galilean Idealization", *Studies in the History and Philosophy of Science 16*: 247–73.

Maddy, P. (1997): *Naturalism in Mathematics*. Oxford: Clarendon Press.

Massimi, M. (2005): *Pauli's Exclusion Principle: The Origin and Validation of a Scientific Principle*. Cambridge: Cambridge University Press.

Maudlin, T. (2014): *New Foundations for Physical Geometry: The Theory of Linear Structures*. Oxford: Oxford University Press.

Mayer, J. (1937): "The Statistical Mechanics of Condensing Systems, I", *Journal of Chemical Physics 5*: 67–73.

Mehra, J. (1994): *The Beat of a Different Drum*. Oxford: Oxford University Press.

Melia, J. (1995): "On What There's Not", *Analysis 55*: 223–9.

Melia, J. (2000): "Weaseling Away the Indispensability Argument", *Mind 109*: 455–80.

Menon, T., and Callender, C. (2013): "Turn and Face the Strange . . . Ch-Ch-Changes: Philosophical Questions Raised by Phase Transitions", in Batterman, R.W. (ed.), *The Oxford Handbook of Philosophy of Physics*. Oxford: Oxford University Press, pp. 189–223.

Mikenberg, I., da Costa, N.C.A., and Chuaqui, R. (1986): "Pragmatic Truth and Approximation to Truth", *Journal of Symbolic Logic 51*: 201–21.

Miller, A. (1987): "Symmetry and Imagery in the Physics of Bohr, Einstein and Heisenberg", in Doncel, M.G., Hermann, A., Michel, L., and Pais, A. (eds.), *Symmetries in Physics (1600–1980): Proceedings of the 1st International Meeting on the History of Scientific Ideas*. Barcelona: Bellaterra, pp. 299–325.

Moore, W. (1989): *Schrödinger: Life and Thought*. Cambridge: Cambridge University Press.

Morgan, M.S., and Morrison, M. (eds.) (1999): *Models as Mediators: Perspectives on Natural and Social Science*. Cambridge: Cambridge University Press.

Morrison, M. (1999): "Models as Autonomous Agents", in Morgan, M.S., and Morrison, M. (eds.), *Models as Mediators: Perspectives on Natural and Social Science*. Cambridge: Cambridge University Press, pp. 38–65.

Morrison, M. (2007): "Spin: All is Not What it Seems", *Studies in History and Philosophy of Modern Physics 38*: 529–57.

Morrison, M. (2015): *Reconstructing Reality: Models, Mathematics, and Simulations*. New York: Oxford University Press.

Mortensen, C. (1995): *Inconsistent Mathematics*. Dordrecht: Kluwer Academic.

Moulines, C.U. (1996): "Structuralism: The Basic Ideas", in Balzer, W., and Moulines, C.U. (eds.), *Structuralist Theory of Science: Focal Issues, New Results*. Berlin: De Gruyter, pp. 1–13.

Muller, F.A. (1997): "The Equivalence Myth of Quantum Mechanics", *Studies in History and Philosophy of Modern Physics 28*: 35–61; 219–47.

Murray, F.J., and von Neumann, J. (1936): "On Rings of Operators", *Annals of Mathematics 37*: 116–229.

Nagel, E. (1961): *The Structure of Science*. Indianapolis: Hackett.

Ne'eman, Y. (1987): "Hadron Symmetry, Classification and Compositeness", in Doncel, M.G., Hermann, A., Michel, L., and Pais, A. (eds.), *Symmetries in Physics (1600–1980): Proceedings of the 1st International Meeting on the History of Scientific Ideas*. Barcelona: Bellaterra, pp. 499–540.

Nersessian, N. (1992): "How Do Scientists Think?", in Giere, R. (ed.), *Cognitive Models of Science*. (Minnesota Studies in the Philosophy of Science, vol. XV.) Minneapolis: University of Minnesota Press, pp. 3–44.

Ney, A., and Albert, D. (eds.) (2013): *The Wave Function: Essays on the Metaphysics of Quantum Mechanics*. Oxford: Oxford University Press.

Nickles, T. (ed.) (1980): *Scientific Discovery, Logic, and Rationality*. Dordrecht: Reidel.

North, J. (2009): "The 'Structure' of Physics: A Case Study", *Journal of Philosophy 106*: 57–88.

Norton, J. (2012): "Approximation and Idealization: Why the Difference Matters", *Philosophy of Science 79*: 207–32.

Orilia, F., and Swoyer, C. (2016): "Properties", in Zalta, E.N. (ed.), *The Stanford Encyclopedia of Philosophy* (Spring 2016 edition). Stanford: The Metaphysics Research Lab, Center for the Study of Language and Information (CSLI), Stanford University. URL: http://plato.stanford.edu/archives/spr2016/entries/properties/.

Pais, A. (1982): *Subtle is the Lord*. Oxford: Oxford University Press.

Pais, A. (1986): *Inward Bound: Of Matter and Forces in the Physical World*. Oxford: Clarendon Press.

Pais, A. (1998): "Paul Dirac: Aspects of His Life and Work", in Goddard, P. (ed.), *Paul Dirac: The Man and His Work*. Cambridge: Cambridge University Press, pp. 1–45.

Parsons, C. (2008): *Mathematical Thought and Its Objects*. Cambridge: Cambridge University Press.

Peressini, A. (1997): "Troubles with Indispensability: Applying Pure Mathematics in Physical Theory", *Philosophia Mathematica 3*: 210–27.

Pexton, M., and Saatsi, J. (2012): "Reassessing Woodward's Account of Explanation: Regularities, Counterfactuals, and Non-Causal Explanations", *Philosophy of Science 80*: 613–24.

Pickering, A. (1995): "Concepts and the Mangle of Practice: Constructing Quaternions", *The South Atlantic Quarterly 94*: 417–65.

Pincock, C. (2005): "Overextending Partial Structures: Idealization and Abstraction", *Philosophy of Science 72*: 1248–59.

Pincock, C. (2012): *Mathematics and Scientific Representation*. New York: Oxford University Press.

Post, H. (1971): "Correspondence, Invariance and Heuristics", *Studies in History and Philosophy of Science 2*: 213–55. (Reprinted in French, S., and Kamminga, H. (eds.) (1993): *Correspondence, Invariance and Heuristics: Essays in Honour of Heinz Post*. Dordrecht: Reidel, pp. 1–44.)

Psillos, P. (1999): *Scientific Realism: How Science Tracks Truth*. Abingdon: Routledge.

Psillos, S. (2010): "Scientific Realism: Between Platonism and Nominalism", *Philosophy of Science 77*: 947–58.

Psillos, S. (2011): "Moving Molecules above the Scientific Horizon: On Perrin's Case for Realism", *Journal for General Philosophy of Science 42*: 339–63.

Putnam, H. (1962): "What Theories Are Not", in Nagel, E., Suppes, P., and Tarski, A. (eds.), *Logic, Methodology and the Philosophy of Science: Proceedings of the 1960 International Congress*. Stanford: Stanford University Press, pp. 240–51.

Putnam, H. (1979): *Mathematics, Matter and Method. Philosophical Papers*, vol. 1 (2nd edition). Cambridge: Cambridge University Press.

Quine, W.V. (1953): "On What There Is", in Quine, W.V., *From a Logical Point of View*. Cambridge, Mass.: Harvard University Press, pp. 1–19.

Quine, W.V. (1960): *Word and Object*. Cambridge, Mass.: The MIT Press.

Quine, W.V. (1976): *The Ways of Paradox and Other Essays* (revised and enlarged edition). Cambridge, Mass.: Harvard University Press.

Rédei, M. (1997): "Why John von Neumann Did Not Like the Hilbert Space Formalism of Quantum Mechanics (and What He Liked Instead)", *Studies in History and Philosophy of Modern Physics 28*: 493–510.

Rédei, M. (1998): *Quantum Logic in Algebraic Approach*. Dordrecht: Kluwer Academic.

Redhead, M. (1975): "Symmetry in Intertheory Relations", *Synthese 32*: 77–112.

Redhead, M. (1980): "Models in Physics", *British Journal for the Philosophy of Science 31*: 145–63.

Redhead, M. (2004): "Discussion: Asymptotic Reasoning", *Studies in History and Philosophy of Modern Physics 35*: 527–30.

Redhead, M., and Teller, P. (1991): "Particles, Particle Labels, and Quanta: The Toll of Unacknowledged Metaphysics", *Foundations of Physics 21*: 43–62.

Redhead, M., and Teller, P. (1992): "Particle Labels and the Theory of Indistinguishable Particles in Quantum Mechanics", *British Journal for the Philosophy of Science 43*: 201–18.

Rescher, N. (1978): *Scientific Progress*. Pittsburgh: University of Pittsburgh Press.

Reutlinger, A. (2017): "Do Renormalization Group Explanations Conform to the Commonality Strategy?", *Journal for General Philosophy of Science 48*: 143–50.

Rosen, G. (2014): "Abstract Objects", in Zalta, E.N. (ed.), *The Stanford Encyclopedia of Philosophy* (Fall 2014 edition). Stanford: The Metaphysics Research Lab, Center for the Study of Language and Information (CSLI), Stanford University. URL: http://plato.stanford.edu/archives/fall2014/entries/abstract-objects/.

Rueger, A. (1990): "Independence from Future Theories: A Research Strategy in Quantum Theory", *PSA 1990*, vol. 1: 203–11.

Rugar, D., Budakian, R., Mamin, H.J., and Chui, B.W. (2004): "Single Spin Detection by Magnetic Resonance Force Microscopy", *Nature 430*: 329–32.

Saatsi, J. (2008): "Eclectic Realism—the Proof of the Pudding", *Studies in History and Philosophy of Science 39*: 273–6.

Saatsi, J. (2011): "The Enhanced Indispensability Argument: Representational vs. Explanatory Role of Mathematics in Science", *British Journal for the Philosophy of Science 62*: 143–54.

Saatsi, J. (2016): "On the 'Indispensable Explanatory Role' of Mathematics", *Mind 125*: 1045–70.

Saatsi, J. (forthcoming): "On Explanations from Geometry of Motion", forthcoming in *British Journal for the Philosophy of Science,* Advance Access, 2016. (DOI: 10.1093/bjps/axw007.)

Saatsi, J., and Reutlinger, A. (forthcoming): "Taking Reduction to the Limit: How to Rebut the Anti-Reductionist Argument from Infinite Limits", forthcoming in *Philosophy of Science.*

Saunders, S. (1993): "To What Physics Corresponds", in French, S., and Kamminga, H. (eds.), *Correspondence, Invariance and Heuristics: Essays in Honour of Heinz Post.* Dordrecht: Reidel, pp. 295–326.

Scheibe, E. (1992): "The Role of Mathematics in Physical Science", in Echeverria, J., Ibarra, A., and Mormann, T. (eds.), *The Space of Mathematics.* Berlin: de Gruyter, pp. 141–55.

Scholz, E. (2007): "Weyl Entering the 'New' Quantum Mechanics Discourse", in Joas, C., Lehner, C., and Renn, J. (eds.), *HQ-1: Conference on the History of Quantum Physics* (Berlin, 2–6 July 2007). Preprint 350. Berlin: MPI History of Science, vol. 2, pp. 253–71.

Schrödinger, E. (1926): "Über das Verhältnis der Heisenberg-Born-Jordanschen Quantenmechanik zu der Meinen", in Schrödinger, E. (1927), *Collected Papers on Wave Mechanics.* (translated by J.F. Shearer.) New York: Chelsea, pp. 45–61.

Schweber, S. (1994): *QED and the Men Who Made It.* Princeton: Princeton University Press.

Shapiro, S. (1983): "Mathematics and Reality", *Philosophy of Science* 50: 523–48.

Shapiro, S. (1997): *Philosophy of Mathematics: Structure and Ontology.* New York: Oxford University Press.

Shech, E. (2013): "What is the Paradox of Phase Transitions?", *Philosophy of Science* 80: 1170–81.

Simons, P. (2001): "Review of: *The Applicability of Mathematics as a Philosophical Problem*", *British Journal for the Philosophy of Science* 52: 181–4.

Simons, P. (unpublished): "Wittgenstein on Surprise in Mathematics". <https://sites.google.com/site/petermsimons/peter-s-files>.

Sklar, L. (1998): "The Language of Nature is Mathematics—But Which Mathematics and What Nature?", *Proceedings of the Aristotelian Society* 98: 241–61.

Skow, B. (2014): "Are There Non-Causal Explanations (of Particular Events)?", *British Journal for the Philosophy of Science* 65: 445–67.

Speiser, D. (1987): "The Principle of Relativity in Euler's Work", in Doncel, M.G., Hermann, A., Michel, L., and Pais, A. (eds.), *Symmetries in Physics (1600–1980): Proceedings of the 1st International Meeting on the History of Scientific Ideas.* Barcelona: Bellaterra, pp. 31–47.

Stegmüller, W. (1976): *The Structure and Dynamics of Theories.* New York: Springer-Verlag.

Steiner, M. (1978): "Mathematics, Explanation, and Scientific Knowledge", *Nous* 12: 17–28.

Steiner, M. (1989): "The Application of Mathematics to Natural Science", *Journal of Philosophy* 86: 449–80.

Steiner, M. (1998): *The Applicability of Mathematics as a Philosophical Problem.* Cambridge, Mass.: Harvard University Press.

Stöltzner, M. (2004): "On Optimism and Opportunism in Applied Mathematics: Mark Wilson Meets John von Neumann on Mathematical Ontology", *Erkenntnis* 60: 121–43.

Strevens, M. (2008): *Depth: An Account of Scientific Explanation.* Cambridge, Mass.: Harvard University Press.

Suárez, M. (1995): "Idealization and the Semantic View of Scientific Theories", paper given at the 1995 Annual Meeting of the British Society for the Philosophy of Science.

Suárez, M. (1999): "Theories, Models and Representations", in Magnani, L., Nersessian, N.J., and Thagard, P. (eds.), *Model-Based Reasoning in Scientific Discovery.* Dordrecht: Kluwer, pp. 75–83.

Suárez, M. (2003): "Scientific Representation: Similarity and Isomorphism", *International Studies in the Philosophy of Science 17*: 225–44.

Suárez, M. (ed.) (2009): *Fictions in Science*. London: Routledge.

Suárez, M., and Cartwright, N. (2008): "Theories: Tools versus models", *Studies in History and Philosophy of Modern Physics 39*: 62–81.

Sudarshan, E.C.G., and Duck, I.M. (2003): "What Price the Spin-Statistics Theorem?", *Pramana: Indian Academy of Sciences 61*: 645–53.

Suppe, F. (ed.) (1977): *The Structure of Scientific Theories* (2nd edition). Urbana: University of Illinois Press.

Suppe, F. (1989): *The Semantic Conception of Theories and Scientific Realism*. Urbana: University of Illinois Press.

Suppes, P. (1961): "A Comparison of the Meaning and Uses of Models in Mathematics and the Empirical Sciences", in Freudenthal, H. (ed.), *The Concept and the Role of the Model in Mathematics and Natural and Social Sciences*. Dordrecht: Reidel, pp. 163–77.

Suppes, P. (1962): "Models of Data", in Nagel, E., Suppes, P., and Tarski, A. (eds.), *Logic, Methodology and the Philosophy of Science: Proceedings of the 1960 International Congress*. Stanford: Stanford University Press, pp. 252–67.

Suppes, P. (1967): *Set-Theoretical Structures in Science*, mimeograph, Stanford University. (Substantially expanded and revised version published as Suppes, P. (2002): *Representation and Invariance of Scientific Structures*. Stanford: Center for the Study of Language and Information (CSLI) Publications.)

Suppes, P. (1968): "The Desirability of Formalization in Science", *Journal of Philosophy 65*: 651–64.

Swoyer, C. (1991): "Structural Representation and Surrogative Reasoning", *Synthese 87*: 449–508.

Tritton, D.J. (1977): *Physical Fluid Dynamics*. New York: Van Nostrand.

Truesdell, C. (1984): "Suppesian Stews", in Truesdell, C., *An Idiot's Fugitive Essays on Science: Methods, Criticism, Training, Circumstances*. Dordrecht: Springer, pp. 503–79.

Tuomela, R. (1973): *Theoretical Concepts*. Dordrecht: Springer-Verlag.

Tuomela, R. (1978): "On the Structuralist Approach to the Dynamics of Theories", *Synthese 39*: 211–31.

Uhlenbeck, G. (1927): *Over Statistische Methoden in de Theorie der Quanta*. PhD thesis. s-Gravenhage, M. Nijhoff.

Vallicella, W.F. (2002): "Relations, Monism, and the Vindication of Bradley's Regress", *Dialectica 56*: 3–35.

van Fraassen, B.C. (1980): *The Scientific Image*. Oxford: Clarendon Press.

van Fraassen, B.C. (1985): "Empiricism in the Philosophy of Science", in Churchland, P.M., and Hooker, C.A. (eds.), *Images of Science: Essays on Realism and Empiricism, with a Reply by Bas C. van Fraassen*. Chicago: The University of Chicago Press, pp. 245–308.

van Fraassen, B.C. (1987): "The Semantic Approach to Scientific Theories", in Nersessian, N. (ed.), *The Process of Science*. Dordrecht: Kluwer, pp. 105–24.

van Fraassen, B.C. (1989): *Laws and Symmetry*. Oxford: Clarendon Press.

van Fraassen, B.C. (1991): *Quantum Mechanics: An Empiricist View*. Oxford: Clarendon Press.

van Fraassen, B.C. (1997): "Structure and Perspective: Philosophical Perplexity and Paradox", in Dalla Chiara, M.L., Doets, K., Mundici, D., and van Bentham, J. (eds.), *Structures and Norms in Science*. Dordrecht: Kluwer Academic, pp. 511–30.

van Fraassen, B.C. (2008): *Scientific Representation: Paradoxes of Perspective*. Oxford: Clarendon Press.

van Fraassen, B.C. (2014): "One or Two Gentle Remarks about Hans Halvorson's Critique of the Semantic View", *Philosophy of Science 81*: 276–83.

Varadarajan, V.S. (1968): *Geometry of Quantum Theory*, vol. 1. Princeton: van Nostrand.

Vickers, P. (2013): *Understanding Inconsistent Science*. Oxford: Oxford University Press.

von Meyenn, K. (1987): "Pauli's Belief in Exact Symmetries", in Doncel, M.G., Hermann, A., Michel, L., and Pais, A. (eds.), *Symmetries in Physics (1600–1980): Proceedings of the 1st International Meeting on the History of Scientific Ideas*. Barcelona: Bellaterra, pp. 331–58.

von Neumann, J. (1932): *Mathematical Foundations of Quantum Mechanics*. (The English translation, by R.T. Beyer, of the original German edition was first published in 1955.) Princeton: Princeton University Press.

von Neumann, J. (1937a): "On Alternative Systems of Logics", manuscript, von Neumann Archives, Library of Congress, Washington, D.C.

von Neumann, J. (1937b): "Quantum Logics. (Strict- and Probability-Logics)", unpublished manuscript, von Neumann Archives, Library of Congress, Washington, D.C. (A brief summary, written by A.H. Taub, can be found in von Neumann, J. (1962): *Collected Works, vol. IV. Continuous Geometry and Other Topics* (edited by A.H. Taub). Oxford: Pergamon Press, pp. 195–7.)

von Neumann, J. (1954): "Unsolved Problems in Mathematics", typescript, von Neumann Archives, Library of Congress, Washington, D.C.

von Neumann, J. (1960): *Continuous Geometry*. Princeton: Princeton University Press.

von Neumann, J. (1981): "Continuous Geometries with a Transition Probability", *Memoirs of the American Mathematical Society 34* (No. 252): 1–210.

Wakil, S., and Justus, J. (2017): "Mathematical Explanation and the Optimization Fallacy", *Philosophy of Science 84*: 916–30.

Wallace, D. (2011): "Taking Particle Physics Seriously: A Critique of the Algebraic Approach to Quantum Field Theory", *Studies in the History and Philosophy of Modern Physics 42*: 116–25.

Weisberg, M. (2007): "Three Kinds of Idealization", *The Journal of Philosophy 104*: 639–59.

Wessels, L. (1979): "Schrödinger's Route to Wave Mechanics", *Studies in History and Philosophy of Science 10*: 311–40.

Wessels, L. (1980): "What Was Born's Statistical Interpretation?", *PSA 1980*: 187–200.

Weyl, H. (1925): "Theorie der Darstellung kontinuierlicher halb-einfacher Gruppen durch lineare Transformationen, I", *Mathematische Zeitschrift 23*: 271–309; II, *ibid. 24* (1926): 328–76; III, *ibid. 24* (1926): 377–95. (Reprinted in Weyl (1968), vol. 2, pp. 543–647.)

Weyl, H. (1927): "Quantenmechanik und Gruppentheorie", *Zeitschrift für Physik 46*: 1–46. (Reprinted in Weyl (1968), vol. 3, pp. 90–135.)

Weyl, H. (1931): *The Theory of Groups and Quantum Mechanics* (translated from the 2nd, revised German edition by H.P. Robertson. 1st edition published 1928). New York: Dover.

Weyl, H. (1952): *Symmetry*. Princeton: Princeton University Press.

Weyl, H. (1968): *Gesammelte Abhandlungen*. Berlin: Springer-Verlag.

Wigner, E.P. (1927): "Über nicht kombinierende Terme in der neueren Quantentheorie", *Zeitschrift für Physik 40*: 883–92.

Wigner, E.P. (1959 [1931]): *Group Theory and Its Application to the Quantum Mechanics of Atomic Spectra*. (The English translation was published in 1959.) New York: Academic Press.

Wigner, E. (1935): "Symmetry Relations in Various Physical Problems", *Bulletin of the American Mathematical Society 41*: 306.

Wigner E.P. (1937): "On the Consequences of the Symmetry of the Nuclear Hamiltonian on the Spectroscopy of Nuclei", *Physical Review 51*: 106–19.

Wigner, E. (1939): "On Unitary Representations of the Inhomogeneous Lorentz Group", *Annals of Mathematics 40*: 149–204.

Wigner, E.P. (1960): "The Unreasonable Effectiveness of Mathematics in the Natural Sciences", *Communications in Pure and Applied Mathematics 13*: 1–14.

Wigner, E.P. (1963): "Oral History Transcript", interviews held at the Niels Bohr Library and Archives, Niels Bohr Library and Archives, Sessions II and III. URL: https://www.aip.org/history-programs/niels-bohr-library/oral-histories/4963-2.

Wigner, E.P. (1964): "Events, Laws of Nature, and Invariance Principles", in *The Nobel Prize Lectures*. Amsterdam: Elsevier, pp. 6–17. (Reprinted in Wigner, E.P. (1967): *Symmetries and Reflections: Scientific Essays of Eugene P. Wigner* (edited by W.J. Moore and M. Scriven). Bloomington: Indiana University Press, pp. 38–50.)

Wilson, K. (1971): "Renormalization Group and Critical Phenomena: I. Renormalization Group and the Kadanoff Scaling Picture", *Physical Review B 4*: 3174–83.

Wilson, K. (1974): "Confinement of Quarks", *Physical Review D 10*: 2445.

Wilson, K. (2008): "The Renormalization Group and Critical Phenomena", Nobel Prize Lecture. URL: http://www.nobelprize.org/nobel_prizes/physics/laureates/1982/wilson-lecture.pdf.

Wilson, M. (1992): "Frege: The Royal Road from Geometry", *Noûs 26*: 149–80.

Wilson, M. (2000): "The Unreasonable Uncooperativeness of Mathematics in Natural Science", *The Monist 83*: 296–314.

Wilson, M. (2002): "On the Mathematics of Spilt Milk", in Grosholz, E., and Breger, H. (eds.), *The Growth of Mathematical Knowledge*. Dordrecht: Springer, pp. 143–52.

Wilson, M. (2006): *Wandering Significance*. Oxford: Oxford University Press.

Wilson, M. (2009): "Determinism and the Mystery of the Missing Physics", *British Journal for the Philosophy of Science 60*: 173–93.

Wilson, M. (2014): "What is 'Classical Mechanics' Anyway?". URL: http://www.philosophy.pitt.edu/sites/default/files/whatisclassicalmechanicsanyway.pdf (an abridged version can be found in Batterman, R.W. (ed.), *The Oxford Handbook of Philosophy of Physics*. Oxford: Oxford University Press, pp. 43–106.)

Woodward, J. (2003): *Making Things Happen*. New York: Oxford University Press.

Name Index

Achinstein, P. 25–6
Azzouni, J. xv, 2, 20, 142–3, 152, 153–4

Baker, A. 131, 160, 161, 163, 166, 167–8
Balzer, W. 30
Bangu, S. 3, 98, 146, 222–6
Barrantes, M. xv, 10 n. 32, 14 n. 36, 19 n. 42,
    160 n. 15, 161 n. 16, 164 n. 20, 167 n. 21,
    202 n. 27, 219 n. 49
Batterman, R. x, 164, 183, 185–93, 195–7, 199,
    201–14, 218–21, 222
Belot, G. 189, 204–5, 218
Bergia, S. 77
Berry, M. 186–7, 189, 190
Beth, E. 25–6, 30
Bird, A. 177
Birkhoff, G. 116, 123, 125–6
Bohr, N. 4, 7, 58, 60, 146
Bokulich, A. 199, 200–2, 203, 204
Boltzmann, L. 4, 76–7
Bonolis, L. 73–6
Boolos, G. 44
Born, M. 5, 9, 79, 105, 117
Bose, S. 77
Braithwaite, R. 25, 178
Brezis, H. 136
Browder, F. 136
Brush, S. 103
Bub, J. 116, 123
Bueno, O. xv–xvi, 32, 35, 43–4, 50, 52–3, 60, 64,
    65, 67, 72, 90, 102, 103, 106, 111, 113, 148,
    153, 159, 171, 183, 188, 193, 218, 220
Bunge, M. 26

Callender, C. 57–60, 214
Cao, T. 171
Carlton, C. 160, 163
Carnap, R. 25
Carson, C. 107, 170
Cartwright, N. 35, 106, 111, 113, 231–2
Chakravartty, A. 63, 64–5, 66, 71, 177
Chayut, M. 79
Chuaqui, R. 40, 42
Cohen, J. 57–60
Coleman, A. 89, 95
Collier, J. 46
Colyvan, M. xv, 52, 64, 131, 136, 137, 141–2,
    144, 150, 152, 153, 160, 165, 168, 169, 171,
    183, 186, 188, 198, 218, 223–4
Contessa, G. 58–60

Cox, R. 160, 163
Cummins, R. 69
Curd, M. 5
Curiel, E. 228

da Costa, N. xv, 23, 26, 32, 33, 35, 40, 41–2, 46,
    49, 60, 65, 68, 70, 71, 111–12, 121, 155, 185,
    191, 225, 232
Daly, C. 160, 167
Davis, P. 11
Dawid, R. 155
de Broglie, L. 6–7, 77, 105
de Regt, H. 207
Demopoulos, W. 46
Diederich, W. 29–30
Dirac, P. ix–x, 4, 11, 12, 15, 20, 22, 38, 77, 79,
    80, 84, 89, 105, 117, 118, 119, 127–8, 129,
    130–50, 151–2, 153, 156, 157–8, 167, 174,
    175, 176, 223–4, 229, 234
Doncel, M. 76, 78, 79, 91, 92, 94, 95
Donini, E. 79
Douven, I. 4
Downes, S. 26
Duck, I. 77, 104, 174

Eco, U. 2
Eddington, A. 149, 173, 177, 178
Ehrenfest, P. 4, 77, 104, 109
Einstein, A. 4, 6, 8, 14, 39, 67, 76–7, 104–6,
    165, 228

Fermi, E. 77
Feyerabend, P. 37
Feynman, P. 110–11, 115, 197
Field, H. 148–9, 198
Franklin, J. 182
Fraser, D. 230
Fraser, J. 4, 205
French, S. 19, 26, 32, 33, 35, 41, 43, 46, 47,
    49–50, 51, 53, 55, 57, 60, 64, 65, 66, 67, 68,
    70–1, 72, 76, 79, 80, 87, 90, 92, 94, 99–100,
    102, 103, 105, 106, 107, 111, 112, 113, 121,
    152, 155, 171–2, 175, 177, 178, 179, 180, 185,
    188, 191, 199, 209, 211, 220, 222, 225, 231, 232
Friederich, S. 97
Friedman, M. 46, 204
Frigg, R. 55, 57, 62–3, 67

Gavroglu, K. 88, 103–7, 112
Gell-Man, M. 91, 94, 223–4, 226

Subject Index